Computational Methods in the Design of Anticancer Drugs

Computational Methods in the Design of Anticancer Drugs

Editors

Marialuigia Fantacuzzi
Mariangela Agamennone

MDPI

Basel • Beijing • Wuhan • Barcelona • Belgrade • Novi Sad • Cluj • Manchester

Editors

Marialuigia Fantacuzzi
Department of Pharmacy
"G. d'Annunzio" University of Chieti-Pescara
Chieti
Italy

Mariangela Agamennone
Department of Pharmacy
"G. d'Annunzio" University of Chieti-Pescara
Chieti
Italy

Editorial Office

MDPI
St. Alban-Anlage 66
4052 Basel, Switzerland

This is a reprint of articles from the Special Issue published online in the open access journal *Pharmaceuticals* (ISSN 1424-8247) (available at: www.mdpi.com/journal/pharmaceuticals/special_issues/computational_anticancer).

For citation purposes, cite each article independently as indicated on the article page online and as indicated below:

Lastname, A.A.; Lastname, B.B. Article Title. *Journal Name* **Year**, *Volume Number*, Page Range.

ISBN 978-3-7258-0814-4 (Hbk)
ISBN 978-3-7258-0813-7 (PDF)
doi.org/10.3390/books978-3-7258-0813-7

Contents

Preface

Cancer remains a major threat to human health and one of the leading causes of death worldwide. In recent years, there has been continuous progress in the development of new anticancer drugs, and several compounds (small molecules, engineered antibodies, immunomodulators, etc.) have been approved for the treatment of cancer.

In recent decades, computational methods have become an essential tool in the drug design process, as they are able to reduce research costs and accelerate the development process. The application of computational methods has been proven to be very effective in the design of anticancer drugs. Given the wide variety of tumor types and the large number of possible pharmacological targets, this area of research is very fruitful.

This Special Issue on "Computational Methods in the Design of Anticancer Drugs" collected nine articles and two reviews covering a large number of targets and several computational approaches, ranging from ligand- to structure-based methods, representing the latest discoveries in the field of computational anticancer drug design. The Special Issue is addressed to all scientists working in the anticancer field, especially those interested in computer-aided drug design.

As guest editors, we would like to thank all the authors for their excellent contributions and Kate Zhou from MDPI for the editorial support.

Marialuigia Fantacuzzi and Mariangela Agamennone
Editors

pharmaceuticals

Editorial

Computational Methods in the Design of Anticancer Drugs

Marialuigia Fantacuzzi * and Mariangela Agamennone *

Department of Pharmacy, University "G. d'Annunzio" of Chieti-Pescara, Via Dei Vestini, 31, 66100 Chieti, Italy
* Correspondence: marialuigia.fantacuzzi@unich.it (M.F.); mariangela.agamennone@unich.it (M.A.)

In recent years, continuous progress has been made in the development of new anticancer drugs, and several compounds (small molecules, engineered antibodies, immunomodulators, etc.) have been approved, dramatically changing the landscape of tumor treatment [1]. Despite these efforts, cancer remains a major threat to human health and one of the leading causes of death worldwide [2]. This underscores the need for an even better understanding of the molecular mechanisms behind cancer initiation and progression. To date, the advent of immunotherapy, gene therapy and molecular targeted therapy has revolutionized the treatment of most cancers. In targeted therapy, the genetic signature of each type of cancer is targeted with drugs designed to act against actionable driver genes, avoiding the side effects of conventional chemotherapy and improving treatment efficacy. The advent of cancer immunotherapy and gene therapy has enriched the available armamentarium in the fight against this pathology, even with some limitations [3]. The efforts of the scientific community against cancer can also be seen in the number of drugs approved by the FDA for cancer treatment in 2023 (15 out of a total of 55 new drugs) [4].

Over the past few decades, computational methods have become an essential tool in the drug design process because they can reduce research costs and accelerate the development process [5]. Several factors have been contributing to the expansion of in silico applications. The increasing availability of 3D macromolecule structures through experimental (X-ray or cryo-EM) or computational methods (AlphaFold) [6] allows us to study most of the genome. The development of supercomputers enables the atom-based simulation of even larger systems. The availability of ultra-large compound libraries, which expand the explorable chemical space through screening campaigns, is another important factor. Furthermore, the growing application of artificial intelligence algorithms to drug discovery is having an increasing impact: AI algorithms are being applied in a variety of areas, such as the aforementioned protein structure prediction, QSAR/QSPR, structure-based modeling, and the prediction of AD-ME/toxicity profiles [7].

The application of computational methods in the design of anticancer drugs has proven to be very effective [8]. Given the wide variety of tumor types and the large number of possible pharmacological targets, this is a challenging area of research [2].

For the Special Issue on "Computational Methods in the Design of Anticancer Drugs", we aimed to collect the most recent discoveries in the field of anticancer drug design using computational methods. The 11 articles (8 papers and 3 reviews) cover a wide range of topics, from pharmacophore modeling to molecular docking, molecular dynamics, and ADMET prediction, and focus on many different targets, highlighting the diverse target landscape in cancer treatment.

Bülbül et al. (contribution 1) focused on the development of novel selective HDAC3 (Histon DeACetylase 3) inhibitors containing the alkylhydrazide zinc-binding group. They generated and evaluated pharmacophore and atom-based QSAR models, and the binding mode of compounds was determined using molecular docking and molecular dynamics simulations. The developed models provide a clear explanation for the in vitro data.

Moreover, Córdova-Bahena et al. (contribution 2) generated a pharmacophore model using a set of well-known Casein Kinase 1 isoform epsilon (CK1ε) inhibitors. The resulting

Citation: Fantacuzzi, M.; Agamennone, M. Computational Methods in the Design of Anticancer Drugs. *Pharmaceuticals* **2024**, *17*, 404. https://doi.org/10.3390/ph17040404

Received: 14 March 2024
Accepted: 20 March 2024
Published: 22 March 2024

model was used to screen a library of FDA-approved drugs for repositioning purposes. Molecular docking and molecular dynamics were used to analyze new compounds. The antineoplastic drug Etravirine, which activates the WNT pathway in osteosarcoma cells by increasing the expression of the cyclin-dependent kinase (CDK) inhibitor p21, emerged as a CK1ε inhibitor.

In addition, Bhujbal et al. (contribution 3) focused on Polo-like kinase 1 (PLK1) inhibitors, which can be used to treat various types of cancer, such as lung, colon, prostate, ovarian, breast, melanoma, and AML. They performed hybrid 3D-QSAR and molecular docking to design potent and selective inhibitors. Two compounds showed good IC_{50} values.

Crocetti and coworkers (contribution 4) used a ligand-based technique to develop more potent fatty acid binding protein 4 (FABP4) inhibitors, starting with a known pyrimidine ligand and applying bioisosteric replacements and scaffold hopping in the pyrimidine skeleton. They synthesized and biologically tested novel 4-amino and 4-ureido-pyridazinone-based compounds as FABP4 inhibitors. The molecular docking study confirmed the ability of the most active molecules to better interact inside the FABP4 binding pocket.

Al-Zahrani et al. (contribution 5) virtually screened 1289 flavonoids using molecular docking to the mitogen-activated protein kinase (MAPK) MEK1. ADMET prediction and 100 ns molecular dynamics (MD) simulations were then applied to the top five docked compounds, revealing them as promising potent inhibitors.

Franco et al. (contribution 6) focused on the inhibition of nicotinic acid phosphoribosyl transferase (NAPRT), the rate-limiting enzyme of the Preiss–Handler NAD biosynthetic pathway, which can overcome resistance to nicotinamide phosphoribosyl transferase (NAMPT) inhibition and lead to better anti-tumor effects. Selected hits from the virtual screening were tested in a cellular assay using the ovarian cell line OVCAR-5, and the recombinant hNAPRT and showed a synergistic effect with the NAMPT inhibitor FK866.

The same research group (contribution 7) performed a structure-based virtual screening on a 537,009 drug-like compound library and identified two additional chemical scaffolds that functioned as NAPRT inhibitors. The new compounds showed comparable anti-cancer activity with respect to the previously discovered NAPRT inhibitor, 2-hydroxynicotinic acid (2-HNA), a better predicted solubility, and favorable drug-like properties.

Bartelink et al. (contribution 8) applied a computational method to develop a physiological pharmacokinetic (PBPK) model to predict the image quality (tumor-to-lung contrast) of three PET radiotracers binding the epidermal growth factor receptor tyrosine kinase (EGFR TKI PET/CT: ^{11}C-erlotinib, ^{18}F-afatinib and ^{11}C-osimertinib), used to assess EGFR overexpression and mutation in NSCLC. The model was also developed to predict the uptake of healthy tissue in three radiolabeled EGFR ligands.

Finally, there are three reviews in this Special Issue. One, written by Wang et al. (contribution 9), focuses on applications of artificial intelligence in the design of anticancer drugs, demonstrating the basic ideas behind these techniques, as well as their advantages and disadvantages. The authors reviewed the literature from the past decade, focusing on all articles presenting computational studies using AI to assist in the identification of effective cancer treatments. In addition, the authors provided a compilation of useful databases (omics, chemical compounds, drugs, etc.) as a valuable tool in the application of AI for drug discovery.

Primavera et al. (contribution 10) focused on small-molecule AKT inhibitors that were validated for anticancer activity using computer-aided drug design methods. The authors provided an introductory analysis of AKT structural features and binding sites. Then, a comprehensive analysis of inhibitors identified via different approaches (pharmacophore screening, docking, QSAR, machine learning) is reported, distinguishing between orthosteric and allosteric binders.

In our review (contribution 11), we examined the most relevant papers that elucidated the binding mechanism of PD-L1 with PD-1 and small molecules through computational analyses. In particular, the hot spot residues involved in the interaction between PD-L1 and PD-1 and the PD-L1 dimerization induced by small molecule binding are described. Virtual screening campaigns, mainly structure-based, that were performed to identify new small-molecule PD-L1 binders are also reported.

As Guest Editors, we hope that the findings included in this Special Issue will inspire further investigations in this challenging field.

Acknowledgments: The Guest Editors thank all the authors for their high-quality contributions and the reviewers for providing critical feedback.

Conflicts of Interest: The authors declare no conflicts of interest.

List of Contributions:

1. Bülbül, E.F.; Robaa, D.; Sun, P.; Mahmoudi, F.; Melesina, J.; Zessin, M.; Schutkowski, M.; Sippl, W. Application of Ligand- and Structure-Based Prediction Models for the Design of Alkylhydrazide-Based HDAC3 Inhibitors as Novel Anti-Cancer Compounds. *Pharmaceuticals* **2023**, *16*, 968. https://doi.org/10.3390/ph16070968.
2. Córdova-Bahena, L.; Sánchez-Álvarez, A.A.; Ruiz-Moreno, A.J.; Velasco-Velázquez, M.A. Repositioning of Etravirine as a Potential CK1ε Inhibitor by Virtual Screening. *Pharmaceuticals* **2022**, *15*, 8. https://doi.org/10.3390/ph15010008.
3. Bhujbal, S.P.; Kim, H.; Bae, H.; Hah, J.-M. Design and Synthesis of Aminopyrimidinyl Pyrazole Analogs as PLK1 Inhibitors Using Hybrid 3D-QSAR and Molecular Docking. *Pharmaceuticals* **2022**, *15*, 1170. https://doi.org/10.3390/ph15101170.
4. Crocetti, L.; Floresta, G.; Zagni, C.; Merugu, D.; Mazzacuva, F.; de Oliveira Silva, R.R.; Vergelli, C.; Giovannoni, M.P.; Cilibrizzi, A. Ligand Growing Experiments Suggested 4-amino and 4-ureido pyridazin-3(2H)-one as Novel Scaffold for FABP4 Inhibition. *Pharmaceuticals* **2022**, *15*, 1335. https://doi.org/10.3390/ph15111335.
5. AlZahrani, W.M.; AlGhamdi, S.A.; Zughaibi, T.A.; Rehan, M. Exploring the Natural Compounds in Flavonoids for Their Potential Inhibition of Cancer Therapeutic Target MEK1 Using Computational Methods. *Pharmaceuticals* **2022**, *15*, 195. https://doi.org/10.3390/ph15020195.
6. Franco, J.; Piacente, F.; Walter, M.; Fratta, S.; Ghanem, M.; Benzi, A.; Caffa, I.; Kurkin, A.V.; Altieri, A.; Herr, P.; et al. Structure-Based Identification and Biological Characterization of New NAPRT Inhibitors. *Pharmaceuticals* **2022**, *15*, 855. https://doi.org/10.3390/ph15070855.
7. Ghanem, M.S.; Caffa, I.; Del Rio, A.; Franco, J.; Parenti, M.D.; Monacelli, F.; Cea, M.; Khalifa, A.; Nahimana, A.; Duchosal, M.A.; et al. Identification of NAPRT Inhibitors with Anti-Cancer Properties by In Silico Drug Discovery. *Pharmaceuticals* **2022**, *15*, 848; https://doi.org/10.3390/ph15070848.
8. Bartelink, I.H.; van de Stadt, E.A.; Leeuwerik, A.F.; Thijssen, V.L.J.L.; Hupsel, J.R.I.; van den Nieuwendijk, J.F.; Bahce, I.; Yaqub, M.; Hendrikse, N.H. Physiologically Based Pharmacokinetic (PBPK) Modeling to Predict PET Image Quality of Three Generations EGFR TKI in Advanced-Stage NSCLC Patients. *Pharmaceuticals* **2022**, *15*, 796. https://doi.org/10.3390/ph15070796.
9. Wang, L.; Song, Y.; Wang, H.; Zhang, X.; Wang, M.; He, J.; Li, S.; Zhang, L.; Li, K.; Cao, L. Advances of Artificial Intelligence in Anti-Cancer Drug Design: A Review of the Past Decade. *Pharmaceuticals* **2023**, *16*, 253. https://doi.org/10.3390/ph16020253.
10. Primavera, E.; Palazzotti, D.; Barreca, M.L.; Astolfi, A. Computer-Aided Identification of Kinase-Targeted Small Molecules for Cancer: A Review on AKT Protein. *Pharmaceuticals* **2023**, *16*, 993. https://doi.org/10.3390/ph16070993.
11. Fantacuzzi, M.; Paciotti, R.; Agamennone, M. A Comprehensive Computational Insight into the PD-L1 Binding to PD-1 and Small Molecules. *Pharmaceuticals* **2024**, *17*, 316. https://doi.org/10.3390/ph17030316.

References

1. Zhang, Z.; Zhou, L.; Xie, N.; Nice, E.C.; Zhang, T.; Cui, Y.; Huang, C. Overcoming cancer therapeutic bottleneck by drug repurposing. *Signal Transduct. Target. Ther.* **2020**, *5*, 113. [CrossRef] [PubMed]
2. Jassim, A.; Rahrmann, E.P.; Simons, B.D.; Gilbertson, R.J. Cancers make their own luck: Theories of cancer origins. *Nat. Rev. Cancer* **2023**, *23*, 710–724. [CrossRef] [PubMed]
3. Zafar, A.; Khan, M.J.; Abu, J.; Naeem, A. Revolutionizing cancer care strategies: Immunotherapy, gene therapy, and molecular targeted therapy. *Mol. Biol. Rep.* **2024**, *51*, 219. [CrossRef] [PubMed]

4. Available online: https://www.fda.gov/drugs/new-drugs-fda-cders-new-molecular-entities-and-new-therapeutic-biological-products/novel-drug-approvals-2023 (accessed on 1 March 2024).
5. Sadybekov, A.V.; Katritch, V. Computational Approaches Streamlining Drug Discovery. *Nature* **2023**, *616*, 673–685. [CrossRef] [PubMed]
6. Available online: https://alphafold.ebi.ac.uk/ (accessed on 1 March 2024).
7. Jiménez-Luna, J.; Grisoni, F.; Weskamp, N.; Schneider, G. Artificial intelligence in drug discovery: Recent advances and future perspectives. *Expert Opin. Drug Discov.* **2021**, *16*, 949–959. [CrossRef] [PubMed]
8. Zhu, H.; Zhang, Y.; Li, W.; Huang, N. A Comprehensive Survey of Prospective Structure-Based Virtual Screening for Early Drug Discovery in the Past Fifteen Years. *Int. J. Mol. Sci.* **2022**, *23*, 15961. [CrossRef] [PubMed]

pharmaceuticals

Article

Application of Ligand- and Structure-Based Prediction Models for the Design of Alkylhydrazide-Based HDAC3 Inhibitors as Novel Anti-Cancer Compounds

Emre F. Bülbül [1], Dina Robaa [1], Ping Sun [1], Fereshteh Mahmoudi [1], Jelena Melesina [1], Matthes Zessin [2], Mike Schutkowski [2] and Wolfgang Sippl [1,*]

[1] Department of Medicinal Chemistry, Institute of Pharmacy, Martin-Luther University of Halle-Wittenberg, 06120 Halle (Saale), Germany; efbulbul@gmail.com (E.F.B.); dina.robaa@pharmazie.uni-halle.de (D.R.); sunpingmc@gmail.com (P.S.); fereshteh.mahmoudi85@gmail.com (F.M.); jelenamelesina@gmail.com (J.M.)

[2] Department of Enzymology, Institute of Biotechnology, Martin-Luther University of Halle-Wittenberg, 06120 Halle (Saale), Germany; matthes.zessin@gmail.com (M.Z.); mike.schutkowski@biochemtech.uni-halle.de (M.S.)

* Correspondence: wolfgang.sippl@pharmazie.uni-halle.de

Abstract: Histone deacetylases (HDAC) represent promising epigenetic targets for several diseases including different cancer types. The HDAC inhibitors approved to date are pan-HDAC inhibitors and most show a poor selectivity profile, side effects, and in particular hydroxamic-acid-based inhibitors lack good pharmacokinetic profiles. Therefore, the development of isoform-selective non-hydroxamic acid HDAC inhibitors is a highly regarded field in medicinal chemistry. In this study, we analyzed different ligand-based and structure-based drug design techniques to predict the binding mode and inhibitory activity of recently developed alkylhydrazide HDAC inhibitors. Alkylhydrazides have recently attracted more attention as they have shown promising effects in various cancer cell lines. In this work, pharmacophore models and atom-based quantitative structure–activity relationship (QSAR) models were generated and evaluated. The binding mode of the studied compounds was determined using molecular docking as well as molecular dynamics simulations and compared with known crystal structures. Calculated free energies of binding were also considered to generate QSAR models. The created models show a good explanation of in vitro data and were used to develop novel HDAC3 inhibitors.

Keywords: docking; binding free energy; pharmacophore; atom-based QSAR; alkylhydrazide; histone deacetylases (HDAC)

check for updates

Citation: Bülbül, E.F.; Robaa, D.; Sun, P.; Mahmoudi, F.; Melesina, J.; Zessin, M.; Schutkowski, M.; Sippl, W. Application of Ligand- and Structure-Based Prediction Models for the Design of Alkylhydrazide-Based HDAC3 Inhibitors as Novel Anti-Cancer Compounds. *Pharmaceuticals* **2023**, *16*, 968. https://doi.org/10.3390/ph16070968

Academic Editors: Marialuigia Fantacuzzi and Mariangela Agamennone

Received: 15 May 2023
Revised: 30 June 2023
Accepted: 4 July 2023
Published: 6 July 2023

1. Introduction

Epigenetics refers to reversible alterations in the gene expressions that do not modify the DNA sequence [1]. Post-translational modifications such as methylation, acetylation, and others introduce changes on the N-terminal tails of histones [2]. Histone acetylation and deacetylation are controlled by different classes of enzymes, namely histone acetyltransferases (HAT) and histone deacetylases (HDAC) [3,4]. Thus, chemical modification is reversible [5,6].

To date, 18 human HDACs have been characterized. HDACs are separated into two groups and four classes depending on their sequence similarity to yeast HDAC [7]. Zinc-dependent HDACs are class I, class II, and class IV HDACs, while nicotinamide adenine dinucleotide (NAD+)-dependent enzymes are class III HDACs which are also known as sirtuins [8–10]. Class I HDACs (HDAC1, 2, 3, and 8) are located in the nucleus [11]. HDAC1 and HDAC2 interact with the nucleosome remodeling and deacetylase complex (NuRD), transcriptional regulatory protein sin3A, corepressor of REST (CoREST), and mitotic deacetylase complex (MIDAC) [12–16], while HDAC3 forms a complex only with the silencing mediator for retinoid and thyroid receptors (SMRT) and nuclear receptor corepressor (NCoR) [17,18]. HDAC8 does not need to form a complex and works alone [19,20].

HDACs are involved in signal transduction, cell growth, and cell death [21]. So far, several inhibitors including SAHA, FK228, belinostat, and panobinostat have been approved by the FDA against T-cell lymphoma [22–25]. However, due to reported side effects and unfavorable pharmacokinetics, much effort has been made to develop novel selective and better bioavailable HDAC inhibitors against several diseases such as cancer, parasitic diseases, inflammation, and others [26–29].

The majority of HDAC inhibitors consist of three pharmacophore features: a zinc-binding group (ZBG), which chelates the zinc ion at the bottom of the catalytic pocket, a linker group, which is located at the lysine binding tunnel, and a cap group, which is solvent-exposed at the rim of the pocket [7,30]. Some HDACs can be selectively inhibited by compounds addressing available subpockets of HDACs such as the side-pockets, lower pocket, and foot pocket (FP) [31–39]. For example, class I HDACs show an additional foot pocket [40]. Targeting this foot pocket resulted in class I HDAC-selective inhibitors [31,34,35,41–44]. However, the development of selective inhibitors for the class I members of HDACs, particularly HDAC1-3, remains a critical challenge to overcome. The zinc-binding group is an integral part of most HDAC inhibitors. Until now, hydroxamic acid, 2-aminobenzamide, 2-substituted benzamide, alkyl/arylketone, and thiol groups have often been used as warheads in the inhibitors of class I HDACs [35,37,38,41–45].

Recently, Wang et al. discovered a lead compound containing a benzoylhydrazide moiety that selectively inhibits HDAC1, HDAC2, and HDAC3 [46]. These compounds showed a fast-on/slow off binding mechanism [46]. Consequently, the alkylhydrazide scaffold attracted attention for the development of HDAC3 inhibitors, and some alkylhydrazide derivatives were found to show high potency, increased selectivity, and good bioavailability [46–52]. Therefore, the alkylhydrazide zinc-binding group represents a promising alternative to classical hydroxamic acids. The general structures of thealkylhydrazides are shown in Figure 1.

Interestingly, increasing the length of the *N*-alkyl group (from *n*-propyl to *n*-hexyl) resulted in a shift of the selectivity towards HDAC8 and provided substrate-competitive and highly potent inhibitors [53].

HDAC3 deacetylates various histone and non-histone proteins [54]. The catalytic activity of HDAC3 is dependent on the formation of a complex with silencing of mediator co-repressor 1 (NCoR1) and retinoic acid and thyroid hormone receptor (SMRT3) [55]. As a class I HDAC member, HDAC3 deletes the acetyl mark from histone tails, resulting in a tightly packed and transcriptionally inactive chromatin structure [56]. HDAC3 has hence been implicated in several pathophysiological processes and disorders including different cancer types, inflammatory conditions such as rheumatoid arthritis, neurodegenerative disorders like Huntington's and Alzheimer's disease, diabetes, kidney diseases, as well as cardiovascular diseases [54,56–63]. The exact role of HDAC3 in the various pathological conditions remains poorly understood, as potent and selective HDAC3 inhibitors have been scarce. Often, the described HDAC3 inhibitors in cells are also able to inhibit the structurally very similar HDAC1 and HDAC2 [41–46]. Therefore, it is a promising task to develop effective and selective HDAC3 inhibitors.

In the current study, we performed docking and molecular dynamics studies of alkylhydrazides as HDAC3 inhibitors. In order to understand the structure–activity relationship of this class of inhibitors, available data were compiled to apply ligand-based and structure-based methods. Various quantitative structure–activity relationship (QSAR) methods were evaluated for this purpose, including pharmacophore models, atom-based 3D QSAR models, and binding-free-energy-based QSAR models. In addition, we tested the models on novel designed alkylhydrazides. The workflow followed in this study is shown in Figure 2.

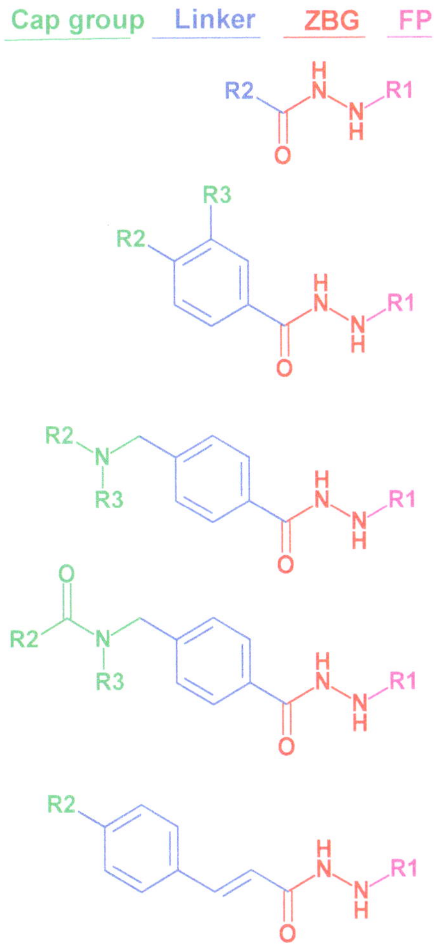

Figure 1. Structures of alkylated-hydrazide-based HDAC inhibitors that were used in this article. (The molecular structures of all compounds under study are shown in Table S1, Supplement). ZBG: zinc-binding group; FP: foot pocket.

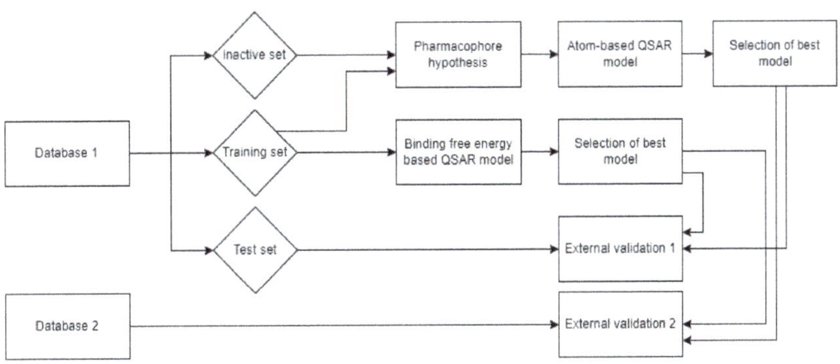

Figure 2. Workflow for the generation of ligand- and structure-based models for alkylhydrazide-based HDAC3 inhibitors.

2. Results and Discussion

2.1. Diversity Analysis of the Datasets

A dataset containing 63 compounds with an *N*-monosubstituted hydrazide scaffold was collected from the literature [47,49,51]. The 2D structures and IC_{50} values of all alkylhydrazides are shown in Table S1 (Supplement). The selected compounds cover a reasonable biological activity range (Table 1). All compounds were measured in vitro using recombinant human HDAC3 and the peptidic substrate (Boc-Lys(acetyl)-AMC). The fluorescence intensity was measured at excitation and emission wavelengths of 360 and 460 nm, respectively [47,49,51].

Table 1. Distribution of inhibitors in the training and test sets according to their HDAC3 IC_{50} values.

HDAC3 Dataset	Number of Compounds	$7 < pIC_{50}$ Highly Active	$5.3 < pIC_{50} < 7$ Moderately Active	$pIC_{50} < 5.3$ Inactive
Training	**39**	30	9	-
Test	**17**	11	6	-
Inactive	**7**	-	-	7
Total	**63**	41	15	7

We first grouped the compounds into three activity classes according to their HDAC3 inhibitory data (Table 1, Table S1 Supplement):

1. Highly active inhibitors showing $pIC_{50} > 7$
2. Moderately active inhibitors showing pIC_{50} between 5.30 and 7
3. Inactive inhibitors showing $pIC_{50} < 5.30$

The compounds were randomly divided into a training set (70%; 39 compounds) and a test set (30%; 17 compounds) using the "RAND" function in the MOE program (MOE–Molecular database calculator–RAND) [64]. The compounds either having no exact IC_{50} values or showing an IC_{50} value higher than 5 μM were classified as inactive (Table 1). The same training and test sets were used for the ligand- and structure-based model development studies. The training set was used to generate the models, while the independent test set was utilized to evaluate the predictive accuracy of the selected best models. The inactive set was only utilized for the validation of the pharmacophore models by calculating the inactive-survival score (detail in Section 2.2).

The applicability domains of the training and external test sets were analyzed by plotting the three most important principal components (PCA1, PCA2, and PCA3) [60,65] of the calculated descriptors (PEOE_VSA_HYD, GCUT_SLOGP_0, TPSA, b_single, lip_acc, lip_don, and vsa_hyd—explained in Table 2). The most important PCA of the molecular descriptors can explain about 100% of the original space. The PCA analysis showed that the training set and external test set were homogeneously distributed in the chemical space (Figure 3).

Table 2. List of selected molecular descriptors for the PCA analysis.

Abbreviations	Molecular Descriptors
PEOE_VSA_HYD	The partial equalization of orbital electronegativity (PEOE). Total hydrophobic van der Waals surface area
GCUT_SLOGP_0	The GCUT descriptors using atomic contribution to logP
TPSA	Polar surface area
b_single	Number of single bonds (including implicit hydrogens). Aromatic bonds are not considered to be single bonds
lip_acc	The number of O and N atoms
lip_don	The number of OH and NH atoms
vsa_hyd	Approximation of the sum of VDW surface areas of hydrophobic atoms.

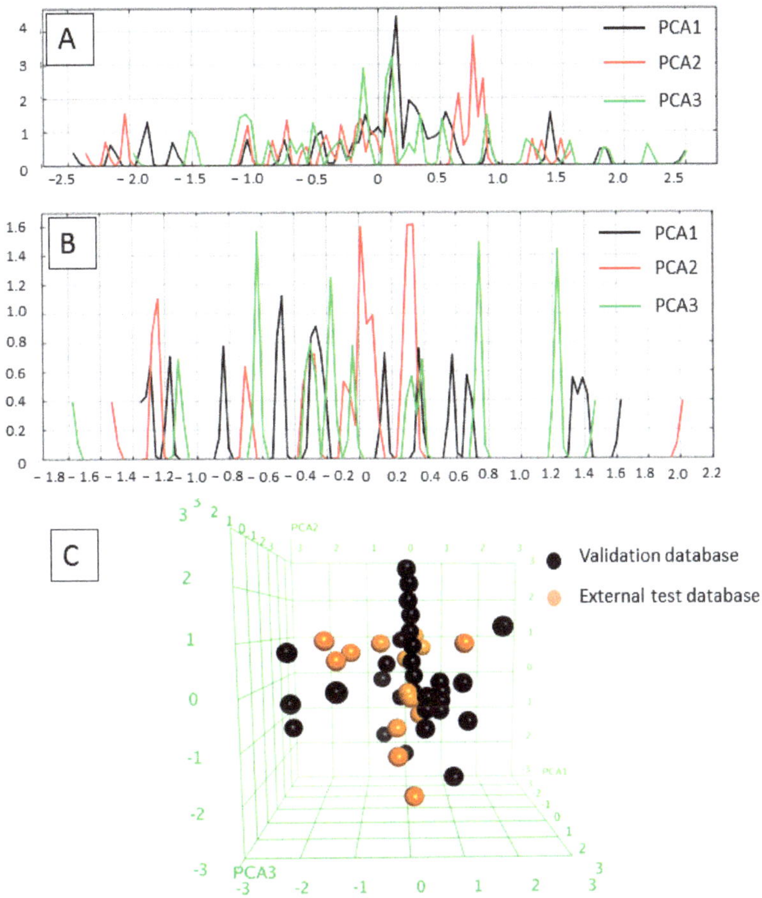

Figure 3. (**A**) Field histogram to visualize the variation of the three most important principal components for the training set. (**B**) Field histogram to visualize the variation of the three most important principal components for the test set. (**C**) Three-dimensional plot of the first three PCAs. The training set is colored black; the test set is colored orange.

2.2. Analysis of the Pharmacophore Model

An important step in establishing a 3D-QSAR model is the development of the correct alignment, usually based on a generated pharmacophore model. In the current work, the pharmacophore model was generated using the Phase module implemented in Schrödinger considering 30 active compounds ($pIC_{50} > 7$) [66]. Then, seven inactive compounds were used to analyze the ability of the generated models to discriminate between the active and inactive compounds.

The pharmacophore model shows the 3D (three-dimensional) structural features which might be essential for the biological activity [67,68]. Hence, the pharmacophore features that are common to the 30 active compounds showing a pIC_{50} more than 7 were investigated. In total, 38 pharmacophore hypotheses were generated and scored according to the survival score. The survival score was generated by evaluating how well the selected pharmacophore hypothesis fits to the most active inhibitors. Additionally, the Phase module penalizes the generated pharmacophore hypothesis that cannot discriminate the actives from inactives. Thus, the developed hypotheses were mapped onto the inactive compounds and scored to yield the inactive-survival score. Pharmacophore hypotheses which showed a better inactive score than survival score was discarded since it cannot discriminate between active and inactive compounds. For the selected pharmacophore hypothesis, all inactive compounds should show low fitness to the pharmacophore hypothesis.

After scoring the generated pharmacophore hypotheses, the best-scored pharmacophore model consisting of seven pharmacophore features (ADDDHRR–Figure 4A) was selected. The survival score (6.923) and the inactive score (1.688) of the hypothesis are shown in Table 3. The pharmacophore features were specified as the hydrogen-bond acceptor (A), the hydrogen bond donor (D), the hydrophobic (H), the negative ionic (N), the positive ionic (P), and the aromatic ring (R). It is worth noting that the less feature-based pharmacophores show weak discrimination between actives and inactives. Most inactive compounds showed a high fitness to the established pharmacophore features which led to an increase in the inactive score as shown for the DDDHRR and DDHR hypotheses (Table 3).

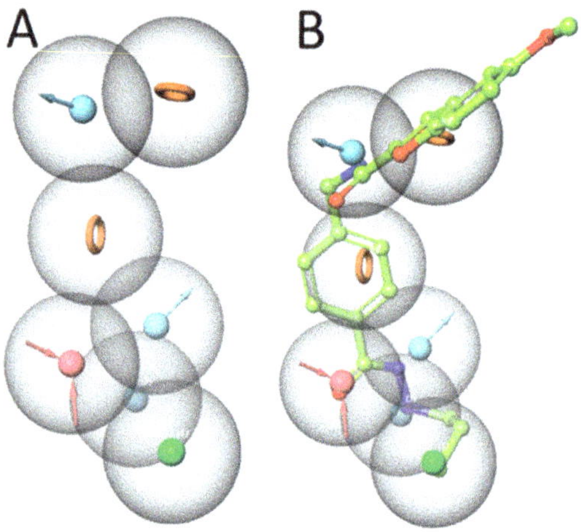

Figure 4. (**A**) Developed pharmacophore model (ADDDHRR) for HDAC3 inhibitors. (**B**) Superposition of compound **1** and the pharmacophore hypothesis (hydrogen bond acceptor—red color, hydrogen bond donor—cyan color, hydrophobic—green color, and ring—orange color).

Table 3. Calculated scores of the best performing pharmacophore hypotheses.

HYPO ID	Survival Score	Inactive Score
ADDDHRR	6.923	1.688
DDDHRR	6.464	1.711
DDHR	5.405	2.069

The generated pharmacophore model (ADDDHRR) was mapped onto the most active compound **1**. This pharmacophore model shows the importance of the hydrogen bond donor and acceptor functions of the hydrazide moiety (Figure 4B). The carbonyl oxygen of the hydrazide serves as a hydrogen bond acceptor, while the two nitrogen atoms serve as hydrogen bond donor groups. The alkyl chain shows hydrophobic features while the two aromatic ring systems are assigned as aromatic features. The amide moiety between the linker acts as a hydrogen bond donor via the amide-*NH* (details shown in the docking part). Accordingly, the selected ADDDHRR pharmacophore model shows the important structural features which can interact with the HDAC3 pocket.

In conclusion, the common pharmacophore features were determined using the active compounds in this step. The established pharmacophore model shows the required features for the binding to HDAC3. Since there is no reported X-ray structure of HDAC3 with an alkylhydrazide, the pharmacophore model gives an insight into the possible binding mode of alkylhydrazide derivatives.

2.3. Analysis of the Atom-Based 3D-QSAR Model

The atom-based 3D-QSAR model was built in Schrödinger19 using the 39 compounds in the training set [66–68]. Atom-based QSAR treats molecules as a set of overlapping van der Waals spheres. The spheres are divided into six categories: hydrogen bond donors; hydrophobic/non-polar; negative ionic; positive ionic; electron withdrawing; and miscellaneous [67,68]. The 3D-QSAR model enables us to consider all relevant structural features such as steric clashes as well as pharmacophores which play a role in the HDAC3 activity of the compounds. In this step, the previously selected seven-featured pharmacophore model (ADDDHRR) was used as an alignment rule for the generation of an atom-based QSAR model. First, 39 compounds were aligned to the pharmacophore model and then the atom-based 3D-QSAR models were generated and cross-validated. The best atom-based 3D-QSAR model was obtained with a good correlation coefficient (R^2: 0.95) and cross-validated correlation coefficient (Q^2: 0.88) (Table 4).

Table 4. The best performing atom-based 3D_QSAR model.

HDAC3 Model	N	SD	R^2	RMSE	Q^2
1	39	0.27	0.95	0.39	0.88

Abbreviations: SD (standard deviation of the regression), R^2 (correlation coefficient of the regression), RMSE (root mean square error of test set prediction), and Q^2_{LOO} (leave one-out cross-validation for the prediction values).

The atom-based 3D-QSAR techniques visualize the compounds as 3D (three-dimensional) based on the non-covalent protein–ligand interactions such as the hydrogen bond acceptor and donor, hydrophobic, and positive and negative ionic interactions. The model marks the favorable structural features with green cubes and unfavorable structural features with red cubes. To understand the most favorable and less favorable interactions, we analyzed all compounds from the training set. As examples, three compounds with low activity (compounds **35**, **36**, and **38**) and three compounds with good activity (compounds **1**, **2**, and **3**) from the training set were chosen to analyze the atom-based QSAR model (Figure S1, Supplement). According to the atom-based QSAR model, compound **35** exhibited poor activity due to its heptyl alkyl chain (Figure S1A, Supplement). As shown in Figure S1B (Supplement), the meta-substituent on the phenyl linker, as exemplified with compound **36**, showed an unfavorable effect on the HDAC3 activity. In the case of compound **38**, the thiophene ring showed unfavorable structural features, decreasing the HDAC3 activity (Figure S1C, Supplement). On the other hand, the propyl alkyl chain attached to the hydrazide group is favored for three active compounds (Figure S1D–F, Supplement). In addition to that, para-substituted cap groups are also favored and covered by green cubes. According to the model visualization, the least active compounds (Figure S1A–C, Supplement) are mainly covered by red cubes, while the more active compounds, especially the cap groups (Figure S1D–F, Supplement), are mostly covered by green cubes.

The external validation was performed using a test set which was not used for model generation, with the aim of evaluating the predictive accuracy and reliability of the generated atom-based QSAR model. The scatter plot of the training and test set is shown in Figure 5. The prediction results of the training, test, and inactive databases are shown in Table S2.

Analysis of the test set revealed that the atom-based QSAR model predicts the active inhibitors well, with differences less than 1 log unit. However, several of the moderately active inhibitors (compounds **51**, **53**, **55**, and **56**) in the external test set as well as the seven inactive compounds were predicted, with differences of more than 1 log unit (Table 5, Table S2 Supplement). The atom-based QSAR model classified most of the moderately active and inactive inhibitors into the active class. Due to the limited accuracy of the atom-based models in correctly predicting the inactives/weakly actives, we tried to overcome this by generating structure-based prediction models. For this, we applied the docking and binding free energy calculation techniques discussed in the next section.

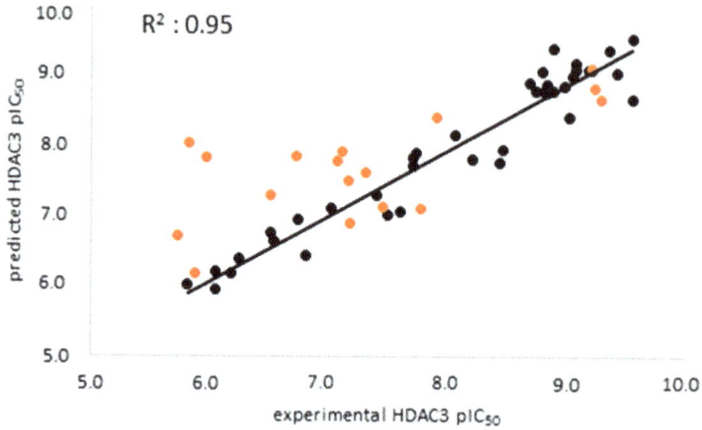

Figure 5. Scatter plot of the best atom-based QSAR model for the training set (black) and the test set (orange).

Table 5. Prediction results of the test set compounds (atom-based QSAR model).

Compound Number	pIC_{50} HDAC3	Prediction by Atom-Based QSAR	Difference (Experimental—Predicted Activity) Atom-Based	References
40	9.29	8.69	0.60	[47]
41	9.24	8.87	0.37	[47]
42	9.21	9.14	0.07	[47]
43	7.90	8.43	0.53	[49]
44	7.78	7.13	0.64	[49]
45	7.46	7.17	0.29	[49]
46	7.31	7.65	0.34	[49]
47	7.17	6.92	0.24	[51]
48	7.16	7.54	0.38	[51]
49	7.11	7.96	0.85	[49]
50	7.07	7.81	0.74	[51]
51	6.73	7.88	1.16	[51]
52	6.51	7.32	0.81	[51]
53	5.96	7.85	1.89	[51]
54	5.87	6.19	0.33	[51]
55	5.81	8.08	2.26	[51]
56	5.72	6.74	1.01	[51]

2.4. Analyzing the Binding Mode of Alkylhydrazides in HDAC3

We started with docking all inhibitors to HDAC3 (PDB ID: 4A69 [69]) (Figure S2, Supplement). We used a docking set-up in Glide which we previously validated for HDAC inhibitors from different chemical series [31,34,53]. To date, there is no crystal structure of an HDAC in complex with an alkylhydrazide, but we have recently shown that alkylhydrazides similar to inhibitors 1 and 47 (Table S1, Supplement) [47,51] are reversible and substrate competitive inhibitors of HDACs [53]. Thus, we docked the alkylhydrazides into the catalytic pocket of HDAC3 and analyzed whether they are able to chelate the catalytic zinc ion. The analysis of the docking results of the active inhibitors, as exemplified by compounds **1** and **2** from the training set (Figure 6), showed that the hydrazide moiety chelates the zinc ion in a bidentate manner through its nitrogen and carbonyl oxygen and exhibits conserved hydrogen bond interactions with H134, H135, and Y298 at the bottom of the catalytic pocket. The aromatic linker group was placed into the hydrophobic tunnel consisting of F144, H172, F200, and L266, where it undergoes aromatic pi–pi interactions with F144 and F200. The cap group interacts with residues at the surface by forming hydrogen bond interactions with D93 as well hydrophobic interactions with H22 and P23 in HDAC3. A structural difference which influences the potency and selectivity on HDAC3 is observed in the foot pocket region. According to the docking results, the propyl and butyl chains of the alkylhydrazides fit well into the foot pocket of HDAC3. However, replacing the propyl or butyl chains by pentyl or longer side chains resulted in a dramatic decrease in HDAC3 activity due to the steric hindrance observed in HDAC3. The Y107 residue pushes L133, resulting in a narrower foot pocket region [31,34]. Hence, the pentyl and longer alkyl chains in the foot pocket region show steric clashing with M24 and L133, causing a decrease in or loss of HDAC3 activity.

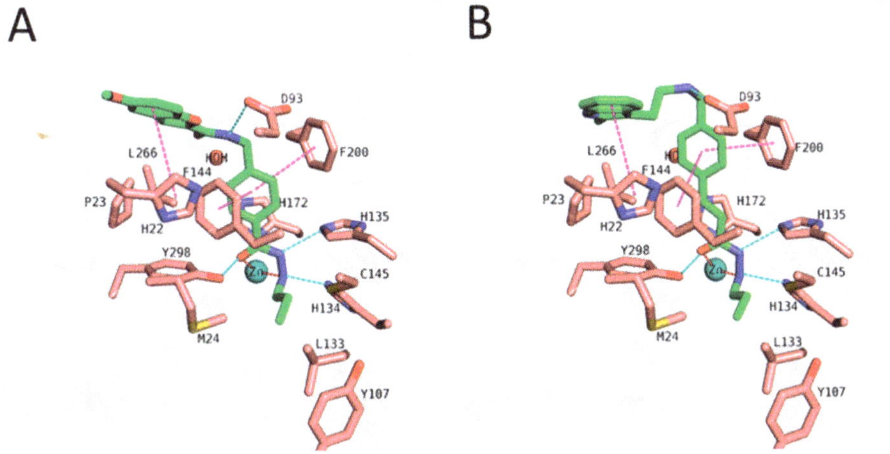

Figure 6. Docking poses of **1** ((**A**), green-colored sticks), **2** ((**B**), orange-colored sticks) in HDAC3 (PDB ID: 4A69). The hydrogen bonds (cyan dashed lines), hydrophobic interactions (magenta dashed lines), and metal coordination (red dashed lines) between the inhibitors and the protein are shown. Relevant residues are shown in stick representation with salmon carbon atoms in HDAC3. The zinc ion is shown as a cyan-colored sphere. The conserved water molecule is shown as a red sphere.

Although the docking poses show reasonable binding modes in the HDAC3 active site, the correlation between the docking scores and pIC_{50} values was poor ($R^2 = 0.28$ for HDAC3). Thus, we rescored the docking poses by calculating the binding free energies.

In addition to the docking results, we checked the stability of the predicted interactions of the potent inhibitors 1 and 2 with the binding site using 100 ns MD simulation (Figures 7, 8, S3 and S4, Supplement). MD simulation analysis of compounds **1** and **2** revealed that the *n*-propyl chain attached to the hydrazide fit into the foot pocket consisting of M24 and L133. Notably, M24 and L133 play a key role as a gate keeper in the foot pocket region of HDAC3. M24 and L133 closed the foot pocket and made the volume narrower where only a propyl or butyl side chain favorably fit. This conformational change of M24 and L133 might explain the decrease in or loss of HDAC3 activity of the compounds with longer alkyl chains than butyl and propyl. The zinc-binding group which is the common part of compounds **1** and **2** preserves its bidentate coordination and undergoes hydrogen bond interactions with H134, H135, and Y298 throughout the MD simulation. Furthermore, the linker groups of compounds **1** and **2** remain sandwiched between F144 and F200. Besides these similar protein–ligand interactions of compounds **1** and **2**, the MD simulation analysis displayed some differences in the cap region of compounds **1** and **2**.

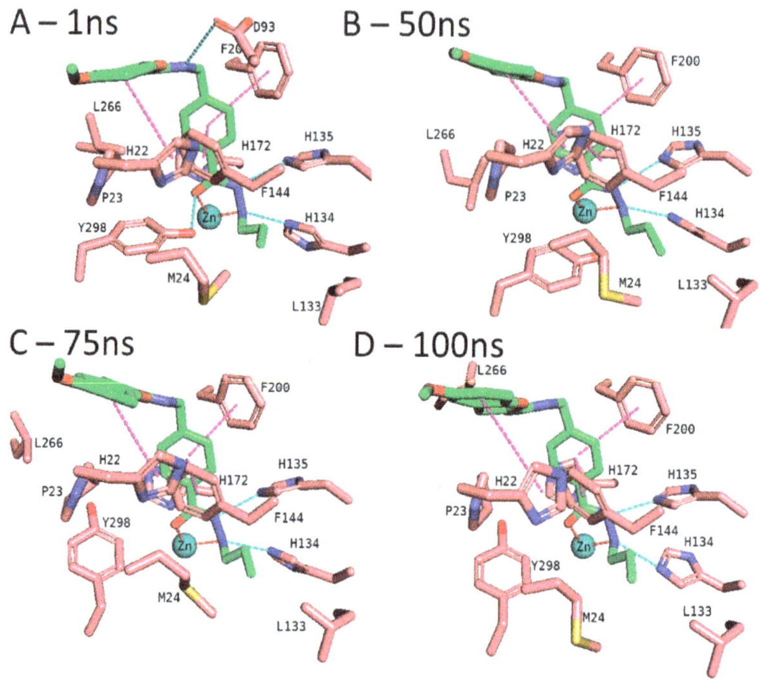

Figure 7. MD frames of the HDAC3-1 complex. (**A**) The frame at 1 ns MD simulation, (**B**) the frame at 50 ns MD simulation, (**C**) the frame at 75 ns MD simulation, and (**D**) the frame at 100 ns MD simulation. The hydrogen bonds (cyan dashed lines), hydrophobic interactions (magenta dashed lines), and metal coordination (red dashed lines) between the inhibitors and the protein are shown. Relevant residues are shown in stick representation with salmon carbon atoms in HDAC3. The ligand is shown in stick representation with green carbon atoms. The zinc ion is shown as a cyan-colored sphere.

According to the MD simulation of compound **1**, the selected docking pose was stable during the 100 ns MD simulation (Figures 7 and S3, Supplement). Throughout the MD simulation, the ligand maintained the predicted binding conformation, albeit two of the predicted interactions were lost, namely the hydrogen bond interaction between the amide group and D93 as well as the interaction between the hydrazide-carbonyl-O and Y298, due to the flexibility of the latter residue (Figure 7). The hydrogen bond distances of the HDAC3-inhibitor 1 complex throughout the 100 ns MD simulations were analyzed and plotted in Figure S4 (Supplement). No significant fluctuation was observed for the benzofuran cap group of compound **1**, which remains embedded in a hydrophobic pocket and undergoes aromatic interaction with H22 at the surface of the protein.

In the case of compound **2** (Figure 8), the flexible 2-methylindole cap group showed conformational changes. Hence, the ligand RMSD of compound **2** showed higher fluctuations (Figure S5, Supplement). During the MD simulation, the 2-methylindole group showed two different orientations: between 40–60 ns of the MD simulation, the cap group adopts an orientation where it undergoes edge-to-face interaction with F144. For the rest of the simulation time, the cap group showed the initially observed position and interacted with H22. In contrast to compound **1**, Y298 showed less fluctuation and maintained its interaction with compound **2**. Throughout the 100 ns MD simulation, compound **2** showed stable binding and maintained its bidentate chelation with the zinc ion. The hydrogen bond distances of the HDAC3-inhibitor 2 complex throughout the 100 ns MD simulations were analyzed and plotted in Figure S6 (Supplement).

In conclusion, since so far no X-ray structure has been released of an HDAC with an alkylhydrazide inhibitor complex, we docked the compounds to HDAC3 to examine the putative binding mode and to rationalize the observed SAR. Additionally, the observed proteinligand interactions were analyzed by MD simulations. The interaction at the catalytic pocket was found to be highly stable whereas some fluctuation was observed for the flexible capping groups that are located at the solvent-exposed part of the binding pocket. To provide further support for the predicted binding poses of the alkyl hydrazides, we previously examined the substrate competition of two alkyl hydrazides for the related class I member HDAC8 and confirmed that they reversibly inhibit and exhibit competitive substrate binding. However, cocrystal structures of HDACs with alkylhydrazide-based inhibitors have to be obtained to confirm the modeling results.

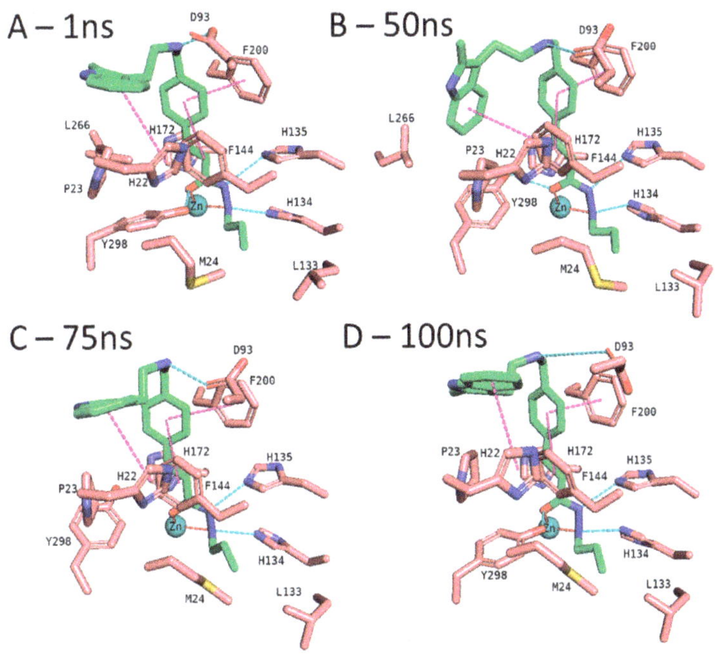

Figure 8. MD frames of the HDAC3-2 complex. (**A**) The frame at 1 ns MD simulation, (**B**) the frame at 50 ns MD simulation, (**C**) the frame at 75 ns MD simulation, and (**D**) the frame at 100 ns MD simulation. The hydrogen bonds (cyan dashed lines), hydrophobic interactions (magenta dashed lines), and metal coordination (red dashed lines) between the inhibitors and the protein are shown. Relevant residues are shown in stick representation with salmon carbon atoms in HDAC3. The ligand is shown in stick representation with green carbon atoms. The zinc ion is shown as a cyan-colored sphere.

2.5. Binding Free Energy Calculation

Due to the low correlation between the docking scores and pIC_{50} values, rescoring of the selected docking poses was performed using the MM/GBSA method in AMBER16 [70]. The total energies of HDAC3–inhibitor complexes were calculated using four different parameter settings (solvation models) and six different frame settings (see the Methods part for details). The same training set including the 39 compounds that was used for the atom-based 3D-QSAR model was also used for model generation based on the calculated binding free energies of the compounds. In total, 24 models were generated. The models were assessed based on the correlation coefficients (R^2) between the biological data and the calculated energy values, taking into account Tropsha's criteria for reliable QSAR models [71] (Figure 9). The prediction results of the training, test set, and inactive compounds are shown in Table S3 (Supplement).

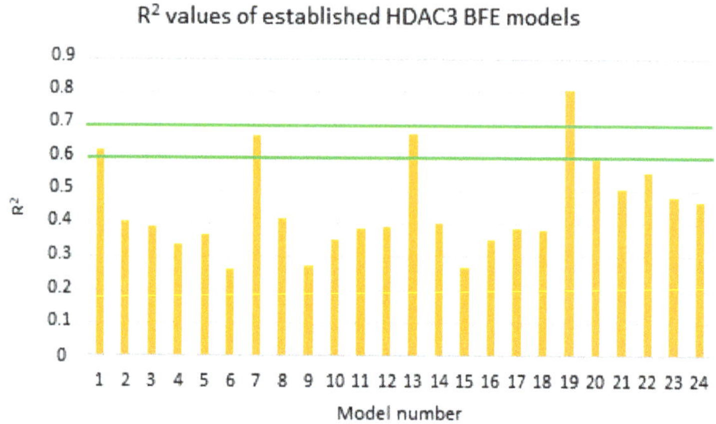

Figure 9. Calculated R^2 values of the established models for alkylhydrazides (training set: orange bars). Green lines highlight the 0.6 and 0.7 R^2 values.

According to the Tropsha criteria [71], a good QSAR model should abide by the following rules; $R^2 > 0.6$ and $Q^2 > 0.5$. Based on the mentioned rule, models showing a calculated R^2 value > 0.6 were considered for further statistical analysis (MODEL1, 7, 13, and 19). Interestingly, the four selected models are all based on the protein–ligand complex obtained with one minimization step (Emin1, Table 6). Further internal validation of the selected models was analyzed using the leave-one-out (LOO) method, 3-fold-cross-validation (cv), and 10-fold cv.

Table 6. Best performing BFE models.

							LOO CV		3-Fold CV		10-Fold CV	
Model Number	**N**	**Method**	**Frame**	**2D Descriptor**	R^2	**RMSE**	Q^2	**QMSE**	Q^2	**QMSE**	Q^2	**QMSE**
MODEL1	39	GB1	Emin1	-	0.63	0.69	0.58	0.73	0.60	0.74	0.60	0.73
MODEL7	39	GB2	Emin1	-	0.66	0.65	0.63	0.69	0.64	0.69	0.64	0.69
MODEL13	39	GB5	Emin1	-	0.67	0.65	0.64	0.68	0.65	0.69	0.65	0.68
MODEL19	39	GB8	Emin1	-	0.81	0.49	0.78	0.52	0.78	0.54	0.77	0.53
MODEL19_1	39	GB8	Emin1	PEOE_VSA_HYD	0.87	0.40	0.84	0.44	0.85	0.45	0.83	0.45

Abbreviations: R^2 (correlation coefficient), RMSE (root mean square error), Q^2_{LOO} (leave one-out cross-validation), QMSE (crossed-root mean square error), and Emin1 (single frame after the first energy minimization step).

The four selected models showed $R^2 > 0.6$ and $Q^2 > 0.5$. Among the selected four models, the GB8 (GBNeck) implicit solvation model outperformed the other methods (GB^{HCT} refers to GB1, and GB^{OBC} refers to GB2-5 in the article). The reason might be that the GBNeck model (referred to GB8) was generated to correct the van der Waals surface that is inaccessible to water [72]. This improvement in GB8 might help to obtain better results for the compounds used in this article. Model 19 based on the GB8 implicit solvation model showed the highest R^2 and Q^2 values (LOO-method) with 0.81 and 0.78, respectively, and the lowest RMSE and QMSE values with 0.49 and 0.52, respectively (Figure 10 and Table 6). In addition, we tested whether the inclusion of a 2D descriptor for the shape/electronic properties of the inhibitors could improve the models. Two-dimensional descriptors were computed for all compounds in MOE [64]. All available 2D descriptors were then assessed for their ability to improve the model. The total hydrophobic van der Waals surface area

(PEOE_VSA_HYD) gave the best improvement. The combination of this 2D descriptor and energy term improved the R^2 from 0.81 to 0.87 and the Q^2 (LOO) from 0.78 to 0.84. In addition, this model (MODEL19_1) exhibited lower RMSE and QMSE values compared to MODEL19 (Table 6).

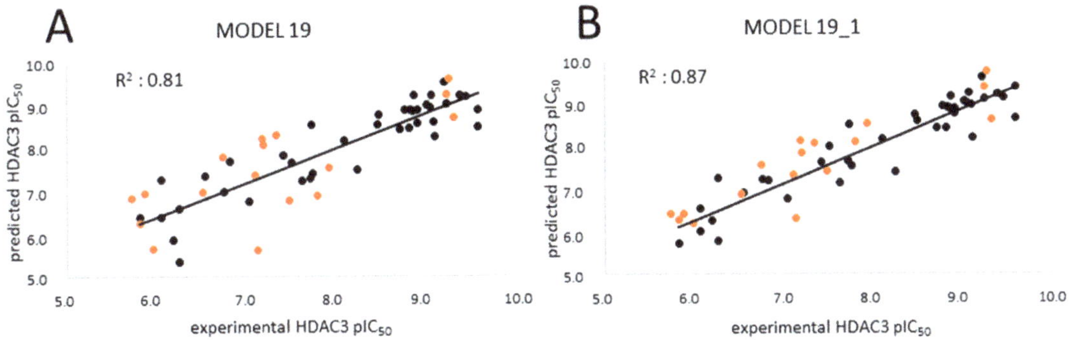

Figure 10. Correlation plots of (**A**) MODEL19 and (**B**) MODEL19_1 showing correlations between the predicted data and experimental data for HDAC3. Training set (black); external test set (orange).

The test set was used to evaluate the accuracy of the best generated model (Model19_1). In the test set (Table 7), all compounds were predicted with less than 1 log unit difference. Additionally, the prediction of the inactive compounds was more satisfying compared to the previously described atom-based QSAR models, with a difference of less than 1 log unit except for compound **61** (Table S3, Supplement). For compound **61**, the docking poses could not explain the incorrect prediction. The scatter plot and prediction results are shown in Figure 10 and Tables 7 and S3 (Supplement), respectively.

Table 7. Prediction results of the test set compounds using the BFE model (MODEL19_1).

Compound Number	pIC_{50} HDAC3	Prediction of BFE	Difference (Experimental—Predicted Activity)	References
40	9.29	8.60	0.70	[47]
41	9.24	9.75	0.51	[47]
42	9.21	9.37	0.17	[47]
43	7.90	8.53	0.63	[49]
44	7.78	8.10	0.33	[49]
45	7.46	7.42	0.03	[49]
46	7.31	8.06	0.75	[49]
47	7.17	7.85	0.68	[51]
48	7.16	8.12	0.96	[51]
49	7.11	6.30	0.81	[49]
50	7.07	7.32	0.25	[51]
51	6.73	7.56	0.83	[51]
52	6.51	6.89	0.38	[51]
53	5.96	6.19	0.22	[51]
54	5.87	6.41	0.55	[51]
55	5.81	6.27	0.46	[51]
56	5.72	6.42	0.69	[51]

2.6. *Evaluation of the Generated Models on Newly Designed Compounds*

The created models, the atom-based 3D QSAR model and Model 19_1, and the generated docking poses were then used to predict alkyl hydrazides with other linkers and capping groups that were synthesized (chemistry and in vitro testing were published elsewhere [53]). These compounds were designed starting from compound **47** (Table S1, Supplement), where several structural modifications were introduced to extend the SAR on this series of HDAC3 inhibitors. In the first series of compounds, the effect of a different length for the alkyl side chain was evaluated. In the second series of compounds, different substitutions on position 3 or 4 of the phenyl ring were introduced. In the next series of compounds, the aminopyrimidine linker group with different *N*-arylmethyl, *N*-arylethyl, or *N*-ethylpiperazinyl moieties were tested, while in the last series of compounds, piperazinyl-piperidine linker groups were attached to the phenylalkylhydrazide core of the compounds. The general structures of the new inhibitors are summarized in Figure 11. The experimentally determined HDAC3 IC_{50} values and the prediction results are shown in Tables 8, S5 and S6 (Supplement, all 2D structures are summarized in Table S4, Supplement).

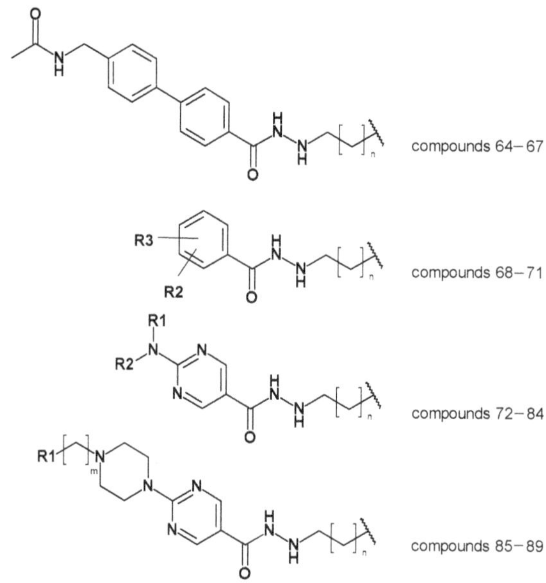

Figure 11. General structures of the newly synthesized compounds [53].

First, the ligand-based models; i.e., the pharmacophore models and 3D_atom-based QSAR models, were tested. The developed pharmacophore model was used to align the 26 new compounds to apply the atom-based QSAR model.

Analyzing the atom-based prediction results of the 26 new compounds revealed interesting results (Table 8). Based on the atom-based QSAR model prediction results, the absolute difference between the experimental and predicted pIC_{50} of the nine compounds, which are moderately active or inactive, was more than >1 log unit, i.e., these compounds were predicted to be more active than experimentally determined. The atom-based QSAR model makes predictions based on the effect of electron-withdrawing groups, electron-donating groups, and hydrophobic groups of compounds considering the created pharmacophore hypothesis. However, in the case of HDAC3, the shape of the foot pocket plays an important role in the inhibitor activity. Due to the smaller volume of the HDAC3 foot pocket, compounds with alkyl groups longer than butyl are moderately active or inactive as determined by the in vitro results. Atom-based QSAR models do not take the pocket volume into account, hence resulting in the observed weak prediction of these derivatives. In conclusion, the atom-based QSAR model showed a weak discriminatory power.

Table 8. Experimental and predicted activities of the BFE-based and atom-based models.

Compound Number	pIC_{50} HDAC3	Prediction by Atom-Based QSAR	Difference (Experimental—Predicted Activity) Atom-Based	Prediction by BFE Model	Difference (Experimental—Predicted Activity) BFE
64	7.04	7.03	0.01	7.16	−0.12
65	6.46	6.59	−0.13	6.89	−0.43
66	5.82	6.57	−0.75	6.36	−0.54
67	<5.00	6.43	<−1.43	5.41	-
68	5.80	7.25	−1.46	6.69	−0.89
69	<5.00	7.05	<−2.05	5.07	-
70	<5.00	6.98	<−1.98	3.92	-
71	9%@1 μM	6.82	-	0.24	-
72	7.37	7.85	−0.49	7.60	−0.23
73	6.70	7.29	−0.59	6.68	0.02
74	41%@1 μM	7.39	-	7.29	-
75	7.09	7.82	−0.73	8.06	−0.97
76	7.22	7.12	0.10	6.86	0.36
77	7.43	7.96	−0.53	7.06	0.38
78	7.24	7.03	0.20	7.98	−0.75
79	6.92	6.80	0.12	6.32	0.60
80	6.96	8.10	−1.14	7.42	−0.46
81	<5.00	6.54	<−1.54	1.65	-
82	<5.00	6.86	<1.86	1.61	-
83	5.52	7.25	−1.73	3.26	2.26
84	<5.00	7.61	<−1.67	2.69	-
85	7.52	7.63	−0.10	7.64	−0.11
86	7.00	7.18	−0.18	7.29	−0.29
87	6.52	6.52	0.00	6.86	−0.33
88	6.00	7.40	−1.40	6.13	−0.13
89	5.85	7.13	−1.28	6.11	−0.26

Then we evaluated the binding-free-energy-based prediction results (Table 8). Initially, all 26 compounds were docked to HDAC3 using the same protocol as for the training set. Similar docking poses in HDAC3 were obtained for all 26 new compounds as obtained for compounds **1** and **2** (exemplified in Figure 12). The docking studies showed that the hydrazide moiety as well as the aromatic linker groups of all 26 compounds exhibited similar interactions as observed for compounds **1** and **2**. A bidentate chelation between the zinc ion and hydrazide moiety was observed for all compounds. In addition, the hydrazide moiety showed hydrogen bonds with H134, H135, and Y298 in the zinc-binding region of the HDAC3 catalytic pocket. The aromatic linker groups were accommodated into the hydrophobic tunnel and interacted with F144 and F200, showing pi–pi interactions. Meanwhile, the cap groups and foot-pocket-targeting groups showed significant differences which has an impact on the HDAC3 activity.

In the first series, exemplified by compound **64** (Figure 12), the acetoamidomethyl cap group was placed at the entrance of the pocket and showed hydrogen bond interactions with D93. The different length of the hydrazide alkyl side chain resulted in a significant difference in HDAC3 activity. Compound **64** possessing a propyl side chain was predicted to be more active than **65**, **66**, and **67**, which is in line with the experimentally determined data. The difference between the experimental and predicted values is indeed less than <1 log unit for this series. Moreover, compound **67** with a hexyl side chain was predicted to be inactive. This result confirms that the BFE model is sensible to the side chain effects on this dataset.

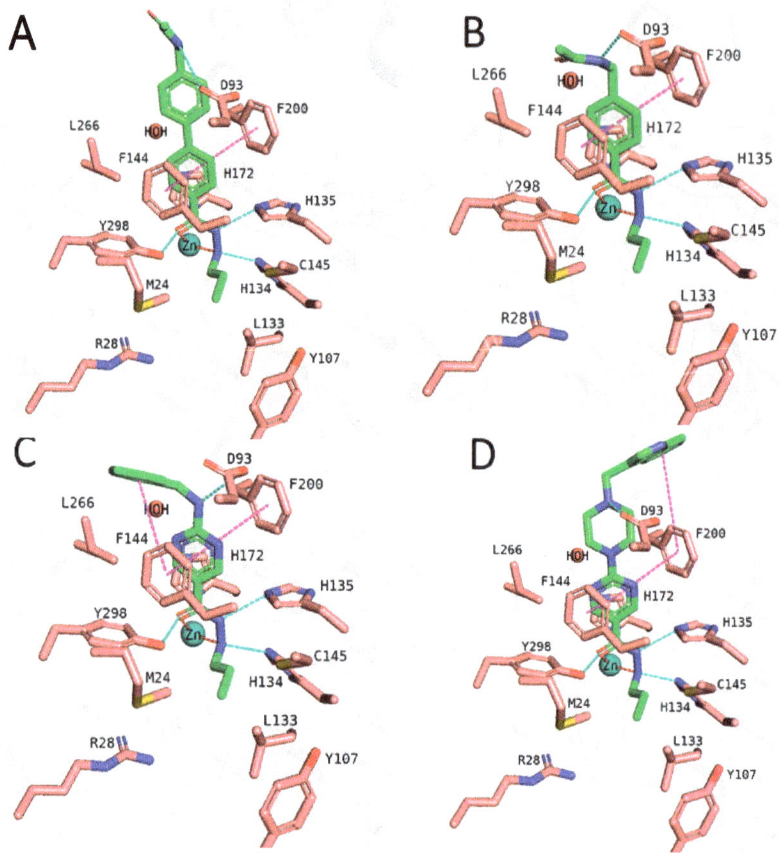

Figure 12. (**A**) Docking pose of compound **64**, (**B**) docking pose of compound **68**, (**C**) docking pose of compound **72**, and (**D**) docking pose of compound **85**. The hydrogen bonds (cyan dashed lines), hydrophobic interactions (magenta dashed lines), and metal coordination (red dashed lines) between the inhibitors and the protein are shown. Relevant residues are shown in stick representation with salmon carbon atoms in HDAC3. The ligand is shown in stick representation with green carbon atoms. The zinc ion is shown as a cyan-colored sphere.

In the second series, only compound **68** (Figure 12) bearing the acetamidomethyl cap group and propyl side chain in the foot-pocket-targeting region showed moderate activity on HDAC3. The other compounds **69**, **70**, and **71** with a hexyl side chain, did not show significant activity on HDAC3. Similar to the experimentally determined activity, the BFE model also predicted compound **68** as a moderate inhibitor, while **69**, **70**, and **71** were predicted as inactive compounds.

In the third series of compounds bearing an aminopyrimidine moiety as a linker group, different *N*-arylmethyl, *N*-arylethyl, or *N*-methylpiperaziniyl moieties as capping groups were tested. This series is exemplified by compound **72** (Figure 12). All compounds bearing a propyl side chain except compound **74** were predicted with less than <1 log unit compared to the experimentally determined activities. Interestingly, compound **74** bearing an *N*-arylethyl cap group did not show significant inhibitory activity (41%@1µM) on HDAC3; however, it was predicted by the model as an active inhibitor. The flexible cap group could be the reason for the reduced activity on HDAC3. Finally, compounds **81–84** possessing a hexyl side chain were predicted as inactive inhibitors which is in line with the experimental findings.

The last series of compounds bearing indole or *N*-methylindole groups attached via a methyl or ethyl linker to the piperazinyl–pyrimidine scaffold were predicted close to the experimental activities. The difference between the experimental and predicted values is less than <1 log unit. The docking poses of this series are exemplified by compound **85** (Figure 12).

In conclusion we applied the validated BFE models to the further development of the alkylhydrazide-based class I HDAC inhibitors. The best inhibitors from this series were also tested for their immunmodulatory effects in Jurkat cells and showed promising cellular effects [53]. As we have recently demonstrated a potent T cell memory response by combined class I HDAC inhibition and immune-checkpoint blockade in hepatocellular carcinoma (HCC) therapy, the new alkylhydrazides represent an interesting class of inhibitors to explore their potential for cancer therapy.

3. Materials and Methods

3.1. Ligand Database Preparation

A ligand dataset of 63 compounds with hydrazide as the zinc-binding group (ZBG) was collected from the literature [47,49,51]. Only compounds having an *N*-monosubstituted hydrazide scaffold were considered. The IC_{50} values of the selected compounds were retrieved from three publications and they all were determined against HDAC3 using the same fluorogenic substrate (Boc-Lys(acetyl)-AMC (amino methyl coumarin)). The same human recombinant HDAC3 enzyme was used for the in vitro studies [47,49,51]. The compounds were prepared in ligprep tool using the OPLS3e forcefield in Schrödinger suite [73]. Subsequently, the output of the ligprep step was submitted the Confgen to generate 64 conformers per ligand while minimizing the output conformers using the OPLS3e forcefield [73,74]. The compounds were automatically divided into a training (70%) and external test set (30%) using the "RAND" function in the MOE program (MOE–Molecular database calculator–RAND) [64]. The same training set and external test set were used for the model development studies. The compounds with no exact IC_{50} values were considered as inactive. The QSAR models were built using the most active and moderate inhibitors for which exact IC_{50} values were available.

The diversity analysis of the compounds was performed by analyzing the three most important principal components using the principal component analysis (PCA) implemented in MOE [60,64,65]. The 2D descriptors were computed in MOE [64]. Several 2D descriptors were selected using the Contingency tool in MOE. The three most important principal components (PCA1, PCA2, and PCA3) were calculated using the selected 2D descriptors. These principal components were used to check the diversity of the compounds.

The 26 compounds were collected from the article published by our group to evaluate the established models and check their reliability in different datasets [53]. The ligands were prepared using the same protocol as used for the validation set.

3.2. Pharmacophore Model

The pharmacophore model was established using 30 inhibitors with IC_{50} values lower than 100 nM in the training set and the 7 inactive compounds in the Phase module of Schrödinger [66]. The compounds were prepared in ligprep using the OPLS3e forcefield in the previous step [73,74]. The conformational search was performed in the Phase module by adjusting 64 conformers per compound and minimizing the output conformers using the "Develop Pharmacophore model" module in Schrödinger [66]. The common pharmacophore hypotheses were developed, scored, and ranked. The selected pharmacophore model was used as an alignment rule for the atom-based 3D-QSAR model.

3.3. Atom-Based 3D-QSAR Model

The ligand-based 3D-QSAR model was generated using the training dataset in the Phase module of Schrödinger [66]. The 39 compounds in the training database were aligned using the selected pharmacophore hypothesis from the previous step. The QSAR models were built with four latent factors and 1.0 Å grid spacing as well as the leave-one-out-cross-validation approach. The generated models were evaluated by means of standard deviation of the regression (SD), R^2 (correlation coefficient of regression), RMSE (root mean square error of test set prediction), and Q^2 (cross-training of test set prediction).

3.4. Docking Study

The hydroxamic acid scaffold and hydrazide scaffold are structurally similar groups. Therefore, the X-ray crystal structures of HDAC2 (PDB ID: 4LXZ [35]) and HDAC3 (PDB ID: 4A69 [69]) were retrieved from the Protein Data Bank (PDB, rcsb.org [(accessed on 20 May 2022) [75]) and analyzed in MOE [64]. SAHA with a hydroxamic acid scaffold in complex with the HDAC2 protein (PDB ID: 4LXZ) was defined as a pan-HDAC inhibitor and showed activity on HDAC3 [35]. First, HDAC2 (PDB ID: 4LXZ) and HDAC3 (PDB ID: 4A69) were superposed in MOE [64]. Then, SAHA was transferred from the HDAC2 protein (PDB ID: 4LXZ) to HDAC3 to mimic the induced fit effect of the zinc-binding group.

The HDAC3–SAHA complex was prepared in the protein preparation wizard of Schrödinger's suite by adding hydrogen bonds and missing side chains and assigning the bond orders [73]. The water molecules (except W2083) and ions (except Zn^{+2} ions) were deleted. The protonation states and tautomers were optimized at pH 7.4 using the PROPKA tool. The optimized complex was minimized using the OPLS3e force field to remove the steric clashes [74].

Molecular docking studies were carried out by applying the standard precision (SP) mode in Glide implemented in Schrödinger Suite [73]. The grid box including the information on the active site coordinates of the proteins was defined with a 10 Å radius around the ligand. Ten docking poses were employed for further post-docking minimization. The other settings were kept as the default. The docking results were visually analyzed in the MOE program [64].

3.5. Molecular Dynamics Simulation

The selected docking poses of compounds **1** and **2** in complex with HDAC3 (PDB ID: 4A69) were subjected to a 100 ns MD simulation in AMBER16 [70]. The Antechamber package was used to prepare the topologies, force field parameters, atom types, and bond types by applying the semi-empirical Austin Model1 with bond charge correction (AM1-BCC) [76,77]. Then, the tLEaP module was employed to prepare the protein–ligand complexes. General amber force field (GAFF), the Duan force field (ff03.r1), and 12-6-4LJ ionic model were used for the ligand, protein, and zinc, respectively [78–81]. The system was solvated by the TIP3P water model and a margin of 10 Å. Two minimization steps including the two sub-steps in each minimization were carried out. In the first step, 4000 iterations (2000 cycles of steepest descent and then 2000 of the conjugate gradient) were performed, while the protein residues, ligand, and zinc ion were restrained to their initial geometries (force constant of 10 kcal*mol^{-1}* $Å^{-2}$) to relieve the bad contacts. In the second step, 4000 iterations (2000 cycles of steepest descent and then 2000 of the conjugate gradient) were performed to remove the steric clashes in the entire complex. The restraint on the protein, ligand, and zinc were removed during the second minimization. Then, the system was heated at 300 K through 100 ps of MD. The protein–ligand complex was restrained to prevent large structural deviations (force constant of 10 kcal*mol^{-1}* $Å^{-2}$). The SHAKE algorithm was activated to constrain bonds involving hydrogens [82]. Finally, the system was equilibrated within a period of 200 ps. Langevin dynamics was applied to keep the temperature at 300 K with a collision frequency of 2 ps [83]. The pressure was kept at 1 bar using isotropic position scaling with a relaxation time of 2 ps. Afterwards, a 100 ns MD simulation was run with a time step of 2 fs using the same conditions as in the equilibration step. A non-bonded cut-off distance of 10 Å was used. The electrostatic interactions were calculated by applying the particle mesh Ewald (PME) method. After the MD simulation, CPPTRAJ module of AMBER was used to analyze the MD snapshots.

3.6. Binding Free Energy Calculation

The binding free energies (BFE) of the prepared protein–ligand complexes were calculated using the AMBER16 program [70]. The MMPBSA.py script was utilized for the calculations [84]. Different implicit solvent models (GB ^{HCT} (igb = 1), GB ^{OBC} (igb = 2), GB ^{OBC2} (igb = 5), and GBn (igb = 8)) were tested [85–87]. Molecular mechanics (MM) and solvent models were combined for the MMGBSA calculations [88–90]. Short 2 ns MD simulation was performed for all BFE calculations. The results of BFE were analyzed using the following six different methods: (1) a single frame at the first minimization step (Emin1), (2) a single frame at the second minimization step (Emin2), (3) a single frame at the third minimization after MD (Emin3), (4) 1–50 frames during MD (MD-1) with an interval of 5, (5) 51–100 frames during MD (MD-2) with an interval of 5, and (6) 101–500 frames during MD (MD-3) with an interval of 5. The correlation between biological activity and the energy results was measured by using the QSAR tool in the MOE program [64].

4. Conclusions

In the current study, we have evaluated several QSAR models including ligand-based and structure-based techniques to understand the structure–activity relationship of alkylhydrazides developed as HDAC3 inhibitors. Additionally, the binding modes of the two most potent HDAC3 inhibitors (compounds **1** and **2**) were verified through 100 ns MD simulation since there is no X-ray structure crystallized with an alkylhydrazide derivative. With the aid of ligand-based and structure-based approaches, in-house computational models have been developed for the prediction of the HDAC3 inhibitory activity of the alkylhydrazide scaffold.

The ligand-based models enabled us to obtain a general overview of the binding of the compounds in the HDAC3 protein. The established pharmacophore model and atom-based 3D-QSAR model can be used to filter the big databases. Since the shape of the foot pocket of HDAC3 has a crucial impact on the HDAC3 inhibitory activity, the predictive power of the ligand-based models was not satisfactory. These models predicted moderate inhibitors and inactive compounds as active compounds, although they predicted the actives as actives. Thus, these methods should be used to reduce the number of compounds in the big database.

The weakness of the ligand-based methods directed us to generate the structure-based methods. The binding mode of the alkylhydrazide was predicted by docking. The selected binding modes from the validation set (compounds **1** and **2**) were verified by the 100 ns MD simulation. The analysis of the MD simulation revealed that M24 and L133 are gatekeepers in the foot pocket of HDAC3. Additionally, the alkylhydrazides kept their bidentate chelation to the zinc ion during the 100 ns MD simulation. The compounds were rescored by means of BFE calculations. The binding free energies were correlated with the experimentally derived inhibitory activities. The established binding-free-energy-based QSAR model predicted all of the compounds in the test set with less than 1 log difference. Additionally, we tested the established model on a new dataset containing 26 molecules which were designed, synthesized, and tested taking knowledge of the developed BFE models [53]. The structure-based model was able to predict these novel compounds with less than 1 log unit error and showed its value for chemical optimization. For the different test sets, the structure-based model showed better accuracy than the ligand-based models. The combination of structure-based and ligand-based models resulted in predictive QSAR models in the current study. These provide useful tools for the further design and optimization of alkylhydrazide derivatives as HDAC inhibitors.

Supplementary Materials: The following supporting information can be downloaded at: https://www.mdpi.com/article/10.3390/ph16070968/s1. Figure S1: Visualization of the atom-based 3D-QSAR model in the training set; Figure S2: Docking poses of the training set, the external test set, and the inactive set; Figure S3: 100 ns MD results for the HDAC3-1 complex; Figure S4: 100 ns MD results for the HDAC3-2 complex; Figure S5: MD analysis results of HDAC3-1 complex. Figure S6: MD analysis results of HDAC3-2 complex. Table S1: Two-dimensional chemical structures and biological data of compounds in the training set, external test set, and inactive set; Table S2: Prediction values of the best atom-based QSAR model generated for the training set, test set, and inactive set in HDAC3; Table S3: Prediction values of the best BFE model generated for the training set, test set, and inactive set in HDAC3; Table S4: Two-dimensional chemical structures and biological data of the test set; Table S5: Prediction values of the best atom-based QSAR model generated for the test set in HDAC3; Table S6: Prediction values of the best BFE model generated for the test set in HDAC3.

Author Contributions: E.F.B. and D.R. carried out the computational studies and wrote the manuscript. J.M. and D.R. supervised the MD simulations and revised the manuscript. P.S., F.M. and M.Z. carried out the data correction and statistical tests and wrote part of the manuscript. M.S. and W.S. supervised the experiments and revised the manuscript. All authors have read and agreed to the published version of the manuscript.

Funding: This work was funded by the Deutsche Forschungsgemeinschaft (DFG) SI868/22-1 project number 46995445 (to W.S.).

Institutional Review Board Statement: Not applicable.

Informed Consent Statement: Not applicable.

Data Availability Statement: Not applicable.

Conflicts of Interest: The authors declare no conflict of interest.

References

1. Weinhold, B. Epigenetics: The science of change. *Environ. Health Perspect.* **2006**, *114*, A160–A167. [CrossRef] [PubMed]
2. Kouzarides, T. Chromatin modifications and their function. *Cell* **2007**, *128*, 693–705. [CrossRef] [PubMed]
3. Fraczek, J.; van Grunsven, L.A.; Vinken, M.; Snykers, S.; Deleu, S.; Vanderkerken, K.; Vanhaecke, T.; Rogiers, V. Histone deacetylase inhibition and the regulation of cell growth with particular reference to liver pathobiology. *J. Cell. Mol. Med.* **2009**, *13*, 2990–3005. [CrossRef]
4. Chen, P.J.; Huang, C.; Meng, X.M.; Li, J. Epigenetic modifications by histone deacetylases: Biological implications and therapeutic potential in liver fibrosis. *Biochimie* **2015**, *116*, 61–69. [CrossRef]
5. Haberland, M.; Montgomery, R.L.; Olson, E.N. The many roles of histone deacetylases in development and physiology: Implications for disease and therapy. *Nat. Rev. Genet.* **2009**, *10*, 32–42. [CrossRef]
6. Seto, E.; Yoshida, M. Erasers of Histone Acetylation: The Histone Deacetylase Enzymes. *Cold Spring Harb. Perspect. Biol.* **2014**, 6. [CrossRef]
7. Melesina, J.; Simoben, C.V.; Praetorius, L.; Bulbul, E.F.; Robaa, D.; Sippl, W. Strategies To Design Selective Histone Deacetylase Inhibitors. *Chemmedchem* **2021**, *16*, 1336–1359. [CrossRef]
8. Bolden, J.E.; Peart, M.J.; Johnstone, R.W. Anticancer activities of histone deacetylase inhibitors. *Nat. Rev. Drug Discov.* **2006**, *5*, 769–784. [CrossRef]
9. Gregoretti, I.V.; Lee, Y.M.; Goodson, H.V. Molecular evolution of the histone deacetylase family: Functional implications of phylogenetic analysis. *J. Mol. Biol.* **2004**, *338*, 17–31. [CrossRef]
10. Hildmann, C.; Riester, D.; Schwienhorst, A. Histone deacetylases--an important class of cellular regulators with a variety of functions. *Appl. Microbiol. Biotechnol.* **2007**, *75*, 487–497. [CrossRef]
11. Barneda-Zahonero, B.; Parra, M. Histone deacetylases and cancer. *Mol. Oncol.* **2012**, *6*, 579–589. [CrossRef]
12. Denslow, S.A.; Wade, P.A. The human Mi-2/NuRD complex and gene regulation. *Oncogene* **2007**, *26*, 5433–5438. [CrossRef]
13. Grozinger, C.M.; Schreiber, S.L. Deacetylase enzymes: Biological functions and the use of small-molecule inhibitors. *Chem. Biol.* **2002**, *9*, 3–16. [CrossRef] [PubMed]
14. Laherty, C.D.; Yang, W.M.; Sun, J.M.; Davie, J.R.; Seto, E.; Eisenman, R.N. Histone deacetylases associated with the mSin3 corepressor mediate Mad transcriptional repression. *Cell* **1997**, *89*, 349–356. [CrossRef] [PubMed]
15. Turnbull, R.E.; Fairall, L.; Saleh, A.; Kelsall, E.; Morris, K.L.; Ragan, T.J.; Savva, C.G.; Chandru, A.; Millard, C.J.; Makarova, O.V.; et al. The MiDAC histone deacetylase complex is essential for embryonic development and has a unique multivalent structure. *Nat. Commun.* **2020**, *11*, 3252. [CrossRef]
16. Xue, Y.T.; Wong, J.M.; Moreno, G.T.; Young, M.K.; Cote, J.; Wang, W.D. NURD, a novel complex with both ATP-dependent chromatin-remodeling and histone deacetylase activities. *Mol. Cell* **1998**, *2*, 851–861. [CrossRef] [PubMed]

17. Li, J.W.; Wang, J.; Wang, J.X.; Nawaz, Z.; Liu, J.M.; Qin, J.; Wong, J.M. Both corepressor proteins SMRT and N-CoR exist in large protein complexes containing HDAC3. *EMBO J.* **2000**, *19*, 4342–4350. [CrossRef]
18. Oberoi, J.; Fairall, L.; Watson, P.J.; Yang, J.C.; Czimmerer, Z.; Kampmann, T.; Goult, B.T.; Greenwood, J.A.; Gooch, J.T.; Kallenberger, B.C.; et al. Structural basis for the assembly of the SMRT/NCoR core transcriptional repression machinery. *Nat. Struct. Mol. Biol.* **2011**, *18*, 177–184. [CrossRef]
19. Hu, E.; Chen, Z.X.; Fredrickson, T.; Zhu, Y.; Kirkpatrick, R.; Zhang, G.F.; Johanson, K.; Sung, C.M.; Liu, R.G.; Winkler, J. Cloning and characterization of a novel human. Class I histone deacetylase that functions as a transcription repressor. *J. Biol. Chem.* **2000**, *275*, 15254–15264. [CrossRef]
20. Park, S.Y.; Kim, J.S. A short guide to histone deacetylases including recent progress on class II enzymes. *Exp. Mol. Med.* **2020**, *52*, 204–212. [CrossRef]
21. Vahid, F.; Zand, H.; Nosrat-Mirshekarlou, E.; Najafi, R.; Hekmatdoost, A. The role dietary of bioactive compounds on the regulation of histone acetylases and deacetylases: A review. *Gene* **2015**, *562*, 8–15. [CrossRef]
22. Chien, W.W.; Lee, D.H.; Zheng, Y.; Wuensche, P.; Alvarez, R.; Wen, D.L.; Aribi, A.M.; Thean, S.M.; Doan, N.B.; Said, J.W.; et al. Growth Inhibition of Pancreatic Cancer Cells by Histone Deacetylase Inhibitor Belinostat Through Suppression of Multiple Pathways Including HIF, NFkB, and mTOR Signaling In Vitro and In Vivo. *Mol. Carcinogen.* **2014**, *53*, 722–735. [CrossRef] [PubMed]
23. Furumai, R.; Matsuyama, A.; Kobashi, N.; Lee, K.H.; Nishiyama, N.; Nakajima, I.; Tanaka, A.; Komatsu, Y.; Nishino, N.; Yoshida, M.; et al. FK228 (depsipeptide) as a natural prodrug that inhibits class I histone deacetylases. *Cancer Res.* **2002**, *62*, 4916–4921. [PubMed]
24. Mann, B.S.; Johnson, J.R.; He, K.; Sridhara, R.; Abraham, S.; Booth, B.P.; Verbois, L.; Morse, D.E.; Jee, J.M.; Pope, S.; et al. Vorinostat for treatment of cutaneous manifestations of advanced primary cutaneous T-cell lymphoma. *Clin. Cancer Res.* **2007**, *13*, 2318–2322. [CrossRef]
25. Sivaraj, D.; Green, M.M.; Gasparetto, C. Panobinostat for the management of multiple myeloma. *Future Oncol.* **2017**, *13*, 477–488. [CrossRef]
26. Fraga, M.F.; Ballestar, E.; Villar-Garea, A.; Boix-Chornet, M.; Espada, J.; Schotta, G.; Bonaldi, T.; Haydon, C.; Ropero, S.; Petrie, K.; et al. Loss of acetylation at Lys16 and trimethylation at Lys20 of histone H4 is a common hallmark of human cancer. *Nat. Genet.* **2005**, *37*, 391–400. [CrossRef] [PubMed]
27. Gryder, B.E. Targeted cancer therapy: Giving histone deacetylase inhibitors all they need to succeed. *Future Med. Chem.* **2012**, *4*, 505–524. [CrossRef]
28. Hailu, G.S.; Robaa, D.; Forgione, M.; Sippl, W.; Rotili, D.; Mai, A. Lysine Deacetylase Inhibitors in Parasites: Past, Present, and Future Perspectives. *J. Med. Chem.* **2017**, *60*, 4780–4804. [CrossRef]
29. Pant, K.; Peixoto, E.; Richard, S.; Gradilone, S.A. Role of Histone Deacetylases in Carcinogenesis: Potential Role in Cholangiocarcinoma. *Cells* **2020**, *9*, 780. [CrossRef]
30. Jung, M.; Brosch, G.; Kolle, D.; Scherf, H.; Gerhauser, C.; Loidl, P. Amide analogues of trichostatin A as inhibitors of histone deacetylase and inducers of terminal cell differentiation. *J. Med. Chem.* **1999**, *42*, 4669–4679. [CrossRef]
31. Bulbul, E.F.; Melesina, J.; Ibrahim, H.S.; Abdelsalam, M.; Vecchio, A.; Robaa, D.; Zessin, M.; Schutkowski, M.; Sippl, W. Docking, Binding Free Energy Calculations and In Vitro Characterization of Pyrazine Linked 2-Aminobenzamides as Novel Class I Histone Deacetylase (HDAC) Inhibitors. *Molecules* **2022**, *27*, 2526. [CrossRef]
32. Burli, R.W.; Luckhurst, C.A.; Aziz, O.; Matthews, K.L.; Yates, D.; Lyons, K.A.; Beconi, M.; McAllister, G.; Breccia, P.; Stott, A.J.; et al. Design, Synthesis, and Biological Evaluation of Potent and Selective Class IIa Histone Deacetylase (HDAC) Inhibitors as a Potential Therapy for Huntington's Disease. *J. Med. Chem.* **2013**, *56*, 9934–9954. [CrossRef] [PubMed]
33. Heimburg, T.; Chakrabarti, A.; Lancelot, J.; Marek, M.; Melesina, J.; Hauser, A.T.; Shaik, T.B.; Duclaud, S.; Robaa, D.; Erdmann, F.; et al. Structure-Based Design and Synthesis of Novel Inhibitors Targeting HDAC8 from Schistosoma mansoni for the Treatment of Schistosomiasis. *J. Med. Chem.* **2016**, *59*, 2423–2435. [CrossRef]
34. Ibrahim, H.S.; Abdelsalam, M.; Zeyn, Y.; Zessin, M.; Mustafa, A.M.; Fischer, M.A.; Zeyen, P.; Sun, P.; Bulbul, E.F.; Vecchio, A.; et al. Synthesis, Molecular Docking and Biological Characterization of Pyrazine Linked 2-Aminobenzamides as New Class I Selective Histone Deacetylase (HDAC) Inhibitors with Anti-Leukemic Activity. *Int. J. Mol. Sci.* **2022**, *23*, 369. [CrossRef]
35. Lauffer, B.E.; Mintzer, R.; Fong, R.; Mukund, S.; Tam, C.; Zilberleyb, I.; Flicke, B.; Ritscher, A.; Fedorowicz, G.; Vallero, R.; et al. Histone deacetylase (HDAC) inhibitor kinetic rate constants correlate with cellular histone acetylation but not transcription and cell viability. *J. Biol. Chem.* **2013**, *288*, 26926–26943. [CrossRef]
36. Luckhurst, C.A.; Breccia, P.; Stott, A.J.; Aziz, O.; Birch, H.L.; Burli, R.W.; Hughes, S.J.; Jarvis, R.E.; Lamers, M.; Leonard, P.M.; et al. Potent, Selective, and CNS-Penetrant Tetrasubstituted Cyclopropane Class IIa Histone Deacetylase (HDAC) Inhibitors. *ACS Med. Chem. Lett.* **2016**, *7*, 34–39. [CrossRef]
37. Marek, M.; Ramos-Morales, E.; Picchi-Constante, G.F.A.; Bayer, T.; Norstrom, C.; Herp, D.; Sales, P.A.; Guerra-Slompo, E.P.; Hausmann, K.; Chakrabarti, A.; et al. Species-selective targeting of pathogens revealed by the atypical structure and active site of Trypanosoma cruzi histone deacetylase DAC2. *Cell Rep.* **2021**, *37*, 110129. [CrossRef] [PubMed]
38. Marek, M.; Shaik, T.B.; Heimburg, T.; Chakrabarti, A.; Lancelot, J.; Ramos-Morales, E.; Da Veiga, C.; Kalinin, D.; Melesina, J.; Robaa, D.; et al. Characterization of Histone Deacetylase 8 (HDAC8) Selective Inhibition Reveals Specific Active Site Structural and Functional Determinants. *J. Med. Chem.* **2018**, *61*, 10000–10016. [CrossRef] [PubMed]

39. Simoben, C.V.; Robaa, D.; Chakrabarti, A.; Schmidtkunz, K.; Marek, M.; Lancelot, J.; Kannan, S.; Melesina, J.; Shaik, T.B.; Pierce, R.J.; et al. A Novel Class of Schistosoma mansoni Histone Deacetylase 8 (HDAC8) Inhibitors Identified by Structure-Based Virtual Screening and In Vitro Testing. *Molecules* **2018**, *23*, 566. [CrossRef] [PubMed]
40. Wang, D.F.; Wiest, O.; Helquist, P.; Lan-Hargest, H.Y.; Wiech, N.L. On the function of the 14 angstrom long internal cavity of histone deacetylase-like protein: Implications for the design of histone deacetylase inhibitors. *J. Med. Chem.* **2004**, *47*, 3409–3417. [CrossRef]
41. Bressi, J.C.; Jennings, A.J.; Skene, R.; Wu, Y.Q.; Melkus, R.; De Jong, R.; O'Connell, S.; Grimshaw, C.E.; Navre, M.; Gangloff, A.R. Exploration of the HDAC2 foot pocket: Synthesis and SAR of substituted N-(2-aminophenyl)benzamides. *Bioorg. Med. Chem. Lett.* **2010**, *20*, 3142–3145. [CrossRef]
42. Liu, J.; Kelly, J.; Yu, W.S.; Clausen, D.; Yu, Y.N.; Kim, H.; Duffy, J.L.; Chung, C.C.; Myers, R.W.; Carroll, S.; et al. Selective Class I HDAC Inhibitors Based on Aryl Ketone Zinc Binding Induce HIV-1 Protein for Clearance. *ACS Med. Chem. Lett.* **2020**, *11*, 1476–1483. [CrossRef] [PubMed]
43. Wagner, F.F.; Weiwer, M.; Steinbacher, S.; Schomburg, A.; Reinemer, P.; Gale, J.P.; Campbell, A.J.; Fisher, S.L.; Zhao, W.N.; Reis, S.A.; et al. Kinetic and structural insights into the binding of histone deacetylase 1 and 2 (HDAC1, 2) inhibitors. *Bioorgan. Med. Chem.* **2016**, *24*, 4008–4015. [CrossRef]
44. Yu, W.S.; Liu, J.; Yu, Y.N.; Zhang, V.; Clausen, D.; Kelly, J.; Wolkenberg, S.; Beshore, D.; Duffy, J.L.; Chung, C.C.; et al. Discovery of ethyl ketone-based HDACs 1, 2, and 3 selective inhibitors for HIV latency reactivation. *Bioorg. Med. Chem. Lett.* **2020**, *30*, 127197. [CrossRef]
45. Liu, J.; Yu, Y.N.; Kelly, J.; Sha, D.Y.; Alhassan, A.B.; Yu, W.S.; Maletic, M.M.; Duffy, J.L.; Klein, D.J.; Holloway, M.K.; et al. Discovery of Highly Selective and Potent HDAC3 Inhibitors Based on a 2-Substituted Benzamide Zinc Binding Group. *ACS Med. Chem. Lett.* **2020**, *11*, 2476–2483. [CrossRef]
46. Wang, Y.F.; Stowe, R.L.; Pinello, C.E.; Tian, G.M.; Madoux, F.; Li, D.W.; Zhao, L.S.Y.; Li, J.L.; Wang, Y.R.; Wang, Y.; et al. Identification of Histone Deacetylase Inhibitors with Benzoylhydrazide Scaffold that Selectively Inhibit Class I Histone Deacetylases. *Chem. Biol.* **2015**, *22*, 273–284. [CrossRef]
47. Jiang, Y.Q.; Xu, J.; Yue, K.R.; Huang, C.; Qin, M.T.; Chi, D.Y.; Yu, Q.X.; Zhu, Y.; Hou, X.H.; Xu, T.Q.; et al. Potent Hydrazide-Based HDAC Inhibitors with a Superior Pharmacokinetic Profile for Efficient Treatment of Acute Myeloid Leukemia In Vivo. *J. Med. Chem.* **2022**, *65*, 285–302. [CrossRef]
48. Kozlov, M.V.; Konduktorov, K.A.; Shcherbakova, A.S.; Kochetkov, S.N. Synthesis of *N*′-propylhydrazide analogs of hydroxamic inhibitors of histone deacetylases (HDACs) and evaluation of their impact on activities of HDACs and replication of hepatitis C virus (HCV). *Bioorg. Med. Chem. Lett.* **2019**, *29*, 2369–2374. [CrossRef] [PubMed]
49. Li, X.Y.; Jiang, Y.Q.; Peterson, Y.K.; Xu, T.Q.; Himes, R.A.; Luo, X.; Yin, G.L.; Inks, E.S.; Dolloff, N.; Halene, S.; et al. Design of Hydrazide-Bearing HDACIs Based on Panobinostat and Their p53 and FLT3-ITD Dependency in Antileukemia Activity. *J. Med. Chem.* **2020**, *63*, 5501–5525. [CrossRef] [PubMed]
50. Li, X.Y.; Peterson, Y.K.; Inks, E.S.; Himes, R.A.; Li, J.Y.; Zhang, Y.J.; Kong, X.J.; Chou, C.J. Class I HDAC Inhibitors Display Different Antitumor Mechanism in Leukemia and Prostatic Cancer Cells Depending on Their p53 Status. *J. Med. Chem.* **2018**, *61*, 2589–2603. [CrossRef] [PubMed]
51. McClure, J.J.; Zhang, C.; Inks, E.S.; Peterson, Y.K.; Li, J.Y.; Chou, C.J. Development of Allosteric Hydrazide-Containing Class I Histone Deacetylase Inhibitors for Use in Acute Myeloid Leukemia. *J. Med. Chem.* **2016**, *59*, 9942–9959. [CrossRef]
52. Xiao, Y.F.; Wang, J.; Zhao, L.S.Y.; Chen, X.Y.; Zheng, G.R.; Zhang, X.; Liao, D.Q. Discovery of histone deacetylase 3 (HDAC3)-specific PROTACs. *Chem. Commun.* **2020**, *56*, 9866–9869. [CrossRef]
53. Sun, P.; Wang, J.; Khan, K.S.; Yang, W.; Ng, B.W.-L.; Ilment, N.; Zessin, M.; Bülbül, E.F.; Robaa, D.; Erdmann, F.; et al. Development of alkylated hydrazides as highly potent and selective class I HDAC inhibitors with T cell modulatory properties. *J. Med. Chem.* **2022**, *65*, 16313–16337. [CrossRef] [PubMed]
54. Adhikari, N.; Jha, T.; Ghosh, B. Dissecting Histone Deacetylase 3 in Multiple Disease Conditions: Selective Inhibition as a Promising Therapeutic Strategy. *J. Med. Chem.* **2021**, *64*, 8827–8869. [CrossRef]
55. You, S.H.; Lim, H.W.; Sun, Z.; Broache, M.; Won, K.J.; Lazar, M.A. Nuclear receptor co-repressors are required for the histone-deacetylase activity of HDAC3 in vivo. *Nat. Struct. Mol. Biol.* **2013**, *20*, 182–187. [CrossRef] [PubMed]
56. Sarkar, R.; Banerjee, S.; Amin, S.A.; Adhikari, N.; Jha, T. Histone deacetylase 3 (HDAC3) inhibitors as anticancer agents: A review. *Eur. J. Med. Chem.* **2020**, *192*, 112171. [CrossRef]
57. Janczura, K.J.; Volmar, C.H.; Sartor, G.C.; Rao, S.J.; Ricciardi, N.R.; Lambert, G.; Brothers, S.P.; Wahlestedt, C. Inhibition of HDAC3 reverses Alzheimer's disease-related pathologies in vitro and in the 3xTg-AD mouse model. *Proc. Natl. Acad. Sci. USA* **2018**, *115*, E11148–E11157. [CrossRef] [PubMed]
58. Jiang, L.P.; Yu, X.H.; Chen, J.Z.; Hu, M.; Zhang, Y.K.; Lin, H.L.; Tang, W.Y.; He, P.P.; Ouyang, X.P. Histone Deacetylase 3: A Potential Therapeutic Target for Atherosclerosis. *Aging Dis.* **2022**, *13*, 773–786. [CrossRef]
59. Zhang, L.; Cao, W. Histone deacetylase 3 (HDAC3) as an important epigenetic regulator of kidney diseases. *J. Mol. Med.* **2022**, *100*, 43–51. [CrossRef]
60. Carey, R.N.; Wold, S.; Westgard, J.O. Principal component analysis: An alternative to "referee" methods in method comparison studies. *Anal. Chem.* **1975**, *47*, 1824–1829. [CrossRef]

61. Eichner, L.J.; Curtis, S.D.; Brun, S.N.; McGuire, C.K.; Gushterova, I.; Baumgart, J.T.; Trefts, E.; Ross, D.S.; Rymoff, T.J.; Shaw, R.J. HDAC3 is critical in tumor development and therapeutic resistance in Kras-mutant non-small cell lung cancer. *Sci. Adv.* **2023**, *9*, eadd3243. [CrossRef]
62. Jia, H.; Wang, Y.; Morris, C.D.; Jacques, V.; Gottesfeld, J.M.; Rusche, J.R.; Thomas, E.A. The Effects of Pharmacological Inhibition of Histone Deacetylase 3 (HDAC3) in Huntington's Disease Mice. *PLoS ONE* **2016**, *11*, e0152498. [CrossRef] [PubMed]
63. Montgomery, R.L.; Potthoff, M.J.; Haberland, M.; Qi, X.; Matsuzaki, S.; Humphries, K.M.; Richardson, J.A.; Bassel-Duby, R.; Olson, E.N. Maintenance of cardiac energy metabolism by histone deacetylase 3 in mice. *J. Clin. Investig.* **2008**, *118*, 3588–3597. [CrossRef]
64. Chemical Computing Group (CCG). *Molecular Operating Environment (MOE), 2019.01*; Chemical Computing Group (CCG): Montreal, QC, Canada, 2019.
65. Jolliffe, I.T.; Cadima, J. Principal component analysis: A review and recent developments. *Philos. Trans. R. Soc. A* **2016**, *374*, 20150202. [CrossRef]
66. Schrödinger LLC. *Release 2019-1: Phase*; Schrödinger LLC: New York, NY, USA, 2019.
67. Dixon, S.L.; Smondyrev, A.M.; Knoll, E.H.; Rao, S.N.; Shaw, D.E.; Friesner, R.A. PHASE: A new engine for pharmacophore perception, 3D QSAR model development, and 3D database screening: 1. Methodology and preliminary results. *J. Comput.-Aid. Mol. Des.* **2006**, *20*, 647–671. [CrossRef] [PubMed]
68. Dixon, S.L.; Smondyrev, A.M.; Rao, S.N. PHASE: A novel approach to pharmacophore modeling and 3D database searching. *Chem. Biol. Drug Des.* **2006**, *67*, 370–372. [CrossRef]
69. Watson, P.J.; Fairall, L.; Santos, G.M.; Schwabe, J.W.R. Structure of HDAC3 bound to co-repressor and inositol tetraphosphate. *Nature* **2012**, *481*, 335–340. [CrossRef]
70. Case, D.A.; Cheatham, T.E.; Darden, T.; Gohlke, H.; Luo, R.; Merz, K.M., Jr.; Onufriev, A.; Simmerling, C.; Wang, B.; Woods, R. The Amber biomolecular simulation programs. *J. Comput. Chem.* **2005**, *26*, 1668–1688. [CrossRef]
71. Tropsha, A. Best Practices for QSAR Model Development, Validation, and Exploitation. *Mol. Inform.* **2010**, *29*, 476–488. [CrossRef] [PubMed]
72. Roe, D.R.; Okur, A.; Wickstrom, L.; Hornak, V.; Simmerling, C. Secondary structure bias in generalized Born solvent models: Comparison of conformational ensembles and free energy of solvent polarization from explicit and implicit solvation. *J. Phys. Chem. B* **2007**, *111*, 1846–1857. [CrossRef]
73. Schrödinger LLC. *Release 2019-1: Maestro, Protein Preparation Wizard, Prime, Epik, Ligprep, Confgen, Glide*; Schrödinger LLC: New York, NY, USA, 2019.
74. Harder, E.; Damm, W.; Maple, J.; Wu, C.J.; Reboul, M.; Xiang, J.Y.; Wang, L.L.; Lupyan, D.; Dahlgren, M.K.; Knight, J.L.; et al. OPLS3: A Force Field Providing Broad Coverage of Drug-like Small Molecules and Proteins. *J. Chem. Theory Comput.* **2016**, *12*, 281–296. [CrossRef]
75. Berman, H.M.; Westbrook, J.; Feng, Z.; Gilliland, G.; Bhat, T.N.; Weissig, H.; Shindyalov, I.N.; Bourne, P.E. The Protein Data Bank. *Nucleic Acids Res.* **2000**, *28*, 235–242. [CrossRef]
76. Jakalian, A.; Jack, D.B.; Bayly, C.I. Fast, efficient generation of high-quality atomic charges. AM1-BCC model: II. Parameterization and validation. *J. Comput. Chem.* **2002**, *23*, 1623–1641. [CrossRef]
77. Jakalian, A.; Bush, B.L.; Jack, D.B.; Bayly, C.I. Fast, efficient generation of high-quality atomic Charges. AM1-BCC model: I. Method. *J. Comput. Chem.* **2000**, *21*, 132–146. [CrossRef]
78. Wang, J.M.; Wolf, R.M.; Caldwell, J.W.; Kollman, P.A.; Case, D.A. Development and testing of a general amber force field. *J. Comput. Chem.* **2004**, *25*, 1157–1174. [CrossRef] [PubMed]
79. Li, P.F.; Song, L.F.; Merz, K.M. Parameterization of Highly Charged Metal Ions Using the 12-6-4 LJ-Type Nonbonded Model in Explicit Water. *J. Phys. Chem. B* **2015**, *119*, 883–895. [CrossRef] [PubMed]
80. Lee, M.C.; Duan, Y. Distinguish protein decoys by using a scoring function based on a new AMBER force field, short molecular dynamics simulations, and the generalized born solvent model. *Proteins-Struct. Funct. Bioinform.* **2004**, *55*, 620–634. [CrossRef] [PubMed]
81. Duan, Y.; Wu, C.; Chowdhury, S.; Lee, M.C.; Xiong, G.M.; Zhang, W.; Yang, R.; Cieplak, P.; Luo, R.; Lee, T.; et al. A point-charge force field for molecular mechanics simulations of proteins based on condensed-phase quantum mechanical calculations. *J. Comput. Chem.* **2003**, *24*, 1999–2012. [CrossRef]
82. Ryckaert, J.-P.; Ciccotti, G.; Berendsen, H.J.C. Numerical integration of the cartesan equations of motion of a system with constraints: Molecular dynamics of n-alkanes. *J. Comput. Phys.* **1977**, *23*, 327–341. [CrossRef]
83. Pastor, R.W.; Brooks, B.R.; Szabo, A. An analysis of the accuracy of Langevin and molecular dynamics algorithms. *Mol. Phys.* **1988**, *65*, 1409–1419. [CrossRef]
84. Miller, B.R., 3rd; McGee, T.D., Jr.; Swails, J.M.; Homeyer, N.; Gohlke, H.; Roitberg, A.E. MMPBSA.py: An Efficient Program for End-State Free Energy Calculations. *J. Chem. Theory Comput.* **2012**, *8*, 3314–3321. [CrossRef]
85. Onufriev, A.; Bashford, D.; Case, D.A. Exploring protein native states and large-scale conformational changes with a modified generalized born model. *Proteins-Struct. Funct. Bioinform.* **2004**, *55*, 383–394. [CrossRef]
86. Hawkins, G.D.; Cramer, C.J.; Truhlar, D.G. Parametrized models of aqueous free energies of solvation based on pairwise descreening of solute atomic charges from a dielectric medium. *J. Phys. Chem.* **1996**, *100*, 19824–19839. [CrossRef]

87. Feig, M.; Onufriev, A.; Lee, M.S.; Im, W.; Case, D.A.; Brooks, C.L. Performance comparison of generalized born and Poisson methods in the calculation of electrostatic solvation energies for protein structures. *J. Comput. Chem.* **2004**, *25*, 265–284. [CrossRef] [PubMed]
88. Karaman, B.; Sippl, W. Docking and binding free energy calculations of sirtuin inhibitors. *Eur. J. Med. Chem.* **2015**, *93*, 584–598. [CrossRef]
89. Genheden, S.; Ryde, U. The MM/PBSA and MM/GBSA methods to estimate ligand-binding affinities. *Expert Opin. Drug Dis.* **2015**, *10*, 449–461. [CrossRef] [PubMed]
90. Cournia, Z.; Allen, B.; Sherman, W. Relative Binding Free Energy Calculations in Drug Discovery: Recent Advances and Practical Considerations. *J. Chem. Inf. Model.* **2017**, *57*, 2911–2937. [CrossRef] [PubMed]

pharmaceuticals

MDPI

Article

Repositioning of Etravirine as a Potential CK1ε Inhibitor by Virtual Screening

Luis Córdova-Bahena [1,2,3], Axel A. Sánchez-Álvarez [1,4], Angel J. Ruiz-Moreno [1,5] and Marco A. Velasco-Velázquez [1,2,*]

1 Departamento de Farmacología, Facultad de Medicina, Universidad Nacional Autónoma de México, Mexico City 04510, Mexico; luisbahena@unam.mx (L.C.-B.); aaalvarez@comunidad.unam.mx (A.A.S.-Á.); angel.j.ruiz.moreno@gmail.com (A.J.R.-M.)
2 Unidad Periférica de Investigación en Biomedicina Traslacional CMN 20 de noviembre ISSSTE, Facultad de Medicina, Universidad Nacional Autónoma de México, Mexico City 04510, Mexico
3 Consejo Nacional de Ciencia y Tecnología (CONACYT), Mexico City 03940, Mexico
4 Facultad de Química, Universidad Nacional Autónoma de México, Mexico City 04510, Mexico
5 Programa de Doctorado en Ciencias Biomédicas, Universidad Nacional Autónoma de México, Mexico City 04510, Mexico
* Correspondence: marcovelasco@unam.mx

Abstract: CK1ε is a key regulator of WNT/β-catenin and other pathways that are linked to tumor progression; thus, CK1ε is considered a target for the development of antineoplastic therapies. In this study, we performed a virtual screening to search for potential CK1ε inhibitors. First, we characterized the dynamic noncovalent interactions profiles for a set of reported CK1ε inhibitors to generate a pharmacophore model, which was used to identify new potential inhibitors among FDA-approved drugs. We found that etravirine and abacavir, two drugs that are approved for HIV infections, can be repurposed as CK1ε inhibitors. The interaction of these drugs with CK1ε was further examined by molecular docking and molecular dynamics. Etravirine and abacavir formed stable complexes with the target, emulating the binding behavior of known inhibitors. However, only etravirine showed high theoretical binding affinity to CK1ε. Our findings provide a new pharmacophore for targeting CK1ε and implicate etravirine as a CK1ε inhibitor and antineoplastic agent.

Keywords: cancer; drug repurposing; pharmacophore model; CK1ε; etravirine; abacavir

Citation: Córdova-Bahena, L.; Sánchez-Álvarez, A.A.; Ruiz-Moreno, A.J.; Velasco-Velázquez, M.A. Repositioning of Etravirine as a Potential CK1ε Inhibitor by Virtual Screening. *Pharmaceuticals* **2022**, *15*, 8. https://doi.org/10.3390/ph15010008

Academic Editors: Marialuigia Fantacuzzi and Mariangela Agamennone

Received: 23 November 2021
Accepted: 19 December 2021
Published: 22 December 2021

1. Introduction

The casein kinase 1 (CK1) family comprises enzymes that regulate signal transduction pathways by reversible phosphorylation of their substrate proteins [1]. They are involved in many cellular processes, including DNA repair, cell differentiation, intracellular trafficking, immune responses, and apoptosis [2]. In mammals, the CK1 family has seven members: α, β1, γ1, γ2, γ3, δ, and ε.

Physiologically, CK1ε participates in circadian clock control via phosphorylation of PER2 and PER3. Phosphorylated PER2/3 translocate into the nucleus, suppressing the activity of the CLOCK/BMAL1 transcriptional complex [3]. However, CK1ε also modulates the transduction of many signals in cancer cells. For example, it phosphorylates p53 and Mdm2, which are important in cell proliferation and the maintenance of genomic integrity [4,5].

Further, CK1ε is a pivotal regulator of the WNT pathways [6], which are commonly altered in various human cancers [7,8]. The upregulation of CK1ε activity that is elicited by WNT ligands [9] leads to the phosphorylation of Dishevelled [10], activating the canonical WNT pathway. Conversely, in the absence of WNT ligands, CK1ε phosphorylates β-catenin, promoting its degradation [11]. This latter activity occurs in glioblastoma cells, in which the of inhibition CK1ε activates β-catenin and induces apoptosis [12]. CK1ε can also interact and phosphorylate the tyrosine-protein kinase WNT co-receptors ROR1/ROR2 in cancer cells [13,14], triggering AKT-mediated signaling and promoting proliferation [14]. Furthermore, CK1ε protects ROR2 protein from degradation [15].

Addition protumoral effects of CK1ε include its control of the expression of the mitochondrial protein adenine nucleotide translocase 2 (ANT2). In ovarian cancer cells, CK1ε interacts with ANT2 to support ATP production [16]. Accordingly, the inhibition of CK1ε suppresses cell proliferation, reduces xenograft growth in vivo, and increases the susceptibility to chemotherapy. Thus, CK1ε is considered an antineoplastic target on several levels [17–19].

Structurally, CK1ε has a highly conserved kinase domain that is organized into a bilobal arrangement [20,21]. The N-terminal lobe is composed primarily of β-sheets, and the C-terminal lobe comprises α-helices [22]. The catalytic site is located between the two lobes. An analysis of the binding mode of ATP has defined five regions [23]: (1) adenine-interacting, (2) sugar-interacting, (3) phosphate-interacting, (4) buried region, and (5) solvent accessible region. The active residues for ATP binding are Ala36, Lys38, and Met80 in the N-lobe; Glu90, Leu91, and Phe95 in the C-lobe; and Met82, Glu83, Leu84, and Leu85 in the linker loop between the two lobes [24].

This extensive characterization of the catalytic pocket of CK1ε has guided the development of several competitive inhibitors with moderate to high biological activity [24–27]. For example, the CK1ε inhibitor PF-4800567 (hereafter referred to as inhibitor 1 [IN1]) is an efficacious inhibitor of circadian rhythms in cycling Rat1 fibroblasts and mice [27]. In February 2021, the Food and Drug Administration (FDA) granted approval to the CK1ε inhibitor umbralisib for the treatment of marginal zone lymphoma and follicular lymphoma [28]. Nevertheless, no other CK1ε inhibitor has attained clinical use, highlighting the need of new inhibitors.

We generated a pharmacophore model by characterizing the binding modes of five reported CK1ε inhibitors. Our model was then used to identify potential molecules that bind the catalytic domain of CK1ε from a database of FDA-approved drugs. We found that the anti-HIV drugs etravirine and abacavir have conformers that match our pharmacophore and were the most likely binding modes in molecular docking experiments. Etravirine emulated the noncovalent interactions in reported CK1ε inhibitors, with a similar theoretical ΔG as IN1. Further, we observed that additional residues outside of the catalytic domain participated in stabilization of the etravirine-CK1ε interaction. Thus, we propose the repurposing of etravirine as a CK1ε inhibitor. Biological validation of our findings will constitute the basis for the development of new clinical CK1ε inhibitors.

2. Results

2.1. Identification of Binding Modes for Noncrystallized CK1ε Inhibitors

The set of inhibitors used in this study is listed in Table 1. The binding mode for inhibitors without available structural data (inhibitors [IN] 2–5) was determined by molecular docking. The protocol was validated by docking IN1 into the catalytic site of CK1ε using 12 combinations of search algorithms and scoring functions. Eight poses were calculated for each combination. Hierarchical clustering of the 96 predicted poses identified subclusters with poses that had a root-mean-square deviation (RMSD) of atomic positions <2.0 Å (Figure 1A). A comparison of a representative pose of the largest subcluster and the bioactive conformer in the structure [22] of PDB 4HNI yielded an RMSD of 0.5 Å (Figure 1B). The largest subcluster included the binding modes with the best scores for various combinations of search algorithms and scoring functions, indicating that the strategy can predict the experimental binding mode.

Table 1. Characteristics of the CK1ε ATP-competitive inhibitors employed for pharmacophore modeling.

Assigned Code	Original Name	Structure	IC50 (μM)	Reference
IN1	PF-4800567	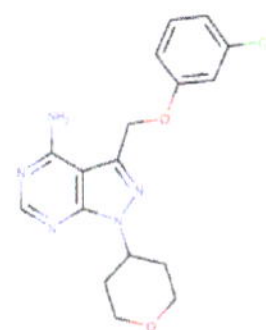	0.034 ± 0.009 [a]	[27]
IN2	IC261	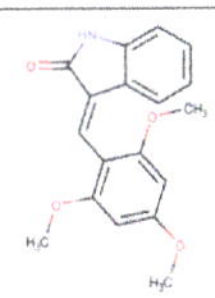	1.0 ± 0.4 [c]	[25]
IN3	compound No. 2	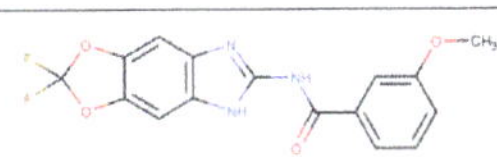	0.52 ± 0.05 [b] 0.16 ± 0.06 [c]	[26]
IN4	compound No. 6	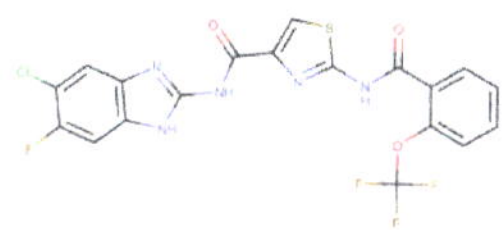	0.033 ± 0.01 [b]	[24]
IN5	compound No. 9	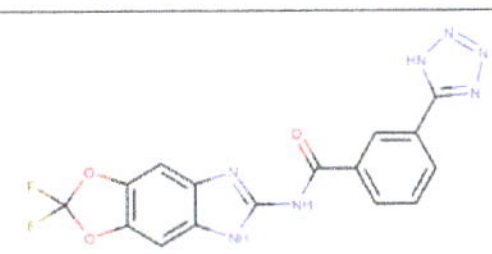	0.62 ± 0.01 [b]	[26]

IC50s were obtained using peptide PLSRTLpSVASLPGL [a], GST-p53 [b], or α-Casein [c] as substrates.

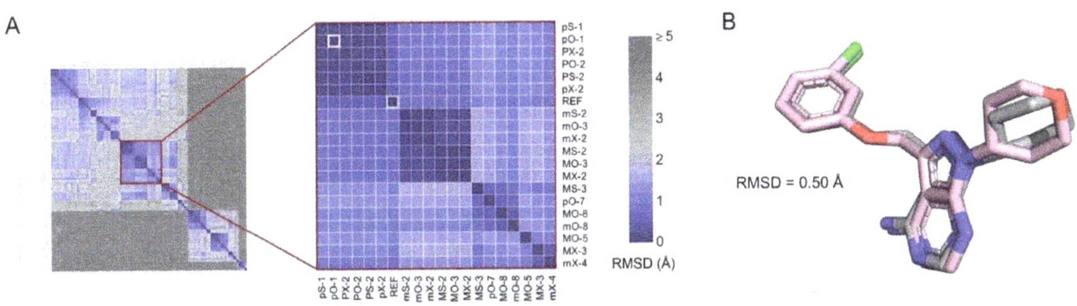

Figure 1. Docking protocol reproduces the active conformer of IN1. (A) Hierarchical clustering of 96 poses obtained by molecular docking. Inset shows the largest subcluster. Poses are labeled using letters that indicate the combination of search algorithm/scoring function employed (see "Methods") and a number indicating the ranked position. The crystal pose of IN1 (REF) and the best scored pose calculated using PLANTS scoring function and Optimizer search algorithm (pO-1) are highlighted in gray and pink, respectively. Color scale shows the RMSD between poses. (B) Superimposition of co-crystallized (gray) and pO-1 docked (pink) poses.

Four additional CK1ε inhibitors with unknown binding mode were docked using the validated protocol. For these inhibitors, the largest subclusters that were generated by hierarchical clustering included poses with RMSD < 2.0 Å (Figure S1). From the representative poses of these subclusters, we identified the binding regions and the intermolecular interactions formed (Figure 2). For all inhibitors, the binding modes were mainly driven by hydrogen bonds with the backbone of residues Glu83, and Leu85. Additional hydrogen bonds were found with Glu52 and Ser88, for IN2 and IN4 respectively. The fused rings of IN2, IN3, IN4, and IN5 were oriented toward the buried region of the catalytic pocket, and the groups on the side opposite to the fused rings were oriented to the solvent accessible region. In contrast, the fused rings of IN1 remained directed toward the adenine region in the middle area of the catalytic pocket. In addition, for all inhibitors the binding was mediated largely by hydrophobic interactions, although charged amino acids also participated.

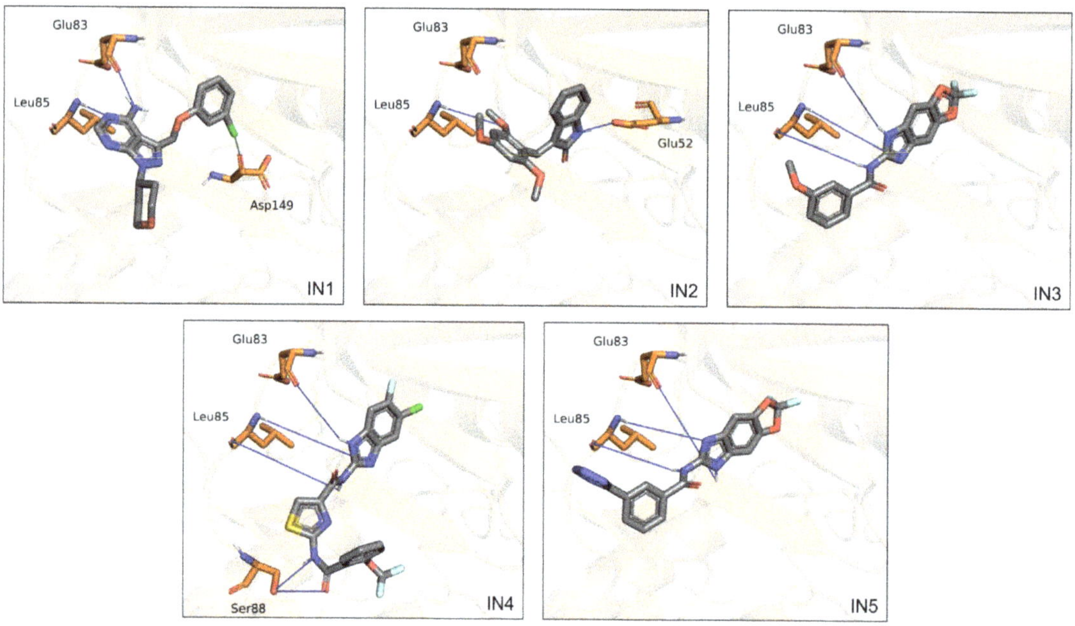

Figure 2. Binding modes calculated by molecular docking for the five inhibitors analyzed. Cartoon representation of CK1ε with residues Glu52, Glu83, and Leu85, and Ser88 on licorice model. Hydrogen bonds are shown as blue lines.

2.2. *Characterization of Relevant CK1ε-Inhibitor Interactions by Molecular Dynamics*

To examine the conformational dynamics of the ligands in the catalytic site of CK1ε, we performed simulations by molecular dynamics (MD). As expected for compounds with proven biological activity, all systems showed stable binding, although IN1 and IN2 conformers had greater stability (average RMSD < 2 Å) than IN3 and IN5 (average RMSD < 4 Å). The greatest changes for IN3 and IN5 occurred in the anisole and tetrazole groups, respectively, in the solvent accessible region. For IN4, the ligand showed transient variability at the beginning of MD but eventually stabilized.

RMSDs of the protein backbone in the apo enzyme and CK1ε-inhibitor complexes were used to analyze structural changes in the target protein. The complete trajectory for the apo enzyme showed an RMSD value of 1.85 ± 0.26 Å (average ± standard deviation). For the CK1ε-inhibitor complexes, the average RMSD was below 2.5 Å, indicating that no significant structural changes occur in CK1ε when bound to inhibitors (Figure 3A).

The root-mean-square fluctuation (RMSF) values from atomic positions of the alpha carbons in the backbone protein showed that all systems behaved similarly, except for a few residues. The loops that comprised residues 42–48 and 74–76 were particularly flexible in the complex with IN3 compared with other inhibitors. Similarly, the 139–140 loop and 216–226 loop-helix, which lie outside of the catalytic site, showed increased mobility in the complexes with IN4 and IN1, respectively (Figure 3B).

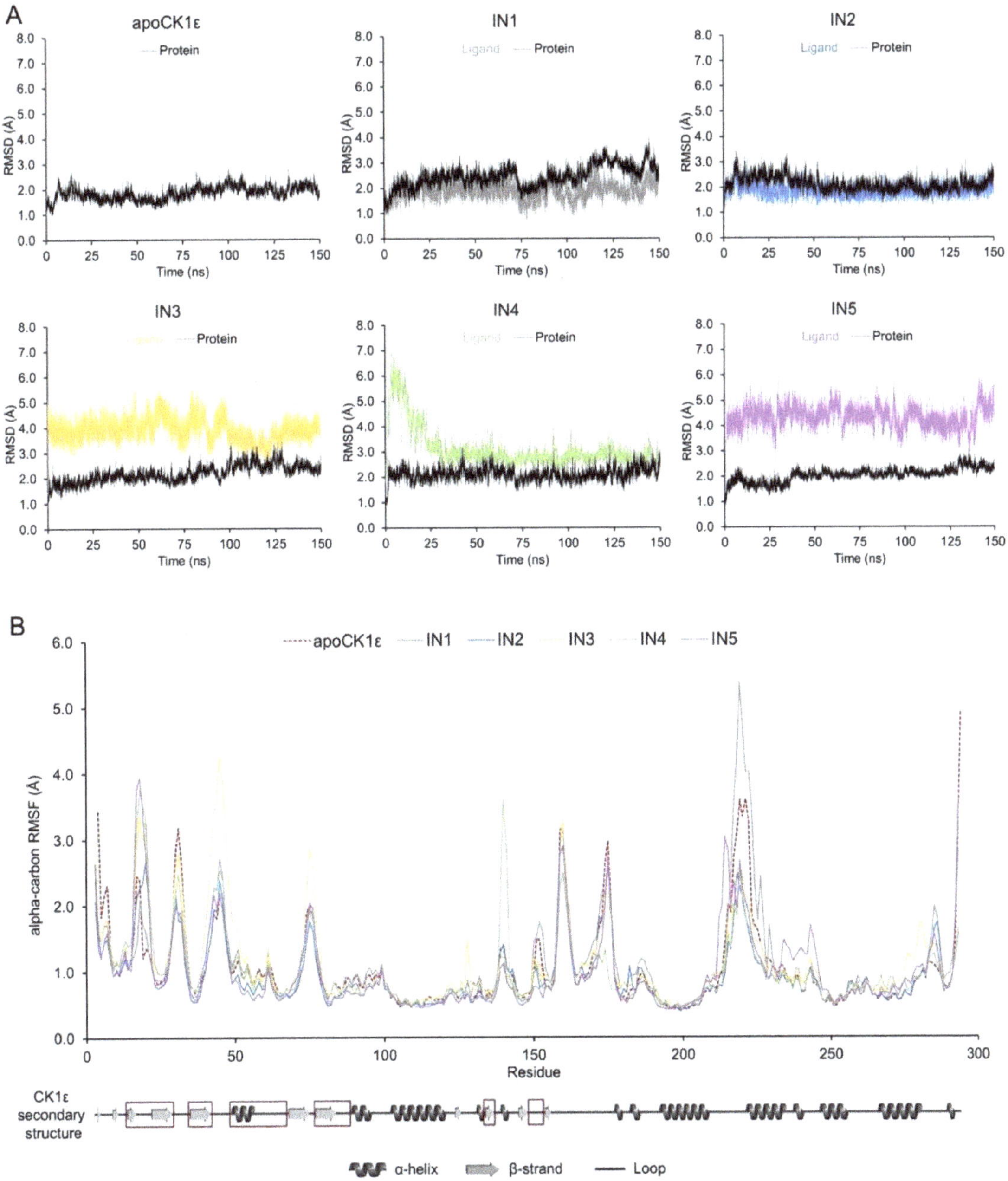

Figure 3. *Cont.*

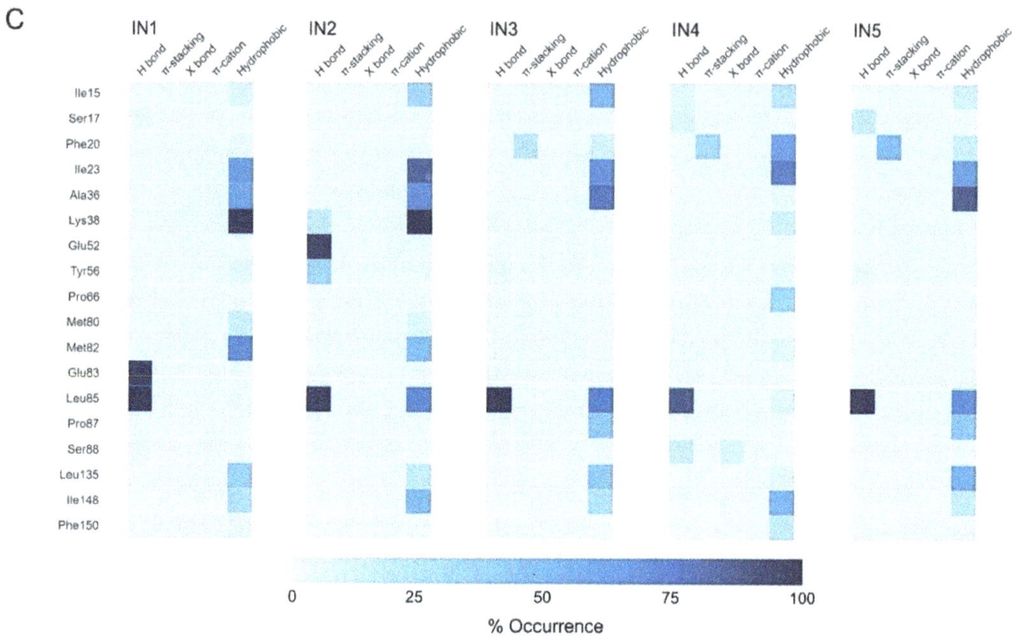

Figure 3. Molecular dynamics analysis identified key inhibitor-CK1ε interactions. (**A**) Protein backbone and ligand RMSDs for apoCK1ε and CK1ε-inhibitor complexes. (**B**) Protein backbone RMSF for all systems. Secondary structure of CK1ε is shown at the bottom with regions comprising the catalytic domain in red squares. (**C**) Heatmap of the non-covalent interaction profile occurrence. For clarity, only residues with occurrence higher than 20% are shown.

An analysis of the 15,000 bioactive conformations that were accessible for each inhibitor allowed us to create dynamic noncovalent interaction profiles for each residue (Figure 3C). Hydrogen bonds with the Leu85backbone in the adenine region were noted, ranging from 13,204 to 14,998 conformations (88.02% to 99.99% occurrence) for all inhibitors, indicating that such an interaction occurs independently of the inhibitor. Similar but less frequent interactions were identified for residues Ile15, Ile23, Leu135, and Ile148 for all inhibitors. Conversely, Phe20 in the phosphate-binding region of the catalytic domain formed 3239 pi-stacking interactions in the simulation for IN3, 4713 for IN4, and 7421 for IN5 (occurrence of 21.59%, 31.42%, and 49.47%, respectively); thus, pi-stacking that is mediated by Phe20 favors the binding of certain compounds.

Finally, we identified residues that mediate only the binding of a particular inhibitor. Hydrophobic interactions with Phe20 and Pro66 occurred frequently for IN4 (63.36% and 33.79% of the time, respectively). Hydrogen bonds with Glu83 appeared for IN1 (97.48% of the time). Several hydrogen bonds with Lys38 and Glu52 were found with IN2; however, Lys38 participated in binding primarily through hydrophobic interactions, and Glu52 had no function in any of the other inhibitors.

2.3. Generation of Pharmacophore Model

Based on their frequency in our MD analyses, we selected eight intermolecular interactions to generate a 3D pharmacophore model (Figure 4):

1. a hydrogen bond acceptor (HBA) from the interactions of all inhibitors with the Leu85 backbone;
2. a hydrogen bond donor (HBD1) from the interactions of all inhibitors with the Leu85 backbone;

3. a second hydrogen bond donor (HBD2) from the interaction of IN1 with Glu83;
4. an aromatic (Aro1) feature from the pi-stacking interactions of IN3, IN4, and IN5 with Phe20;
5. a hydrophobic (Hyd1) feature—at the same position of Aro1—from the hydrophobic interaction of IN4 with Phe20;
6. a second hydrophobic element (Hyd2) from the interactions of inhibitors with Ala36, Pro66, Met82, Leu135, and Ile148;
7. a second aromatic element (Aro2)—at the same position as Hyd2—because a ring can fix the hydrogen bond elements to each other;
8. a third hydrophobic element (Hyd3) from the interactions of IN1 and IN2 with the lateral chain of Lys38.

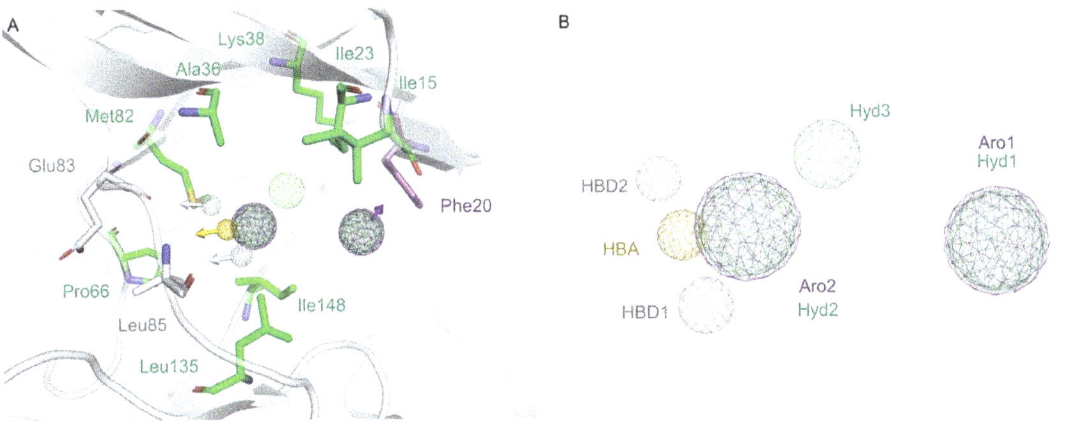

Figure 4. Pharmacophore modeling. (**A**) CK1ε catalytic site representation with selected amino acids showed on licorice model. Residues forming hydrophobic interactions, hydrogen bonds, or stacking interactions are colored green, yellow and white, or purple, respectively. (**B**) The generated pharmacophore model included eight elements. Hydrogen bond donors (HBD) are represented as sphere grid colored white, hydrogen bond acceptor (HBA) on yellow, aromatic (Aro) on purple, and hydrophobic (Hyd) on green.

2.4. Virtual Screening

The generated pharmacophore was used to search a virtual library of FDA-approved drugs. When the complete pharmacophore model was queried, only etravirine appeared as a match. Thus, we performed a second search with a submodel without the Aro1/Hyd1 dual element. These elements were eliminated because Hyd1 makes a less important energetic contribution and because the position of the element remains solvent-exposed. Seven additional compounds matched this simplified pharmacophore model. The set of candidate drugs is presented on Table S1.

Of the eight identified drugs, four were discarded because they have predominant species at physiological pH with protonation states different to the species matching the pharmacophore. The remaining four compounds were docked into the catalytic site of CK1ε to analyze their binding mode. Two compounds bound to the target in modes that clearly differed from that predicted by the pharmacophore models (RMSD > 5.1 Å) and thus were not further studied. In contrast, etravirine and abacavir bound to CK1ε per the model that was used for their identification (Figure 5).

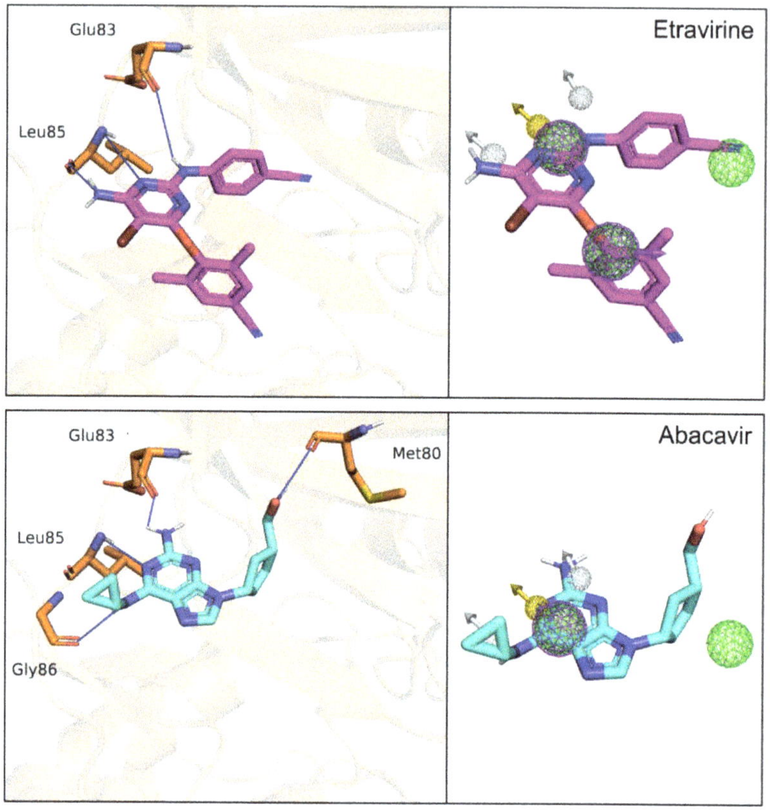

Figure 5. Docking poses of etravirine and abacavir. Overlap of pharmacophore features with the calculated binding modes of etravirine (magenta) and abacavir (cyan). Hydrogen bonds are shown as blue lines. Note that the pharmacophoric submodel used for abacavir lacks the dual element Aro1/Hyd1 (see text for details).

2.5. MD Analyses Support the Repurposing of Etravirine as a CK1ε Inhibitor

Complexes of etravirine or abacavir with CK1ε were examined by MD. The CK1ε-etravirine complex had an RMSD of 2.4 ± 0.5 Å for the enzyme backbone and 3.1 ± 0.5 Å for the ligand, suggesting that the system remains stable during the MD simulation (Figure 6A). Further, etravirine reproduced the pattern of interactions in the known inhibitors (Figure 6B). Etravirine formed a trident of hydrogen bonds with Glu83 and Leu85, with a prevalence of 87.7% and 98.3%, respectively, playing a major role in binding of the compound. Similarly, stacking interactions with Phe20 were 38.2% of the time. In addition, several hydrogen bonds were noted with Ser17, Lys38, and Tyr56 13.7%, 13.1%, and 7.8% of the time, respectively, but hydrophobic interactions were predominant.

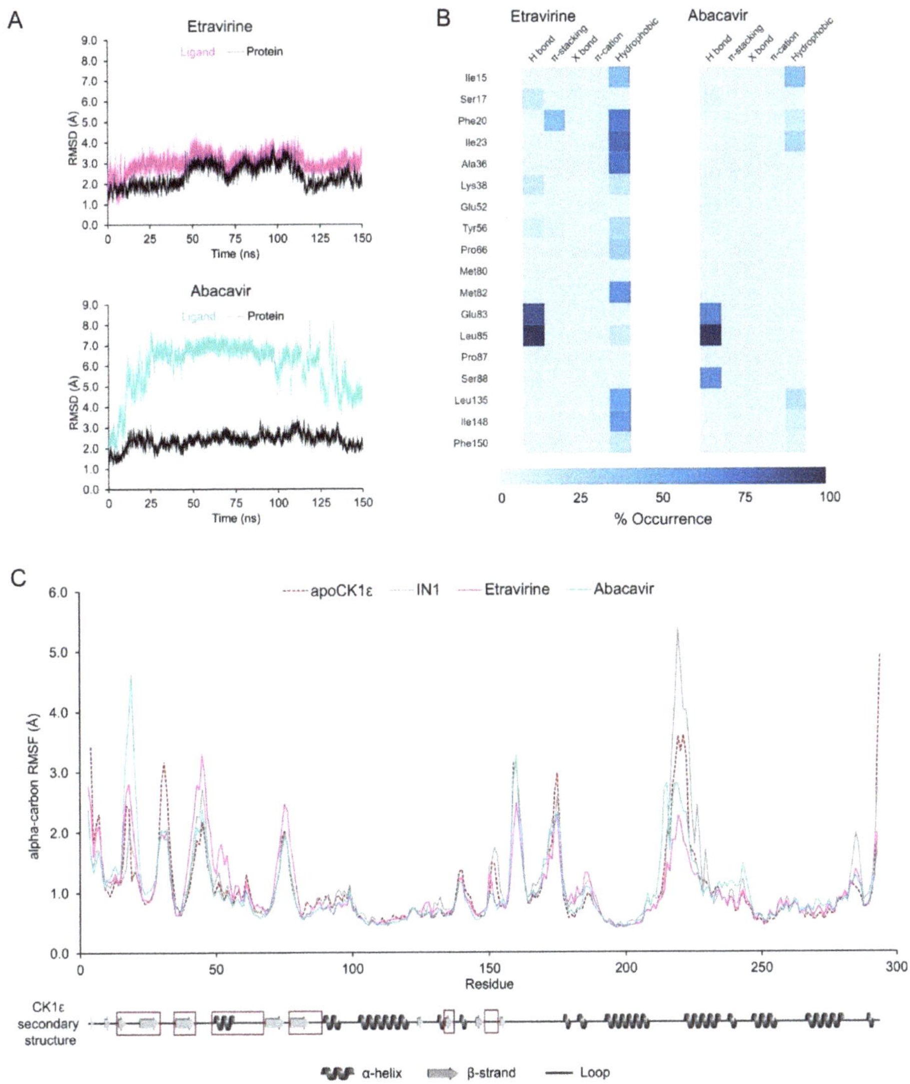

Figure 6. MD analyses for CK1ε-etravirine and CK1ε-abacavir complexes. (**A**) Protein backbone (black) and ligands (colored) RMSDs. (**B**) Heatmap of the non-covalent interactions occurrence for selected residues. (**C**) Protein backbone RMSF. The apo enzyme and the system CK1ε-IN1 are shown for comparison. Secondary structure of CK1ε is shown at the bottom with the catalytic domain in red squares.

The CK1ε-abacavir complex had an RMSD of 2.4 ± 0.3 Å for the enzyme backbone and 6.0 ± 1.5 Å for the ligand (Figure 6A), indicating that although the protein remained stable, abacavir undergoes significant conformational changes. A dynamic noncovalent interaction profile analysis (Figure 6B) identified the predicted hydrogen bonds with Glu83 and Leu85 at a prevalence of 67.7% and 98.8%, respectively, and an additional one with Ser88 at 63.4%. Thus, abacavir partially reproduces the interactions in the pharmacophore, lacking stacking interactions and participating in fewer hydrophobic interactions.

The RMSF analysis (Figure 6C) showed that fluctuation in Ser31, which lies adjacent to the catalytic domain, is reduced by both drugs, compared with the apo enzyme. Similar behavior occurred with the control compound IN1. The CK1ε-abacavir complex showed high fluctuation in Phe20 versus systems with IN1 or etravirine, in which Phe20 was rigid due to stacking interactions with the ligand. Finally, the binding of both drugs restricted the conformational dynamics of the loop conformed by residues 217–226. Given that such a loop showed increased fluctuation with IN1 and it resides outside of the catalytic site, the relevance of the changes that are induced by etravirine and abacavir remain to be determined.

The binding energy of the identified drugs was calculated by MM-PBSA for the entire MD simulation (Figure 7A). The contribution of van der Waals interactions and energy that was associated with a solvent-accessible surface were similar in magnitude for etravirine and IN1, but the electrostatic contribution was slightly higher for etravirine. Thus, the total binding energy of etravirine to CK1ε approximates that of IN1. Conversely, the total binding energy for abacavir was lower by three-fold, suggesting that this drug should not be prioritized in experimental assays.

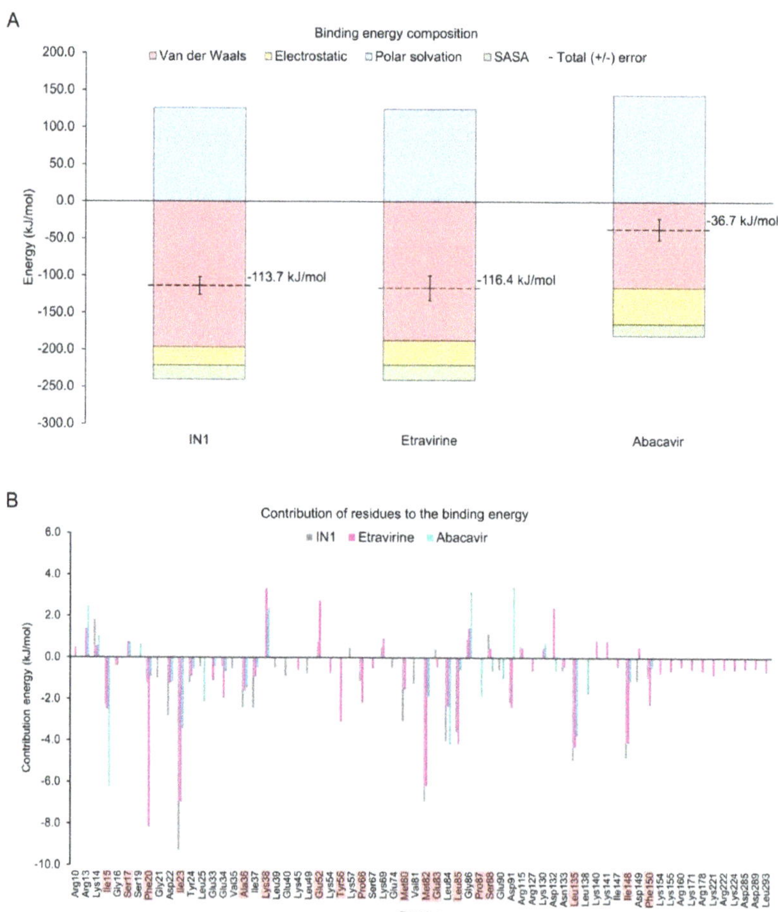

Figure 7. Binding energy-analysis. (**A**) Energy decomposition from systems with etravirine or abacavir were calculated from MD simulations. The system with IN1 is included for comparison. (**B**) Per-residue energy decomposition. Highlighted residues are part of the catalytic domain.

Analysis of the energetic contribution of each residue to the binding energy (Figure 7B) showed that residues Asp22, Ile23, Tyr24, Ala36, Met82, Leu135, and Ile148 contributed more to IN1 but still cooperated in the binding of etravirine and abacavir. Phe20 was crucial in binding etravirine but was minor in interactions with abacavir and IN1. Similar behavior was observed for Glu34, Lys45, Lys54, Tyr56, Pro66, Leu85, and Phe150. Notably, the large and positively charged residues Lys155, Lys171, Arg178, Lys221, Arg222, and Lys224, located in front of the catalytic domain, contributed uniquely to the binding mode of etravirine. However, Lys38, Glu52, Lys69, Gly86, Arg115, Lys130, Asp132, Lys140, Lys141, and Asp149 negatively affected the stability of the CK1ε-etravirine complex. To determine the function of Lys38 and Glu52, the prevalence of a salt bridge between such residues in the CK1ε-etravirine complex was evaluated. Our findings suggest that etravirine competes with Glu52 for Lys38 (Figure S2).

3. Discussion

CK1ε has been implicated as a biological target due its importance in the initiation and progression of various types of cancer [1,4–6,12,17,29]. In this study, we aimed to identify potential CK1ε inhibitors by repurposing FDA-approved drugs by virtual screening. Drug repurposing is an effective strategy for identifying new activities of approved drugs [30–32]. This approach has several advantages, including its accelerated clinical translation, given the known pharmacokinetics and safety profiles of the candidate compounds [33].

Although several ATP-competitive inhibitors of CK1ε have been reported [22,24–26], only 1 has a known binding mode [22]. Thus, we determined the most likely binding modes of four additional CK1ε inhibitors, observing that a fraction of each molecule remains solvent-exposed in the limits of the catalytic pocket, as has been reported for IN1 [22]. The binding of ligands is mediated primarily by hydrophobic residues into the catalytic site. The inhibitors form nonpolar interactions with the side chains of Ile23 and Ala36 and the aliphatic chain of Lys38, located in the buried region of the catalytic pocket. In addition, the backbone of Leu85 forms hydrogen bonds with all inhibitors, as has been reported for the adenine moiety in the binding mode of ATP [20]. These results are in good agreement with the reported binding modes of CK1δ inhibitors and other inhibitors of kinases with bilobal stucture [34–36]. Our findings suggest that common key residues of CK1ε mediate the binding of various inhibitors.

To generate information on the prevalence of such interactions, we performed MD experiments for all five inhibitors. As suggested by our previous experiments, the backbone of Leu85 plays a major role in the binding of the inhibitors, forming hydrogen bonds most of the time. Further, the Leu85 and Glu83 backbones remain highly synchronized, making hydrogen bonds with IN1. We also identified new, frequent hydrophobic interactions with Met82, Leu135, and Ile148, in addition to those by Ile23, Ala36, and Lys38. Finally, Phe20, in the phosphate-binding region, forms stacking interactions that stabilize the regions of the inhibitors that remain exposed to the solvent.

These identified patterns of intermolecular interactions drove the generation of a target-based pharmacophore. A similar strategy has been used successfully in the identification of kinase inhibitors [37–39]. Our model is disposed on a triangle conformation, wherein the first one of the vertexes is situated in the buried region, the second one is situated interacting with the adenine region, and the third one is situated at the phosphate binding region. This model is compatible with the one proposed by Bolcato et al. for CK1δ inhibitors [40], but implies improvement since our analysis revealed additional hotspots.

The pharmacophore, and a simplified submodel in which we removed the stacking interaction with the phosphate-binding region, allowed us to identify the non-nucleoside reverse-transcriptase inhibitors etravirine and abacavir from a library of approved drugs. The pharmacophore that we have described can also be used to search for hits in other chemotheques to identify additional compounds that can be developed into CK1ε inhibitors. Given that our study aimed to repurpose approved drugs, additional screens are outside of its scope, constituting a limitation of this report.

Our MD analysis showed that both drugs stably bind CK1ε, reproducing the interactions that conform the pharmacophore(s). Although both compounds remained fixed to the interconnection loop between lobes, important differences between their binding modes were noted. For etravirine, we corroborated the key role of Phe20. There was also cooperation of additional hydrogen bonds between the ammonium group of Lys38 and the nitrile substituent of the benzonitrile moiety. However, in apo CK1ε, Lys38 and Glu52 frequently form a salt bridge, but etravirine competes with Glu52 for Lys38, driving the energetic contribution of Lys38 to become unfavorable. Conversely, additional hydrogen bonds that formed between the benzonitrile moiety of etravirine and Tyr56 were beneficial for the binding mode. Further, the aliphatic residues Ile15, Ile23, Ala36, Pro66, Leu85, Pro87, Leu135, Ile148, and Phe150 contributed favorably to the binding mode. Notably, many charged residues in the C-lobe contribute favorably to the binding energy, except for Glu52 and Asp132, which interacted with the two nitrile substituents of etravirine, having a prejudicial effect. Finally, the theoretical binding affinity of etravirine approximated that of the reference compound IN1.

In contrast, the calculated binding affinity of abacavir to CK1ε was significantly reduced. As expected for a hit that was obtained with a pharmacophore that lacked the stacking interaction with Phe20, this residue was irrelevant to the interaction. However, we identified several hydrogen bonds with Glu83, Leu85, and Ser88; thus, these three residues fix the fused rings of abacavir. Yet, the cyclopropane moiety remained in movement without strong interactions. In contrast to etravirine and IN1, abacavir received little contribution from van der Waals interactions. Hydrogen bonds appeared in the adenine region, pulling the compound from the pocket and preventing the interaction between the hydroxymethyl cyclopentene and the buried region. Thus, abacavir is not a promising candidate for inhibiting CK1ε.

In summary, our results support the repurposing of etravirine as a CK1ε inhibitor and antineoplastic agent. Notably, etravirine activates the WNT pathway in osteosarcoma cells, increasing the expression of the cyclin-dependent kinase (CDK) inhibitor p21 [41]. This effect correlates with our findings and encourages further studies. Our data suggest that etravirine inhibits CK1ε at similar concentrations as IN1 [22,27].

4. Materials and Methods

4.1. Protein Preparation

The structures of CK1ε apo (PDB ID 4HOK), alone and cocrystallized with IN1 (PDB ID 4HNI) [22], were obtained from the RCSB Protein Data Bank. The unresolved fragments of the structures were built by homology modeling using the GapRepairer server [42].

4.2. Selection of Inhibitors and Ligand Preparation

Five potent CK1ε inhibitors were selected for this study. The ligand 3-[(3-chlorophenoxy) methyl]-1(oxan-4-yl)-1H-pyrazolo[3,4-d]pyrimidin-4-amine (IN1), was considered the reference compound. The compounds (3Z)-3-[(2,4,6-trimethoxyphenyl) methylidene]-2,3-dihydro-1H-indol-2-one (IN2), N-(2,2-difluoro-5H-[1,3]dioxolo[4,5-f]benzimidazol-6-yl)-3-methoxybenzamide (IN3), N-(5-chloro-6-fluoro-1,3-benzodiazol-2-yl)-4-[2-(trifluoromethoxy) benzamido]-4,5-dihydro-1,3-thiazole-2-carboxamide (IN4), and N-(2,2-difluoro-5H-[1,3] dioxolo[4,5-f]benzimidazol-6-yl)-3-(2H-tetrazol-5-yl)benzamide (IN5) were selected because they have reported IC50 values that ranged between 16–1000 nM [24–26].

The two-dimensional chemical structures of the ligands were drawn manually using Marvin Sketch ChemAxon, and the protonation states were calculated at pH > 7.2 (the reported intracellular pH in cancer cells) [43]. Subsequently, three-dimensional structures were built, and their geometries were optimized at the PM6 semiempirical level [44] using Spartan software, and the output files were exported in pdb format.

4.3. Molecular Docking

Docking simulations were performed with the Molegro Virtual Docker suite [45]. The exploration region was delimited by a 10-Å-radius sphere that was centered on the catalytic site of CK1ε, with 0.2 Å grid spacing. All rotatable bonds of ligands were set free in the experiments, and protonation states were adjusted as discussed. We used 12 combinations of 3 search algorithms [Iterated Simplex (X), MolDock Simplex Evolution (S), and MolDock Optimizer (O)] and four scoring functions [MolDock Score (m), MolDock Score [GRID] (M), Plants Score (p), and Plants Score [GRID] (P)]. For every combination, eight runs were performed, with a maximum of 1500 iterations and an initial population of 50 poses. To find the most probable binding mode, all poses were analyzed by clustering, as reported [46]. A representative pose of the largest subcluster was selected as the input file for MD simulations.

4.4. MD Simulations

MD simulations of the CK1ε–IN complexes were performed in GROMACS 4.5.5 [47] using the CHARMM36m force field [48]. The ligand parameters were generated with the CHARMM graphical user interface module in the force field framework [49]. Each complex was in a periodic 75 Å × 75 Å × 75 Å cubic box and solvated using a three-point (TIP3P) model for water molecules. Further, Na+ and Cl− atoms were added to neutralize the charge of the system and achieve an ionic concentration of 0.15 M. All simulations were carried out at 1 bar and 310.15 K. Before production of MD, the energy of the system was minimized using the steepest-descent algorithm, followed by equilibration in an NVT ensemble using a modified Berendsen thermostat. MD simulations were generated for 150 ns with an integration time frame of 2 fs, and the trajectories were saved after every 10 ps. Analysis of ligand-target interactions was carried out by a python tailored-made script (https://github.com/AngelRuizMoreno/Scripts_Notebooks/blob/master/Scripts/plipMD_V3.1.py; last accessed 12 December 2021) implementing MDAnalysis [50] and PLIP [51]. The complete trajectories were visualized in VMD [52].

4.5. Pharmacophore Modeling

We calculated the occurrence of intermolecular interactions between INs and CK1ε. We considered an interaction to be relevant only if it occurred > 20% of the total simulation time; thus, only interactions with higher frequency were analyzed to define the pharmacophoric elements. The generated pharmacophore comprised eight elements (see Results).

4.6. Virtual Screening

4.6.1. Pharmacophore-Based Selection

Pharmacophore-matching compounds were selected from a database of 21,850 conformers from 1856 FDA-approved drugs using Pharmit [53]. Compounds with a molecular volume that was greater than 270 Å^3 were eliminated, due to the size restrictions of the catalytic site of CK1ε. Compounds with at least six elements of the pharmacophore model were candidates for further study.

4.6.2. In Silico Molecular Docking

The candidates were docked using the protocol above. Only compounds for which the most likely binding mode did not differ from the pharmacophore-predicted binding mode (RMSD < 1.5 Å) were selected.

4.6.3. Binding Free Energy

The binding free energy for the etravirine-, abacavir-, and IN1-CK1ε complexes was evaluated using the complete trajectories of the MD simulations (generated as described above). We used the Poisson–Boltzmann surface area (MM/PBSA) method [54] in the g_mmpbsa v1.6 package [55]. van der Walls, electrostatic, polar solvation, and solvent-accessible surface area energies were determined to calculate the average binding energy.

Per-residue decomposition analysis was performed to demonstrate the primary amino acids that were involved in stabilizing the systems.

5. Conclusions

In this study, two drugs that have been approved by the FDA for the treatment of HIV were identified as potential ATP-competitive CK1ε inhibitors. Both emulate the binding mode of known CK1ε inhibitors, but etravirine binds to CK1ε with greater stability and affinity. Our data encourage the evaluation of etravirine as an antineoplastic agent against CK1ε.

Supplementary Materials: The following are available online at https://www.mdpi.com/article/10.3390/ph15010008/s1, Figure S1: Docking protocol for selected inhibitors set; Figure S2: Salt bridge between Lys38 and Glu52 in CK1ε-Etravirine; Table S1: Hits from virtual screening.

Author Contributions: Conceptualization, L.C.-B. and M.A.V.-V.; methodology, A.A.S.-Á. and A.J.R.-M.; software, A.J.R.-M.; formal analysis, L.C.-B., A.A.S.-Á. and M.A.V.-V.; data curation, A.A.S.-Á.; writing—original draft preparation, L.C.-B. and M.A.V.-V.; writing—review and editing, L.C.-B., A.J.R.-M. and M.A.V.-V.; supervision, M.A.V.-V.; funding acquisition, L.C.-B. and M.A.V.-V. All authors have read and agreed to the published version of the manuscript.

Funding: This research was funded by PAPIIT-UNAM IV200121, LANCAD-UNAM-DGTIC-386, and CONACYT (project Cátedras CONACYT 639). The APC was funded by UNAM.

Institutional Review Board Statement: Not applicable.

Informed Consent Statement: Not applicable.

Data Availability Statement: Analyzed data are contained in the main text of the article. Raw data are available from the authors upon request.

Acknowledgments: A.J.R.-M. recognizes the received fellowship 584534 from CONACYT.

Conflicts of Interest: The authors declare no conflict of interest. The funders had no role in the design of the study; in the collection, analyses, or interpretation of data; in the writing of the manuscript, or in the decision to publish the results.

References

1. Schittek, B.; Sinnberg, T. Biological functions of casein kinase 1 isoforms and putative roles in tumorigenesis. *Mol. Cancer* **2014**, *13*, 231. [CrossRef] [PubMed]
2. Fulcher, L.J.; Sapkota, G.P. Functions and regulation of the serine/threonine protein kinase CK1 family: Moving beyond promiscuity. *Biochem. J.* **2020**, *477*, 4603–4621. [CrossRef]
3. Cao, X.; Yang, Y.; Selby, C.P.; Liu, Z.; Sancar, A. Molecular mechanism of the repressive phase of the mammalian circadian clock. *Proc. Natl. Acad. Sci. USA* **2021**, *118*, e2021174118. [CrossRef]
4. Knippschild, U.; Wolff, S.; Giamas, G.; Brockschmidt, C.; Wittau, M.; Würl, P.U.; Eismann, T.; Stöter, M. The role of the casein kinase 1 (CK1) family in different signaling pathways linked to cancer development. *Oncol. Res. Treat.* **2005**, *28*, 508–514. [CrossRef]
5. Knippschild, U.; Krüger, M.; Richter, J.; Xu, P.; García-Reyes, B.; Peifer, C.; Halekotte, J.; Bakulev, V.; Bischof, J. The CK1 family: Contribution to cellular stress response and its role in carcinogenesis. *Front. Oncol.* **2014**, *4*, 96. [CrossRef]
6. Janovská, P.; Normant, E.; Miskin, H.; Bryja, V. Targeting Casein Kinase 1 (CK1) in Hematological Cancers. *Int. J. Mol. Sci.* **2020**, *21*, 9026. [CrossRef] [PubMed]
7. Vinyoles, M.; Del Valle-Pérez, B.; Curto, J.; Padilla, M.; Villarroel, A.; Yang, J.; de Herreros, A.G.; Duñach, M. Activation of CK1ε by PP2A/PR61ε is required for the initiation of Wnt signaling. *Oncogene* **2017**, *36*, 429–438. [CrossRef] [PubMed]
8. Lin, S.-H.; Lin, Y.-M.; Yeh, C.-M.; Chen, C.-J.; Chen, M.-W.; Hung, H.-F.; Yeh, K.-T.; Yang, S.-F. Casein kinase 1 epsilon expression predicts poorer prognosis in low T-stage oral cancer patients. *Int. J. Mol. Sci.* **2014**, *15*, 2876–2891. [CrossRef]
9. Swiatek, W.; Tsai, I.-C.; Klimowski, L.; Pepler, A.; Barnette, J.; Yost, H.J.; Virshup, D.M. Regulation of casein kinase Iϵ activity by Wnt signaling. *J. Biol. Chem.* **2004**, *279*, 13011–13017. [CrossRef]
10. Klimowski, L.K.; Garcia, B.A.; Shabanowitz, J.; Hunt, D.F.; Virshup, D.M. Site-specific casein kinase 1ε-dependent phosphorylation of Dishevelled modulates β-catenin signaling. *FEBS J.* **2006**, *273*, 4594–4602. [CrossRef]
11. Amit, S.; Hatzubai, A.; Birman, Y.; Andersen, J.S.; Ben-Shushan, E.; Mann, M.; Ben-Neriah, Y.; Alkalay, I. Axin-mediated CKI phosphorylation of β-catenin at Ser 45: A molecular switch for the Wnt pathway. *Genes Dev.* **2002**, *16*, 1066–1076. [CrossRef]

12. Varghese, R.T.; Young, S.; Pham, L.; Liang, Y.; Pridham, K.J.; Guo, S.; Murphy, S.; Kelly, D.F.; Sheng, Z. Casein kinase 1 epsilon regulates glioblastoma cell survival. *Sci. Rep.* **2018**, *8*, 13621. [CrossRef]
13. Kani, S.; Oishi, I.; Yamamoto, H.; Yoda, A.; Suzuki, H.; Nomachi, A.; Iozumi, K.; Nishita, M.; Kikuchi, A.; Takumi, T. The receptor tyrosine kinase Ror2 associates with and is activated by casein kinase Iε. *J. Biol. Chem.* **2004**, *279*, 50102–50109. [CrossRef] [PubMed]
14. Zhang, S.; Chen, L.; Cui, B.; Chuang, H.-Y.; Yu, J.; Wang-Rodriguez, J.; Tang, L.; Chen, G.; Basak, G.W.; Kipps, T.J. ROR1 is expressed in human breast cancer and associated with enhanced tumor-cell growth. *PLoS ONE* **2012**, *7*, e31127. [CrossRef]
15. Curto, J.; Del Valle-Pérez, B.; Villarroel, A.; Fuertes, G.; Vinyoles, M.; Peña, R.; García de Herreros, A.; Duñach, M. CK 1ε and p120-catenin control Ror2 function in noncanonical Wnt signaling. *Mol. Oncol.* **2018**, *12*, 611–629. [CrossRef]
16. Rodriguez, N.; Yang, J.; Hasselblatt, K.; Liu, S.; Zhou, Y.; Rauh-Hain, J.A.; Ng, S.; Choi, P.; Fong, W.; Agar, N.Y.R. Casein kinase I epsilon interacts with mitochondrial proteins for the growth and survival of human ovarian cancer cells. *EMBO Mol. Med.* **2012**, *4*, 952–963. [CrossRef]
17. Janovska, P.; Verner, J.; Kohoutek, J.; Bryjova, L.; Gregorova, M.; Dzimkova, M.; Skabrahova, H.; Radaszkiewicz, T.; Ovesna, P.; Vondalova Blanarova, O. Casein kinase 1 is a therapeutic target in chronic lymphocytic leukemia. *Blood J. Am. Soc. Hematol.* **2018**, *131*, 1206–1218. [CrossRef]
18. Brockschmidt, C.; Hirner, H.; Huber, N.; Eismann, T.; Hillenbrand, A.; Giamas, G.; Radunsky, B.; Ammerpohl, O.; Bohm, B.; Henne-Bruns, D. Anti-apoptotic and growth-stimulatory functions of CK1 delta and epsilon in ductal adenocarcinoma of the pancreas are inhibited by IC261 in vitro and in vivo. *Gut* **2008**, *57*, 799–806. [CrossRef] [PubMed]
19. Cheong, J.K.; Hung, N.T.; Wang, H.; Tan, P.; Voorhoeve, P.M.; Lee, S.H.; Virshup, D.M. IC261 induces cell cycle arrest and apoptosis of human cancer cells via CK1δ/ε and Wnt/β-catenin independent inhibition of mitotic spindle formation. *Oncogene* **2011**, *30*, 2558–2569. [CrossRef]
20. Xu, R.-M.; Carmel, G.; Sweet, R.M.; Kuret, J.; Cheng, X. Crystal structure of casein kinase-1, a phosphate-directed protein kinase. *EMBO J.* **1995**, *14*, 1015–1023. [CrossRef] [PubMed]
21. Johnson, L.N.; Noble, M.E.M.; Owen, D.J. Active and inactive protein kinases: Structural basis for regulation. *Cell* **1996**, *85*, 149–158. [CrossRef]
22. Long, A.M.; Zhao, H.; Huang, X. Structural basis for the potent and selective inhibition of casein kinase 1 epsilon. *J. Med. Chem.* **2012**, *55*, 10307–10311. [CrossRef] [PubMed]
23. Vulpetti, A.; Bosotti, R. Sequence and structural analysis of kinase ATP pocket residues. *Farmaco* **2004**, *59*, 759–765. [CrossRef] [PubMed]
24. Bischof, J.; Leban, J.; Zaja, M.; Grothey, A.; Radunsky, B.; Othersen, O.; Strobl, S.; Vitt, D.; Knippschild, U. 2-Benzamido-N-(1 H-benzo [d] imidazol-2-yl) thiazole-4-carboxamide derivatives as potent inhibitors of CK1δ/ε. *Amino Acids* **2012**, *43*, 1577–1591. [CrossRef] [PubMed]
25. Mashhoon, N.; DeMaggio, A.J.; Tereshko, V.; Bergmeier, S.C.; Egli, M.; Hoekstra, M.F.; Kuret, J. Crystal structure of a conformation-selective casein kinase-1 inhibitor. *J. Biol. Chem.* **2000**, *275*, 20052–20060. [CrossRef]
26. Richter, J.; Bischof, J.; Zaja, M.; Kohlhof, H.; Othersen, O.; Vitt, D.; Alscher, V.; Pospiech, I.; Garcia-Reyes, B.; Berg, S. Difluoro-dioxolo-benzoimidazol-benzamides as potent inhibitors of CK1δ and ε with nanomolar inhibitory activity on cancer cell proliferation. *J. Med. Chem.* **2014**, *57*, 7933–7946. [CrossRef] [PubMed]
27. Walton, K.M.; Fisher, K.; Rubitski, D.; Marconi, M.; Meng, Q.-J.; Sladek, M.; Adams, J.; Bass, M.; Chandrasekaran, R.; Butler, T. Selective inhibition of casein kinase 1ε minimally alters circadian clock period. *J. Pharmacol. Exp. Ther.* **2009**, *330*, 430–439. [CrossRef] [PubMed]
28. Dhillon, S.; Keam, S.J. Umbralisib: First Approval. *Drugs* **2021**, *81*, 857–866. [CrossRef]
29. Menck, K.; Heinrichs, S.; Baden, C.; Bleckmann, A. The WNT/ROR Pathway in Cancer: From Signaling to Therapeutic Intervention. *Cells* **2021**, *10*, 142. [CrossRef]
30. Antoszczak, M.; Markowska, A.; Markowska, J.; Huczyński, A. Old wine in new bottles: Drug repurposing in oncology. *Eur. J. Pharmacol.* **2020**, *866*, 172784. [CrossRef]
31. Sleire, L.; Førde, H.E.; Netland, I.A.; Leiss, L.; Skeie, B.S.; Enger, P.Ø. Drug repurposing in cancer. *Pharmacol. Res.* **2017**, *124*, 74–91. [CrossRef] [PubMed]
32. Méndez-Lucio, O.; Naveja, J.J.; Vite-Caritino, H.; Prieto-Martínez, F.D.; Medina-Franco, J.L. One drug for multiple targets: A computational perspective. *J. Mex. Chem. Soc.* **2016**, *60*, 168–181. [CrossRef]
33. Talevi, A.; Bellera, C.L. Challenges and opportunities with drug repurposing: Finding strategies to find alternative uses of therapeutics. *Expert Opin. Drug Discov.* **2020**, *15*, 397–401. [CrossRef]
34. Garcia-Reyes, B.; Witt, L.; Jansen, B.; Karasu, E.; Gehring, T.; Leban, J.; Henne-Bruns, D.; Pichlo, C.; Brunstein, E.; Baumann, U. Discovery of inhibitor of wnt production 2 (IWP-2) and related compounds as selective ATP-competitive inhibitors of casein kinase 1 (CK1) δ/ε. *J. Med. Chem.* **2018**, *61*, 4087–4102. [CrossRef]
35. Halekotte, J.; Witt, L.; Ianes, C.; Krüger, M.; Bührmann, M.; Rauh, D.; Pichlo, C.; Brunstein, E.; Luxenburger, A.; Baumann, U. Optimized 4, 5-diarylimidazoles as potent/selective inhibitors of protein kinase CK1δ and their structural relation to p38α MAPK. *Molecules* **2017**, *22*, 522. [CrossRef] [PubMed]

36. Sato, K.; Padgaonkar, A.A.; Baker, S.J.; Cosenza, S.C.; Rechkoblit, O.; Subbaiah, D.R.C.; Domingo-Domenech, J.; Bartkowski, A.; Port, E.R.; Aggarwal, A.K. Simultaneous CK2/TNIK/DYRK1 inhibition by 108600 suppresses triple negative breast cancer stem cells and chemotherapy-resistant disease. *Nat. Commun.* **2021**, *12*, 4671. [CrossRef]
37. McGregor, M.J. A pharmacophore map of small molecule protein kinase inhibitors. *J. Chem. Inf. Model.* **2007**, *47*, 2374–2382. [CrossRef]
38. Wang, Y.; Sun, Y.; Cao, R.; Liu, D.; Xie, Y.; Li, L.; Qi, X.; Huang, N. In silico identification of a novel hinge-binding scaffold for kinase inhibitor discovery. *J. Med. Chem.* **2017**, *60*, 8552–8564. [CrossRef] [PubMed]
39. Cozza, G.; Gianoncelli, A.; Montopoli, M.; Caparrotta, L.; Venerando, A.; Meggio, F.; Pinna, L.A.; Zagotto, G.; Moro, S. Identification of novel protein kinase CK1 delta (CK1δ) inhibitors through structure-based virtual screening. *Bioorg. Med. Chem. Lett.* **2008**, *18*, 5672–5675. [CrossRef] [PubMed]
40. Bolcato, G.; Cescon, E.; Pavan, M.; Bissaro, M.; Bassani, D.; Federico, S.; Spalluto, G.; Sturlese, M.; Moro, S. A Computational Workflow for the Identification of Novel Fragments Acting as Inhibitors of the Activity of Protein Kinase CK1δ. *Int. J. Mol. Sci.* **2021**, *22*, 9741. [CrossRef]
41. Esposito, V.; Perna, A.; Lucariello, A.; Carleo, M.A.; Viglietti, R.; Sangiovanni, V.; Coppola, N.; Guerra, G.; De Luca, A.; Chirianni, A. Different impact of antiretroviral drugs on bone differentiation in an in vitro model. *J. Cell. Biochem.* **2015**, *116*, 2188–2194. [CrossRef]
42. Jarmolinska, A.I.; Kadlof, M.; Dabrowski-Tumanski, P.; Sulkowska, J.I. GapRepairer: A server to model a structural gap and validate it using topological analysis. *Bioinformatics* **2018**, *34*, 3300–3307. [CrossRef]
43. Persi, E.; Duran-Frigola, M.; Damaghi, M.; Roush, W.R.; Aloy, P.; Cleveland, J.L.; Gillies, R.J.; Ruppin, E. Systems analysis of intracellular pH vulnerabilities for cancer therapy. *Nat. Commun.* **2018**, *9*, 2997. [CrossRef] [PubMed]
44. Kriz, K.; Řezáč, J. Benchmarking of semiempirical quantum-mechanical methods on systems relevant to computer-aided drug design. *J. Chem. Inf. Model.* **2020**, *60*, 1453–1460. [CrossRef] [PubMed]
45. Bitencourt-Ferreira, G.; de Azevedo, W.F. Molegro virtual docker for docking. In *Docking Screens for Drug Discovery*; Springer: Berlin/Heidelberg, Germany, 2019; pp. 149–167.
46. Varela-Salinas, G.; García-Pérez, C.A.; Peláez, R.; Rodríguez, A.J. Visual clustering approach for docking results from vina and autodock. In Proceedings of the International Conference on Hybrid Artificial Intelligence Systems, La Rioja, Spain, 21–23 June 2017; Springer: Berlin/Heidelberg, Germany, 2017; pp. 342–353.
47. Van Der Spoel, D.; Lindahl, E.; Hess, B.; Groenhof, G.; Mark, A.E.; Berendsen, H.J.C. GROMACS: Fast, flexible, and free. *J. Comput. Chem.* **2005**, *26*, 1701–1718. [CrossRef] [PubMed]
48. Lee, J.; Cheng, X.; Swails, J.M.; Yeom, M.S.; Eastman, P.K.; Lemkul, J.A.; Wei, S.; Buckner, J.; Jeong, J.C.; Qi, Y. CHARMM-GUI input generator for NAMD, GROMACS, AMBER, OpenMM, and CHARMM/OpenMM simulations using the CHARMM36 additive force field. *J. Chem. Theory Comput.* **2016**, *12*, 405–413. [CrossRef]
49. Kim, S.; Lee, J.; Jo, S.; Brooks, C.L., III; Lee, H.S.; Im, W. CHARMM-GUI ligand reader and modeler for CHARMM force field generation of small molecules. *J. Comput. Chem.* **2017**, *38*, 1879–1886. [CrossRef]
50. Gowers, R.J.; Linke, M.; Barnoud, J.; Reddy, T.J.E.; Melo, M.N.; Seyler, S.L.; Domanski, J.; Dotson, D.L.; Buchoux, S.; Kenney, I.M. *MDAnalysis: A Python Package for the Rapid Analysis of Molecular Dynamics Simulations*; Los Alamos National Lab.(LANL): Los Alamos, NM, USA, 2019.
51. Adasme, M.F.; Linnemann, K.L.; Bolz, S.N.; Kaiser, F.; Salentin, S.; Haupt, V.J.; Schroeder, M. PLIP 2021: Expanding the scope of the protein-ligand interaction profiler to DNA and RNA. *Nucleic Acids Res.* **2021**, *49*, W530–W534. [CrossRef] [PubMed]
52. Humphrey, W.; Dalke, A.; Schulten, K. VMD: Visual molecular dynamics. *J. Mol. Graph.* **1996**, *14*, 33–38. [CrossRef]
53. Sunseri, J.; Koes, D.R. Pharmit: Interactive exploration of chemical space. *Nucleic Acids Res.* **2016**, *44*, W442–W448. [CrossRef]
54. Wang, C.; Greene, D.; Xiao, L.; Qi, R.; Luo, R. Recent developments and applications of the MMPBSA method. *Front. Mol. Biosci.* **2018**, *4*, 87. [CrossRef] [PubMed]
55. Kumari, R.; Kumar, R.; Open Source Drug Discovery Consortium; Lynn, A. g_mmpbsa-A GROMACS tool for high-throughput MM-PBSA calculations. *J. Chem. Inf. Model.* **2014**, *54*, 1951–1962. [CrossRef] [PubMed]

pharmaceuticals

Article

Design and Synthesis of Aminopyrimidinyl Pyrazole Analogs as PLK1 Inhibitors Using Hybrid 3D-QSAR and Molecular Docking

Swapnil P. Bhujbal [1,2], Hyejin Kim [1,2], Hyunah Bae [1,2] and Jung-Mi Hah [1,2,*]

1 Department of Pharmacy, College of Pharmacy, Hanyang University, Ansan 426-791, Korea
2 Institute of Pharmaceutical Science and Technology, Hanyang University, Ansan 426-791, Korea
* Correspondence: jhah@hanyang.ac.kr; Tel.: +82-31-400-5803

Abstract: Cancer continues to be one of the world's most severe public health issues. Polo-like kinase 1 (PLK1) is one of the most studied members of the polo-like kinase subfamily of serine/threonine protein kinases. PLK1 is a key mitotic regulator responsible for cell cycle processes, such as mitosis initiation, bipolar mitotic spindle formation, centrosome maturation, the metaphase to anaphase transition, and mitotic exit, whose overexpression is often associated with oncogenesis. Moreover, it is also involved in DNA damage response, autophagy, cytokine signaling, and apoptosis. Due to its fundamental role in cell cycle regulation, PLK1 has been linked to various types of cancer onset and progression, such as lung, colon, prostate, ovary, breast cancer, melanoma, and AML. Hence, PLK1 is recognized as a critical therapeutic target in the treatment of various proliferative diseases. PLK1 inhibitors developed in recent years have been researched and studied through clinical trials; however, most of them have failed because of their toxicity and poor therapeutic response. To design more potent and selective PLK1 inhibitors, we performed a receptor-based hybrid 3D-QSAR study of two datasets, possessing similar common scaffolds. The developed hybrid CoMFA (q^2 = 0.628, r^2 = 0.905) and CoMSIA (q^2 = 0.580, r^2 = 0.895) models showed admissible statistical results. Comprehensive, molecular docking of one of the most active compounds from the dataset and hybrid 3D-QSAR studies revealed important active site residues of PLK1 and requisite structural characteristics of ligand to design potent PLK1 inhibitors. Based on this information, we have proposed approximately 38 PLK1 inhibitors. The newly designed PLK1 inhibitors showed higher activity (predicted pIC_{50}) than the most active compounds of all the derivatives selected for this study. We selected and synthesized two compounds, which were ultimately found to possess good IC_{50} values. Our design strategy provides insight into development of potent and selective PLK1 inhibitors.

Keywords: cancer; PLK1; kinase; 3D-QSAR; molecular docking; inhibitors

Citation: Bhujbal, S.P.; Kim, H.; Bae, H.; Hah, J.-M. Design and Synthesis of Aminopyrimidinyl Pyrazole Analogs as PLK1 Inhibitors Using Hybrid 3D-QSAR and Molecular Docking. *Pharmaceuticals* **2022**, *15*, 1170. https://doi.org/10.3390/ph15101170

Academic Editors: Maria Emília De Sousa and Marialuigia Fantacuzzi

Received: 17 August 2022
Accepted: 16 September 2022
Published: 21 September 2022

1. Introduction

The genetic stability of all eukaryotes is governed by the flawless segregation of chromosomes during mitosis. Disruption of this phenomenon can lead to aneuploidy, which is a vital cause of cancer [1]. Cancer continues to be one of the most serious public health problems worldwide. As per the statistics of WHO, cancer is the second major cause of death worldwide, with approximately 9.6 million deaths in 2018. Men are more likely to develop prostate, lung, colorectal, stomach, and liver cancer than women, who are more likely to develop colorectal, breast, lung, thyroid, and cervical cancer. Protein kinase inhibitors developed to treat various cancer types have been of great advantage in recent years [1]. However, these inhibitors are merely efficient and are limited by the emergence of resistant mutations [1,2]. Interestingly, polo-like kinase 1 (PLK1), which is an important type of serine-threonine kinase, controls several phases of mitosis [1,3]. PLK1 is a primary regulator of mitotic progression, the overexpression of which is often related to oncogenesis and is an important therapeutic target for anticancer drug discovery [4].

There are five isoforms of PLKs in humans, referred to as PLK1, PLK2, PLK3, PLK4, and PLK5 [5,6]. PLK1 is highly characterized among them and has the utmost degree of homology with the single PLK gene [5]. PLK1 is structurally defined by the presence of 3 functional domains; a disordered and highly conserved N-terminal domain, C-terminal polo-box domain (PBD) for substrate targeting and implied in its subcellular localization, and a kinase domain (KD) that is regulated through phosphorylation by upstream kinases [7–9]. The N-terminal domain is a Ser/Thr kinase domain that has a T-loop, whose phosphorylation is directly associated with the PLK1's kinase activity [6]. The PBD is C-terminal domain, made of two polo-box structures, which is activated upon ligand binding and separates with T-loop of the kinase domain [6,10].

PLK1 is predominantly related to regulation of the cell cycle, disruption of which is the primary cause of cancer [10]. In addition to its role in the activity of tumor suppressors and oncogenes, PLK1 has a distinctive function in regulating cancer cell metabolism, promoting the growth of cancer cells [5]. It also has an indisputable role in controlling several key transcription factors that promote cell proliferation, transformation, and epithelial-to-mesenchymal transition. PLK1 expression begins to increase from the S/G2 phase and peaks at mitosis [9]. PLK1 activates the cyclin B/cdc2 complex by phosphorylating CDC25, which stimulates cell proliferation [5,6]. A few recent studies have reported that PLK1 also plays role in autophagy, chromosomal stability, and DNA damage response. Considering the crucial role of PLK1 during DNA damage repair and cell-cycle regulation, it is not astonishing that it is upregulated in various types of cancers, such as melanoma, colorectal cancer, non-small cell lung cancer, thyroid carcinoma, esophageal carcinoma, ovarian carcinoma, colorectal cancer, breast cancer, and prostate cancer [3–5]. Moreover, PLK1 is linked to diverse immune and neurological disorders, such as graft versus host disease, liver fibrosis, Huntington's disease, and Alzheimer's disease [11–13].

Consequently, inhibition of PLK1 has been an appealing goal for researchers, who have made substantial efforts to design and develop small molecule PLK1 inhibitors (Figure 1). The FDA has approved at least 55 small molecule anticancer kinase inhibitors as of February 2020 [14]. Researchers from Boehringer Ingelheim developed one of the earliest PLK1 inhibitors, BI2536, a dihydropteridinone derivative. It was the first pioneered PLK1 inhibitor to enter clinical trials for the treatment of cancer, however it was not very effective during monotherapy regimens in clinical trials [5,10,14]. One more Boehringer Ingelheim compound, volasertib (aka BI6727), is an ATP-competitive inhibitor that was developed by modifying the chemical structure of BI2536 and was shown to inhibit the growth of acute myeloid leukemia (AML) cell lines. However, when combined with Trematinib, BI6727 was more effective at arresting dual G1 and G2-M in an NRAS-mutant melanoma cell line [15]. Both of these drugs passed phase I and III clinical trials for breast cancer, B-cell lymphoma and AML but were found to be more effective in combination with other drugs and are no longer used in monotherapy [5,16].

Onvansertib is a 3rd generation, highly selective and oral PLK1 NCD (N-terminal catalytic domain) inhibitor that exhibited promising synergic results during clinical trials as part of a combination regimen treatment of AML and metastatic colorectal cancer [6,15]. Currently, it is the only PLK1 inhibitor under clinical trials for solid tumors. In addition, Rigosertib is presently in Phase III clinical trials for treatment of cancers, including AML. Nevertheless, it is a non ATP-competitive dual inhibitor of PI3K and PLK1 [6,15]. Furthermore, Phase I clinical trials for two PLK1 NCD inhibitors, GSK461364, and Tak960, were finished in 2009 and 2013, respectively, but did not undergo further development [16]. PLK1 inhibitor, NMS-P937 (pyrazoloquinazoline derivative), has shown remarkable reduction of osteosarcoma tumor growth. NMS-P937 is under phase 1 trials for patients having advanced metastatic solid tumors, nevertheless, no findings have yet been published [16,17].

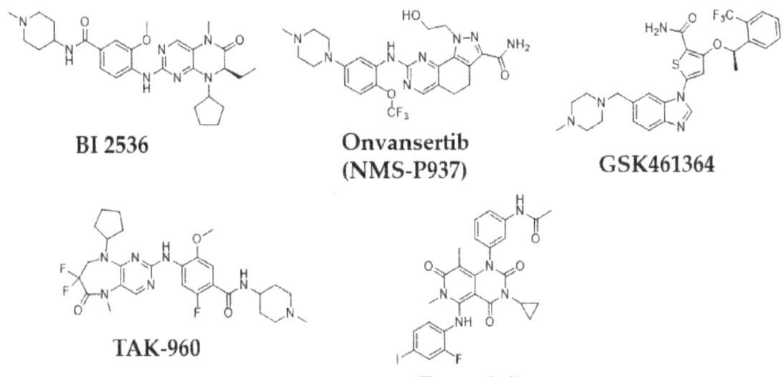

Figure 1. Chemical structures of the previously reported PLK1 inhibitors.

In summary, there are numerous small molecule PLK1 inhibitors in research and in the early stages of clinical progress. Nonetheless, most of these convincing drugs targeting PLK1 have been inefficient in clinical trials due to their toxicity leading to serious side effects or low therapeutic response [15,17]. PLK1 inhibitors that are highly selective are anticipated to overcome side effects impelled by off-target effects. An effective way is to design selective PLK1 inhibitors by using structure-based drug design methods, as in our study. We selected two datasets as PLK1 inhibitors with similar scaffolds, such as pyrimidine derivatives [3] and quinazoline derivatives [4]. We chose compound **17** (Table S1) from dataset 1 as a reference compound (as shown in Figure 2) to combine the remaining compounds from both datasets for hybrid 3D-QSAR because a detailed mechanistic study of compound **17** showed that PLK1 inhibition by **17** enhanced mitotic arrest at the G2/M phase checkpoint and led to apoptosis of cancer cells. Compound **17** is also one of the most active compounds among datasets 1 and 2.

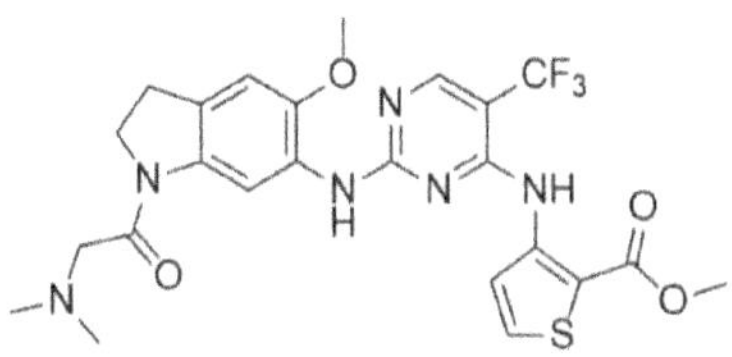

Figure 2. Chemical structure of Compound **17** (The most active compound used in our study).

The docking of compound **17** into the active site of PLK1 revealed important interactions with crucial binding site residues, which were accountable for inhibition of PLK1. Combining the specific structural information from 3D-QSAR contour maps with the overall molecular docking analysis about ligand–protein interactions provided insight toward understanding and modifying pyrimidine scaffold of PLK1 inhibitors for better potency. Hence, we designed few PLK1 inhibitors of aminopyrimidinyl scaffold, which showed better predicted activity (pIC_{50}) than the most active compound in the datasets used in our study. Among our designed compounds, we synthesized two compounds for experimental validation, which showed good inhibitory activity (IC_{50}). Our designed strategy, which was validated by experimental study could be a great start to develop potent and selective PLK1 inhibitors for medicinal chemists and pharmaceutical companies.

2. Results and Discussion

2.1. Molecular Docking

Molecular docking was performed to interpret the binding mode of the most active compound, **17**, inside the active site of PLK1. We utilized an extended sampling protocol to perform induced fit docking that produced 80 binding poses for compound **17**. All 80 poses were checked for a docking score and bonding and non-bonding interactions. One of the poses displayed a docking score of −12.04 and yielded a binding pose with PLK1 that was similar to that witnessed between the co-crystallized ligand (BI6727) and the protein (PDB ID: 3FC2) (Figure 3). This pose was chosen to analyze further interactions. The most active compound **17** docked within the binding pocket, mimicked the binding mode of ATP, and formed three hydrogen bonds (H-bonds) with crucial active site residues of PLK1. The hydrogen and nitrogen atoms from the aminopyrimidine ring of compound **17** formed 2 H-bonds with the key hinge region residue CYS133. This interaction was also observed in previously reported docking studies of several other PLK1 inhibitors including BI6727, which was considered important for inhibition of PLK1 [3,4]. Another H-bond was found between the oxygen atom from the dimethylamino-propan-2-one moiety at R^3 position and the residue SER137. This moiety docked very well within a pocket lined primarily with residues of the kinase C-lobe, including SER137 that formed the floor of the C-lobe [4]. In addition, pyrimidine ring formed pi–pi interaction with GLU131, which is also a part of the hinge region. Overall interactions were alike to those observed between a co-crystallized ligand and PLK1. Hence, the docked pose of compound **17** validated a strong binding conformation.

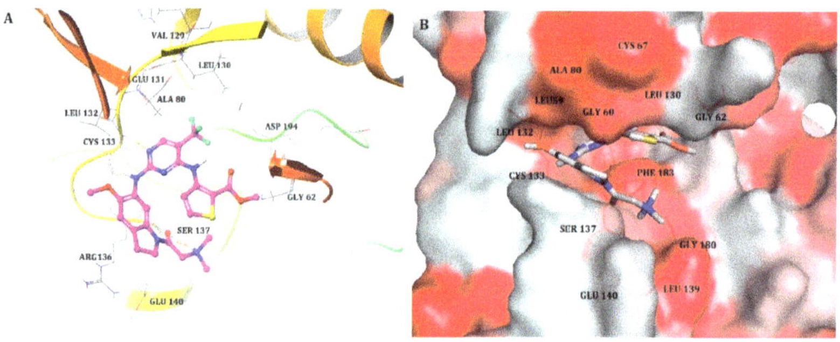

Figure 3. (**A**) Docked pose of the most active compound, **17**, within the active site of PLK1 (Hydrogen bonds are represented as yellow dotted lines, pi-pi interactions are represented as cyan dotted lines); (**B**) The most active compound **17** (shown in stick model) within the hydrophobic pocket of PLK1; the red colored region represents the most hydrophobic surface of the protein, and the white color represents the least hydrophobic surface. Hydrophobic residues are indicated with red lines.

The docked pose of compound **17** was further analyzed to assess hydrophobic interactions. A Python script 'color h' was utilized to color the hydrophobic residues of PLK1 and to identify their interactions with compound **17**. PyMOL uses this script and Eisenberg hydrophobicity scale (Figure 3b) to color the receptor [18]. The most hydrophobic residues were colored red, whereas the least hydrophobic residues were colored white. The substituents, trifluoromethylpyrimidine at the R^2 position and methylthiophene-2-carboxylate at the R^1 position of the ligand, were docked within the hydrophobic pocket lined by hydrophobic residues CYS67, ALA80, LYS82, LEU59, and LEU130. Additionally, the dimethylamine group at R^3 position formed hydrophobic interactions with LEU139 and GLY180. The hydrophobic residues CYS67, ALA80, LYS82, and LEU130 were crucial in the hydrophobic pocket, since their interaction with compound BI6727 (co-crystallized ligand) was observed. Comprehensive docking analysis suggested that the selected docked pose of the most potent compound **17** was appropriate, thus it was employed to further perform 3D-QSAR studies.

2.2. Hybrid 3D-QSAR Models

We obtained the receptor-based hybrid 3D-QSAR models (CoMFA and CoMSIA) after combining two datasets with pyrimidine and quinazoline scaffolds. These two datasets were combined in order to acquire and understand the structural characteristics requisite to propose more potent PLK1 inhibitors and to study structure–activity relationships. The docked pose of the most active compound, compound **17** was selected as a reference compound to align the remaining compounds of the dataset using a common scaffold alignment method (Figure 4), which provided better statistical CoMFA and CoMSIA models in SYBYL-X 2.1. The dataset was separated into a training set of 52 compounds and a test set of 18 compounds using the criteria proposed in activity ranking algorithm 4 in a previously reported article [19]. Accordingly, our test set contained compounds with low, moderate, and high activity (pIC_{50}) values. The test set compounds are specified by * in Table S1 entitled the chemical structures of the selected PLK1 inhibitors with their IC_{50} values.

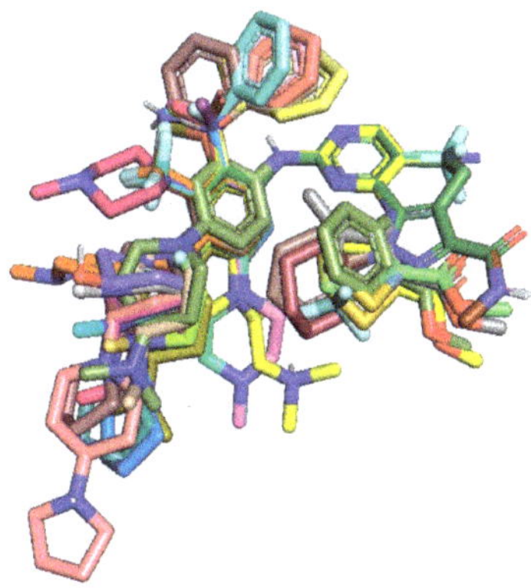

Figure 4. Alignment of the dataset compounds for hybrid 3D-QSAR.

The reliability of the developed hybrid 3D-QSAR models was examined by computing different statistical parameters, such as the non-cross validated correlation coefficient (r^2), cross-validated correlation coefficient (q^2), standard error of estimate (SEE), F-value, and the optimal number of components (ONC) with the help of partial least square (PLS) analysis. Initially, hybrid CoMFA (q^2 = 0.517, ONC = 6, r^2 = 0.847) and CoMSIA (q^2 = 0.540, ONC = 6, r^2 = 0.855) models were obtained for the complete dataset compounds (training + test set), which were called full models. Test set 12 was used to develop CoMFA (q^2 = 0.628, ONC = 6, r^2 = 0.905) and CoMSIA (q^2 = 0.580, ONC = 6, r^2 = 0.895) models. CoMFA models were generated using steric and electrostatic fields, while CoMSIA employed hydrogen bond acceptor, donor, and hydrophobic fields along with steric and electrostatic fields. Thus, CoMSIA models were produced using various combinations of these fields (Table S2). The model with the best q^2 and r^2 values was chosen as the concluding model. The combination of steric, electrostatic, and hydrophobic fields yielded reasonable CoMSIA model. In conclusion, CoMFA and CoMSIA models obtained using external test set 12 were selected for further statistical analysis. Several methods were used to validate these models. Table 1 provides thorough statistical values of the validated CoMFA and CoMSIA models.

Table 1. Detailed statistical values of the selected CoMFA and CoMSIA models.

Parameter	Full Model		Test Set 12	
	CoMFA	**CoMSIA (SEH)**	**CoMFA**	**CoMSIA (SEH)**
q^2	0.517	0.540	0.628	0.580
ONC	6	6	6	6
SEP	0.844	0.824	0.717	0.762
r^2	0.847	0.855	0.905	0.895
SEE	0.475	0.462	0.363	0.381
F value	58.087	61.993	71.401	63.990
LOF	-	-	0.607	0.609
BS-r^2	-	-	0.929	0.936
BS-SD	-	-	0.020	0.020
r^2pred	-	-	0.796	0.783
rm^2	-	-	0.665	0.581
Delta rm^2	-	-	0.181	0.214

q^2: squared cross-validated correlation coefficient; ONC: optimal number of components; SEP: standard error of prediction; r^2: squared correlation coefficient; SEE: standard error of estimation; F value: F-test value; LOF: leave-out-five; BS-r^2: bootstrapping r^2 mean; BS-SD: bootstrapping standard deviation; r^2_{pred}: predictive r^2; rm^2: average rm^2 metric calculation; Delta rm^2: standard error.

Validation of 3D-QSAR Models

The selected hybrid 3D-QSAR models were validated using a number of validation techniques to evaluate predictive ability and robustness. Validation techniques of predictive r^2 (r^2pred), bootstrapping, leave-out-five (LOF), and rm^2 metric calculations presented acceptable statistical values [20,21]. Hence, the generated models were robust and predictive. Table 1 depicts the detailed statistical values. Residual values as well as the experimental and predicted activity values of the selected CoMFA and CoMSIA models can be seen in Table S3 of the Supplementary Materials. The scatter plots for the same are depicted in Figure 5.

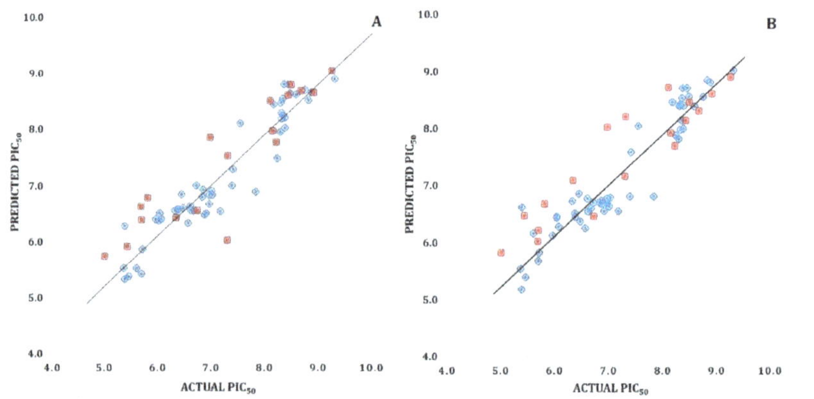

Figure 5. (**A**) Scatter plot for the selected CoMFA model; (**B**) Scatter plot for the selected CoMSIA model; the plot shows the actual pIC_{50} versus predicted pIC_{50} activity of the training and test sets; the training set compounds are represented as blue diamonds; the test set compounds are represented as dark red squares.

2.3. *Contour Map Analysis*

2.3.1. CoMFA Contour Maps

The contour maps of the hybrid CoMFA model are depicted superimposed with those of compound **17** in Figure 6. In the steric contour map (Figure 6A), green and yellow colored contour maps revealed favorable and unfavorable regions for steric substitution, respectively. In Figure 6B (electrostatic contour map), the blue contour denoted a favorable region for electropositive substitution, whereas the red contour denoted unfavorable regions.

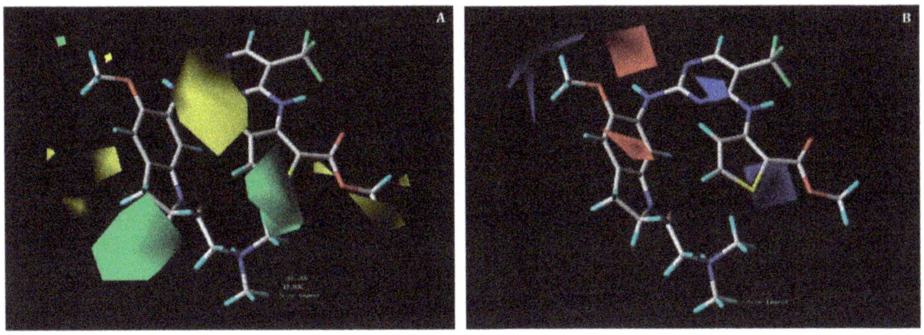

Figure 6. Contour maps for the selected CoMFA model. (**A**) Steric contour map; (**B**) electrostatic contour map; green contours show the region is favorable for bulky substitutions, and yellow contours show the region is unfavorable; blue contours favor electropositive substitutions, whereas red contours favor electronegative substitutions.

Two green colored contours were observed at the R^1 and R^3 positions, illustrating that bulky groups were favorable at these positions to increase potency. The R^1 position that holds a steric group could interact with the hydrophobic residues of PLK1. This could be signified by hydrophobic interactions of methylthiophene-2-carboxylate with residues GLY60 and LEU130, which were also identified in the docking analysis of compound **17**. Conversely, two yellow contour maps were spotted near the indoline ring at the R^3 position, which conveyed the inferiority of bulky groups at this position. There was also a small yellow contour near the methyl acetate moiety at the R^1 position, indicating that addition of bulky groups at this position might not increase potency.

Furthermore, red and blue contours were observed near the aminopyrimidine ring at the R^3 position among which, red contours seemed to be position specific. Therefore, the presence of blue contours at this position showed that electropositive groups were favorable, which could be verified by looking at key hydrogen bond interactions of amine groups with the hinge region residue CYS133 in the docking study of compound **17**. Another red contour was seen near the indoline ring at the R^3 position, which meant that electronegative substitution was favorable at this position. This was verified with a docking study of compound **17** where H-bond interactions were found between the oxygen atom from dimethylamino-propan-2-one moiety and the residue SER137. Compounds **13, 33**, and the most active compound **17**, which have moderate to high activity, could be the result of this.

2.3.2. CoMSIA Contour Maps

The combination of steric, electrostatic and hydrophobic (SEH) fields was utilized to produce CoMSIA contour maps (Figure 7). The steric and electrostatic contour maps are quite similar to the CoMFA steric and electrostatic contours, thus only a hydrophobic contour map is discussed below. The hydrophobic contour map is displayed in the Figure 7C, in which magenta contours represent favorable areas for hydrophobic substitution, however cyan contours denote unfavorable regions.

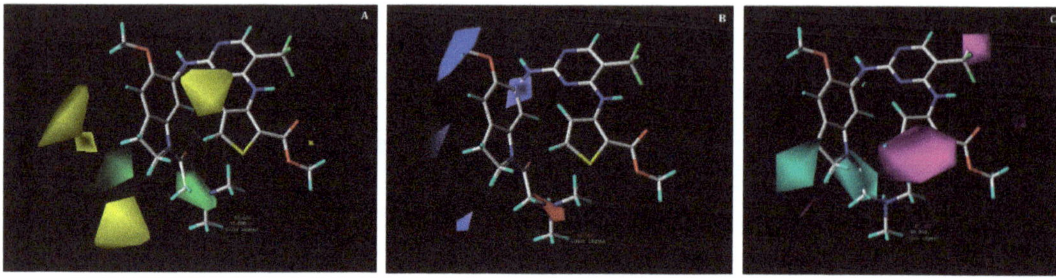

Figure 7. Contour maps for the selected CoMSIA model. (**A**) Steric contour map; (**B**) electrostatic contour map; (**C**) hydrophobic contour map. Green contours show regions that are favorable for bulky substitutions, whereas yellow contours show unfavorable regions; blue contours favor electropositive, while red does not. Magenta contours show the regions favorable for hydrophobic substitutions, whereas cyan contours show unfavorable regions.

Two big magenta contours are present at R^1 and R^2 positions, which illustrates that occurrence of hydrophobic group at this position could increase the activity of the compound. This can be proved by hydrophobic interactions of trifluoromethylpyrimidine and methylthiophene-2-carboxylate with residues GLY60, CYS67, ALA80, LYS82, LEU59, and LEU130 that were seen in our docking analysis of the compound **17**. Moreover, two cyan contour maps are present near the R^3 position, which explain that substituting hydrophobic groups at this location can reduce the activity of a compound.

2.4. Designing New PLK1 Inhibitors

The created 3D-QSAR models revealed crucial structural properties in terms of steric, electrostatic, and hydrophobic fields. These structural characteristics and the important interactions detected in the docking analysis of the most potent compound **17** were utilized to derive a drug design strategy with a new scaffold to design new PLK1 inhibitors (Figure 8). As per this design strategy, we modified the substituent at the R^1 position, while keeping others fixed to assess differences in activity. The same procedure was followed for each position and identified new PLK1 inhibitors. The methoxy group at the R^1 position, dimethylacetamide at the R^2 position, CF_3 at the R^3 position, amide ($CONH_2$) at the R^4 position, and chlorine at the R^6 position showed better activity (pIC_{50}) than the most potent compound **17** in the dataset.

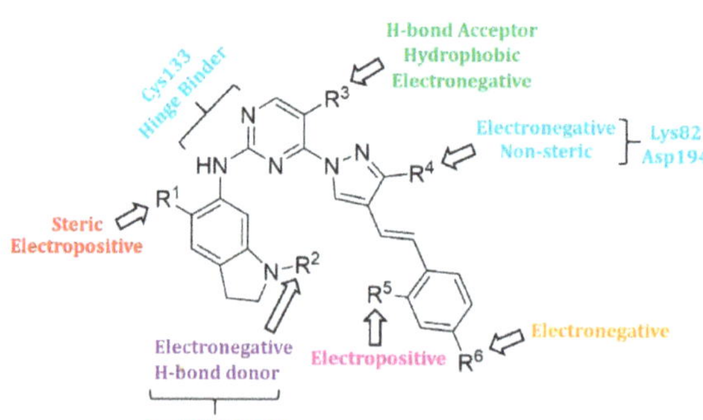

Figure 8. The design strategy for PLK1 inhibitors obtained from hybrid 3D-QSAR models.

The structures and the predicted pIC_{50} values of the newly designed compounds are presented in Table S4 of the Supplementary Materials. A few selected PLK1 inhibitors are shown below in Table 2.

Table 2. The structures and predicted pIC_{50} values of few selected designed PLK1 inhibitors.

Compound Structure	Name	R^1	R^2	R^3	R^4	R^5	R^6	Predicted pIC_{50}
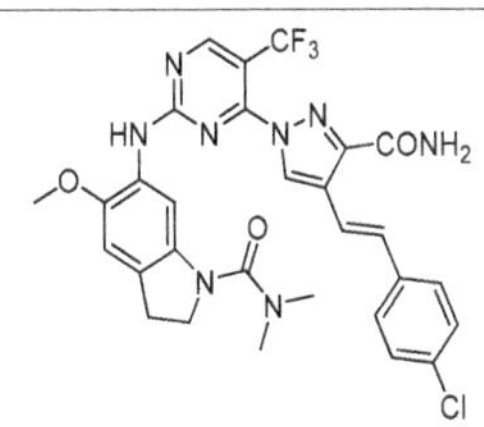	D3	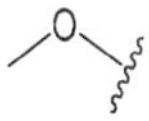	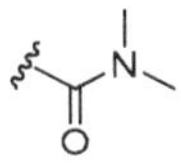	CF_3	$CONH_2$	H	Cl	10.217
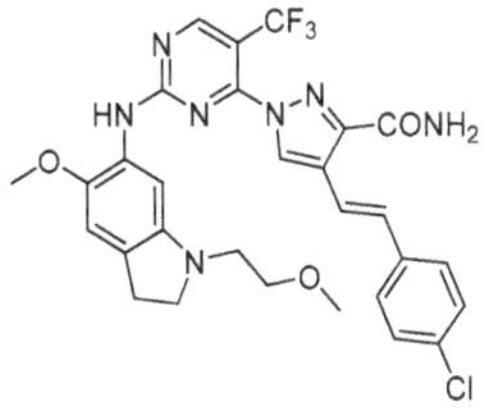	D5		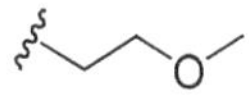					9.715
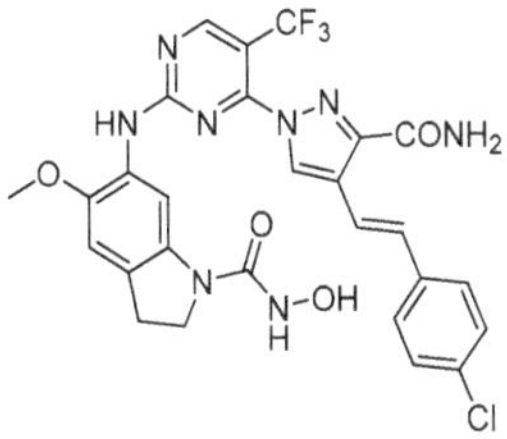	D10		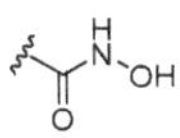					10.272
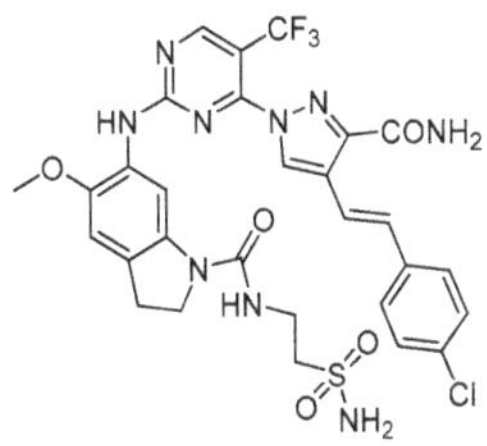	D14		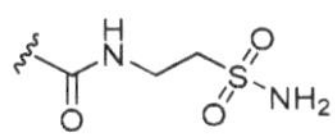					9.811
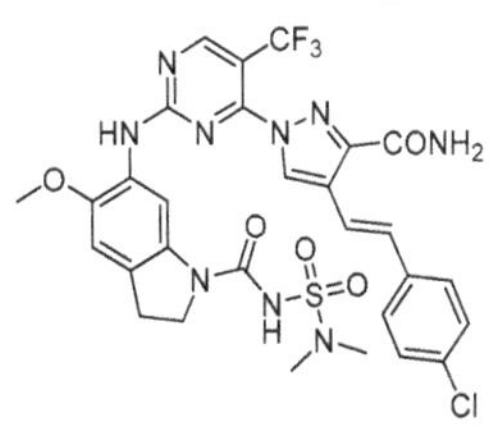	D17		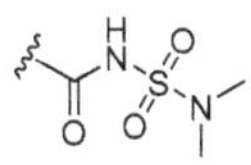					9.903

Synthesis of New PLK1 Inhibitors and Evaluation of IC_{50} Values

Depending on the their fastest synthetic route in order to validate our PLK1 drug design, we selected two compounds for the synthesis: compound D39 [ethyl €-1-(2-((1-(dimethylcarbamoyl)indolin-6-yl)amino)pyrimidin-4-yl)-4-styryl-1H-pyrazole-3-carboxylate] and compound D40 [(E)-6-((4-(3-carbamoyl-4-styryl-1H-pyrazol-1-yl)pyrimidin-2-yl)amino)-*N,N*-dimethylindoline-1-carboxamide]. NMR and HRMS data for these compounds are added in the methodology section. Synthesis of the rest of the compounds is under process, which upon completion will be published as a separate article with detailed procedure and experimental evaluations. Furthermore, we docked these two compounds inside the active site of PLK1 to check their binding mode (Figures S1 and S2 in Supplementary Materials). Both compounds formed two key hydrogen bonds with hinge region residue CYS133 similar to compound **17**. Compound D39 also formed pi–pi and pi–cation interactions with residues PHE183 and LYS82, respectively. Furthermore, a hydrogen from amide group at R^4 position of compound D40 formed additional hydrogen bond with residue ASP194 that is considered to be crucial for the selectivity over other kinases. It also possessed pi–pi interaction with PHE183.

Moreover, IC_{50} evaluation of compounds D39 and D40 revealed that they possess good inhibitory activities of 1.43 μM and 0.359 μM, respectively (Table 3). These results validates our designed compounds and design scheme that could be further utilized to develop more potent PLK1 inhibitors.

Table 3. The comparison of the actual and calculated IC_{50} values of two synthesized PLK1 inhibitors.

Compound Structure	Name	R^1	R^2	R^3	R^4	R^5	R^6	IC_{50} (μM)	Predicted IC_{50} (nM)
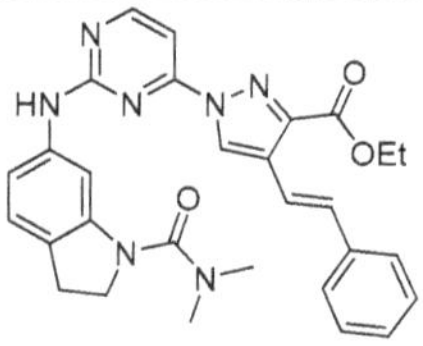	D39	H	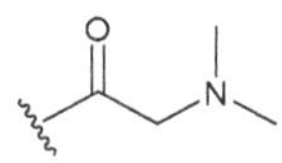	H	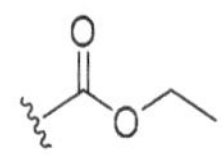	H	H	1.43	0.35
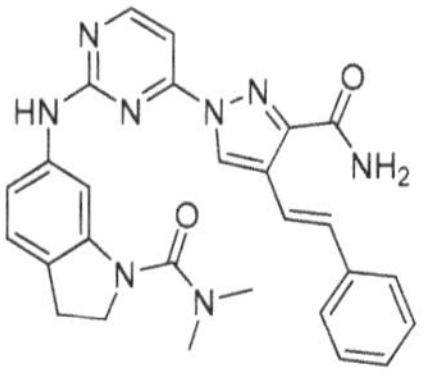	D40	H	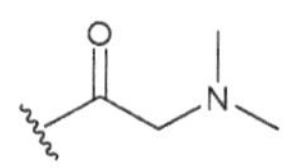	H	$CONH_2$	H	H	0.359	0.13

3. Materials and Methods

3.1. Training Set/Test Set Selection for CoMFA and CoMSIA

We selected two datasets comprising pyrimidine derivatives [3] and quinazoline derivatives [4] as PLK1 inhibitors for the hybrid 3D-QSAR study. Dataset 1 and dataset 2 consisted of 35 and 39 compounds, respectively, exhibiting a log value of greater than 3.5 logarithmic units, which was within the required range [22]. Compound structures were drawn using the sketch module in SYBYL-X 2.1 and were optimized using energy minimization with Tripos force field [23]. Biological activities (IC_{50}) of all compounds in the study were converted into pIC_{50} ($-\log IC_{50}$) values, which were used as dependent variables to develop 3D-QSAR models. The compounds from both the datasets were divided into training set of 52 compounds for model generation and 18 compounds as test set for model validation. The structure and activity of the compounds were considered in order to separate them into training and test sets. The compounds with low, medium, and high activity values were carefully added to the test set as suggested in Algorithm 4 [19]. Compounds with undefined activity values were eliminated as outliers. The chemical structures of the selected dataset compounds with their IC_{50} values are depicted in Table S1 of the Supporting Materials.

3.2. Molecular Docking

Molecular docking of the most active compound **17** from the selected dataset was performed using Schrodinger Maestro 12.8 (Release 2021-2, Schrödinger, LLC, New York, NY, USA) [24]. The structure of compound **17** was drawn using Chemdraw [25] and its 3D conformation was generated using the Schrödinger LigPrep program [26]. LigPrep produced all probable tautomers and states at pH 7.0 using Epik [27] for compound **17**, and specified chiralities were retained following minimization using the OPLS 2005 force field [28]. The crystal structure of PLK1 co-crystallized with BI6727 (PDB ID: 3FC2) was taken from the Protein Data Bank (PDB). The Protein Preparation Wizard was utilized to prepare protein by assigning bond orders, hydrogens at pH 7.0, and removing water molecules [29]. Prime was used to complete missing side chains and loops. Finally, a restrained minimization was performed using the default constraint of 0.30 Å RMSD and the OPLS 2005 force field to finalize the protein preparation. Molecular docking simulations were executed with the help of a Glide induced fit docking module in extended sampling protocol mode [30]. The docked conformations of compound **17** were examined to identify important interactions with the active site residues of PLK1. The selected docked pose of compound **17** was employed as a template to align the rest of the dataset compounds for 3D-QSAR model generation.

3.3. *Receptor-Based Hybrid CoMFA and CoMSIA Models*

SYBYL-X 2.1 [31] was used to develop 3D-QSAR (3-Dimensional Quantitative Structure–Activity Relationship) models utilizing CoMFA (Comparative Molecular Field Analysis) [32] and CoMSIA (Comparative Molecular Similarity Indices Analysis) [33] to correlate 3D structures of the PLK1 inhibitors with the biological activity. The alignment of dataset compounds was performed inside the active site of the receptor using a common scaffold alignment method using the most active compound **17** as a template molecule. In CoMFA, the steric and electrostatic potential energies were estimated using Lennard–Jones and Coulombic potentials, respectively [31].

The application of appropriate partial charge was crucial toward obtaining reasonable 3D-QSAR models. We used Gastegeir Marsili as a partial charge scheme and default parameters to generate 3D-QSAR models [34]. A grid spacing of 2.0 Å and an sp^3 hybridized carbon as a probe atom with +1 charge were used. Statistically acceptable CoMFA and CoMSIA models were obtained using partial least squares (PLS) regression. CoMFA descriptors as independent variables and biological activity values (pIC_{50}) as dependent variables were used in PLS regression. The reliability of the generated models was assessed through PLS analysis with leave-one-out (LOO) cross-validation and to calculate the squared cross-validated correlation coefficient (q^2) value, an optimal number of components (ONC) and the standard deviation of prediction (SEP). A column filtering value of 2.0 and obtained ONC were used in non-cross-validation analysis to compute the squared correlation coefficient (r^2), F-test value (F), and standard error of the estimate (SEE).

Nevertheless, CoMSIA utilizes descriptors, such as steric, electrostatic, hydrophobic, hydrogen bond acceptor, and donor. All of these CoMSIA similarity indices were calculated using a probe atom of radius 1.0 Å and an attenuation factor of 0.30. A Gaussian function was used to calculate the CoMSIA model between the grid point and each atom of the molecule. Various CoMSIA models were derived based on different descriptor combinations using the same lattice box that was used in CoMFA. The model that exhibited acceptable q^2 and r^2 statistical values was chosen as the final model. Additionally, the selected models were statistically validated using the following validation methods.

3D-QSAR Model Validation

Several validation techniques, including bootstrapping, leave-out-five (LOF), rm^2 metric calculation, and external test set validation were implemented to assess the stability, robustness, and predictive ability of the resulting models. To evaluate model's reliability, bootstrapping for 100 runs and progressive scrambling of 10 samplings with 2–10 bins were performed [35]. Last, the predictive ability was calculated as expressed through the predictive correlation coefficient (r^2_{pred}), using the formula given below:

$$r^2_{pred} = (SD - PRESS)/SD$$

where SD is the sum of the squared deviations of each experimental value from the mean and PRESS is the sum of the squared differences between the predicted and actual affinity values.

Standard contour maps were developed for both CoMFA and CoMSIA models. A new design strategy was derived using the structural information from an analysis of the contour maps and molecular docking, and we designed more potent PLK1 inhibitors.

3.4. Synthesized PLK1 Inhibitors

We selected two compounds for synthesis to validate our design of PLK1 inhibitor: compound D39, Ethyl (*E*)-1-(2-((1-(dimethylcarbamoyl)indolin-6-yl)amino)pyrimidin-4-yl)-4-styryl-*1H*-pyrazole-3-carboxylate and compound D40, (*E*)-6-((4-(3-carbamoyl-4-styryl-*1H*-pyrazol-1-yl)pyrimidin-2-yl)amino)-*N,N*-dimethylindoline-1-carboxamide. We state the spectral data of the two synthesized compounds below in Table 4.

Table 4. The spectral data (NMR and HRMS) of the synthesized compounds.

Compound	Structure	NMR	HRMS
D39	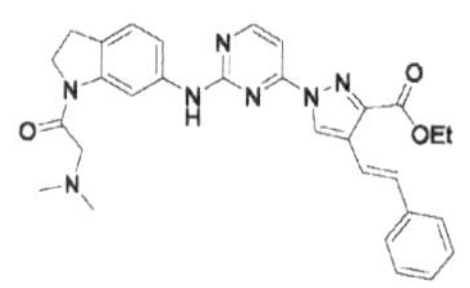	^{1}**H NMR** (400 MHz, DMSO-d_6) δ 10.05–9.95 (m, 1H), 9.40 (s, 1H), 0.16 (d, *J* = 28.1 Hz, 1H), 8.62 (d, *J* = 5.3 Hz, 1H), 7.58 (d, *J* = 7.3 Hz, 1H), 7.48 (d, *J* = 1. 8 Hz, 1H), 7.41 (t, *J* = 7.6Hz, 1H), 7.33 (dd, *J* = 5.0, 3.6 Hz, 1H), 7.31–7.09 (m, 5H), 4.40 (q, *J* = 7.1 Hz, 2H), 4.31–4.23 (m, 3H), 4.10 (q, *J* = 5.2 Hz, 2H), 3.22 (s, 1H), 2.27 (d, *J* = 16.0 Hz, 1H), 2.11 (d, *J* = 23. 6 Hz, 6H), 1.38 (t, *J* = 7.1 Hz, 2H); ^{13}**C NMR** (101 MHz, DMSO-d_6) δ 168.4 (S), 161.7 (s), 161.2 (s), 159.6 (s), 156.5 (s), 143.3 (s), 142.9 (s), 138.9 (s), 137.2 (s), 131.6 (s), 128.8 (s), 127.9 (s), 126.5 (s), 125.9 (s), 125.1 (s), 124.5 (s), 124.3 (s), 117.5 (s), 114.5 (s), 107.1 (s), 99.1 (s), 63.9 (S), 61.0 (s), 47.9 (s), 45.2 (s), 27.1 (s), 14.2 (s)	**HRMS (ESI+)** calculated for $[M + H]^+$ $C_{30}H_{31}N_7O_3$: 538.2561, found 538.2572.
D40	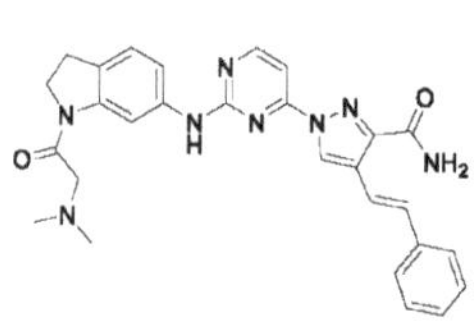	^{1}**H NMR** (400 MHz, DMSO-d_6) δ 9.99 (s, 1H), 9.36 (s, 1H), 9.23 (s, 1H), 8.63 (d, *J* = 5.3 Hz, 1H), 7.97 (s, 1H), 7.63 (d, *J* = 16.7 Hz, 1H), 7.59 (s, 1H), 7.55 (d, *J* = 7.4 Hz, 2H), 7.48 (d, *J* = 2.8 Hz, 1H), 7.47 (d, *J* = 5.3 Hz, 1H), 7.40 (t, *J* = 7.6 Hz, 2H), 7.29 (t, *J* = 7.3 Hz, 1H), 7.17 (d, *J* = 8.1 Hz, 1H), 7.07 (d, *J* = 7.8 Hz, 1H), 4.27 (t, *J* = 8.4 Hz, 2H), 3.24 (s, 2H), 3.09 (t, *J* = 8.3 Hz, 2H), 2.16 (s. 6H); ^{13}**C NMR** (101 MHz, DMSO-d_6) δ 168.2 (s), 163.7 (s), 160.8 (s), 159.6 (s), 156.6 (s), 145.8 (s), 143.3 (s), 139.0 (s), 137.5 (s), 130.7 (s), 128.7 (s), 127.6 (s), 126.4 (s), 125.7 (s), 125.0 (s), 124.5 (s), 123.2 (s), 118.3 (s), 114.4 (s), 107.0 (s), 99.1 (s), 63.8 (s), 47.8 (s), 45.2 (s), 27.1 (s)	**HRMS (ESI+)** calculated for [M + H]+ $C_{28}H_{29}N_8O_2$: 509.2408, found 509.2403.

3.5. Evaluation of IC_{50} Values

We used Reaction Biology Corp. Kinase Hot Spot^{SM} service (www.reactionbiology.com, accessed date 11 January 2022) for screening of D39 and D40 (10 μM) and IC_{50} Profiler Express for IC_{50} measurement. Assay protocol: In a final reaction volume of 25 μL, substrate [Casein], 1 μM, ATP 10 μM, PLK1 (h) (5–10 mU) is incubated with 25 mM Tris pH 7.5, 0.02 mM EGTA, 0.66 mg/mL myelin basic protein, 10 mM Mg acetate, and [^{33}P-ATP] (specific activity approximately 500 cpm/pmol, concentration as required). The reaction is initiated by addition of the Mg-ATP mix. After incubation for 40 min at room temperature, the reaction is stopped by addition of 5 μL of a 3% phosphoric acid solution. Next, 10 μL of the reaction is spotted onto a P30 filtremat and washed three times for 5 min in 75 mM phosphoric acid and once in methanol prior to drying and scintillation counting.

4. Conclusions

One of the most important cell cycle regulators, PLK1, a type of serine-threonine kinase has garnered the attention of the academic research community as well as pharmaceutical companies to develop anticancer inhibitors. PLK1 regulates the cell cycle, and cell cycle dysregulation is the primary cause of cancer. In addition to its role in mitosis, PLK1 has a unique function in regulating cancer cell metabolism, which promotes the growth of cancer cells. Therefore, it is imperative to design and develop new PLK1 inhibitors. Several PLK1 inhibitors were discovered in the past few years; some were unsuccessful due to their low therapeutic response and side effects. Some of these inhibitors, such as volasertib, Onvasertib, etc., were in use for treatment of AML, prostate cancer, and thyroid cancer but were later found to be more effective in combination with other drugs and are no longer used in monotherapy. Hence, we focused on designing more potent PLK1 inhibitors through receptor-based hybrid 3D-QSAR and molecular docking studies. Molecular docking of compound **17** revealed important active site residues of PLK1, such as CYS133, SER137, CYS67, ALA80, LYS82, LEU59, and LEU130. The generated hybrid CoMFA (q^2 = 0.628, ONC = 6, r^2 = 0.905) and CoMSIA (q^2 = 0.580, ONC = 6, r^2 = 0.895) models possessed acceptable statistical results and provided very useful structural information to modify the pyrimidine and quinazoline scaffolds used in this study. Hence, we developed a design strategy and identified additional potent PLK1 antagonists. Our designed compounds showed better predicted activity (pIC_{50}) than the most active compound in the dataset. Altogether, results of our study provided crucial structural insights to develop and synthesize more selective and potent PLK1 inhibitors. Therefore, we synthesized and evaluated IC_{50} values of two designed compounds D39 and D40. The IC_{50} value of compound D39 and D40 was found to be 1.43 μM and 0.359 μM, respectively. Our designed strategy and designed inhibitors can be used as a reference by the drug design community to develop more potent PLK1 antagonists.

Supplementary Materials: The following supporting information can be downloaded at: https://www.mdpi.com/article/10.3390/ph15101170/s1, Table S1: The chemical structures of the selected PLK1 inhibitors with their IC_{50} values; Table S2: CoMSIA models developed using different combinations of fields; Table S3: The predicted pIC_{50} and residual values for both CoMFA and CoMSIA (SEH) models; Table S4: Chemical structures and predicted pIC_{50} activity of designed PLK1 inhibitors.

Author Contributions: Conceptualization, J.-M.H.; methodology and formal analysis, S.P.B., H.K. and H.B.; writing, original draft preparation, S.P.B.; writing-review and editing, J.-M.H. and S.P.B.; project administration, J.-M.H.; funding acquisition, J.-M.H. All authors have read and agreed to the published version of the manuscript.

Funding: This work was financially supported by a National Research Foundation of Korea grant NRF-2019M3A9A8066500 (J.-M.H.) NRF-2020R1A6A1A03042854 (Center for Proteinopathy), NRF-2021R1A2C2007159 (J.-M.H.) and supported by an Institute of Information and Communications Technology Planning and Evaluation (IITP) grant funded by the Korean government (MSIT) (No.2020-0-01343).

Institutional Review Board Statement: Not applicable.

Informed Consent Statement: Not applicable.

Data Availability Statement: Data is contained within the article and Supplementary Material.

Acknowledgments: This work was financially supported by a National Research Foundation of Korea grant NRF-2019M3A9A8066500 (J.-M.H.) NRF-2020R1A6A1A03042854 (Center for Proteinopathy), NRF-2021R1A2C2007159 (J.-M.H.) and supported by an Institute of Information and Communications Technology Planning and Evaluation (IITP) grant funded by the Korean government (MSIT) (No.2020-0-01343).

Conflicts of Interest: The authors declare no conflict of interest.

References

1. Kasahara, K.; Goto, H.; Izawa, I.; Kiyono, T.; Watanabe, N.; Elowe, S.; Nigg, E.A.; Inagaki, M. PI 3-kinase-dependent phosphorylation of Plk1–Ser99 promotes association with 14-3-3γ and is required for metaphase–anaphase transition. *Nat. Commun.* **2013**, *4*, 1882. [CrossRef]
2. Nigg, E.A. Mitotic kinases as regulators of cell division and its checkpoints. *Nat. Rev. Mol. Cell Biol.* **2001**, *2*, 21–32. [CrossRef]
3. Deng, Z.; Chen, G.; Liu, S.; Li, Y.; Zhong, J.; Zhang, B.; Li, L.; Huang, H.; Wang, Z.; Xu, Q.J.E. Discovery of methyl 3-((2-((1-(dimethylglycyl)-5-methoxyindolin-6-yl) amino)-5-(trifluoro-methyl) pyrimidin-4-yl) amino) thiophene-2-carboxylate as a potent and selective polo-like kinase 1 (PLK1) inhibitor for combating hepatocellular carcinoma. *Eur. J. Med. Chem.* **2020**, *206*, 112697. [CrossRef] [PubMed]
4. Beria, I.; Ballinari, D.; Bertrand, J.A.; Borghi, D.; Bossi, R.T.; Brasca, M.G.; Cappella, P.; Caruso, M.; Ceccarelli, W.; Ciavolella, A.J. Identification of 4,5-Dihydro-1 H-pyrazolo [4,3-H] quinazoline derivatives as a new class of orally and selective polo-like kinase 1 inhibitors. *J. Med. Chem.* **2010**, *53*, 3532–3551. [CrossRef] [PubMed]
5. Iliaki, S.; Beyaert, R.; Afonina, I.S. Polo-like kinase 1 (PLK1) signaling in cancer and beyond. *Biochem. Pharmacol.* **2021**, *193*, 114747. [CrossRef]
6. Kothe, M.; Kohls, D.; Low, S.; Coli, R.; Cheng, A.C.; Jacques, S.L.; Johnson, T.L.; Lewis, C.; Loh, C.; Nonomiya, J. Structure of the catalytic domain of human polo-like kinase 1. *Biochemistry* **2007**, *46*, 5960–5971. [CrossRef]
7. Archambault, V.; Glover, D.M. Polo-like kinases: Conservation and divergence in their functions and regulation. *Nat. Rev. Mol. Cell Biol.* **2009**, *10*, 265–275. [CrossRef]
8. Bruinsma, W.; Raaijmakers, J.A.; Medema, R.H. Switching Polo-like kinase-1 on and off in time and space. *Trends Biochem. Sci.* **2012**, *37*, 534–542. [CrossRef]
9. Ruan, H.; Kiselar, J.; Zhang, W.; Li, S.; Xiong, R.; Liu, Y.; Yang, S.; Lai, L. Integrative structural modeling of a multidomain polo-like kinase. *Phys. Chem. Chem. Phys.* **2020**, *22*, 27581–27589. [CrossRef] [PubMed]
10. Liu, Z.; Sun, Q.; Wang, X. PLK1, a potential target for cancer therapy. *Transl. Oncol.* **2017**, *10*, 22–32. [CrossRef]
11. Guan, R.; Tapang, P.; Leverson, J.D.; Albert, D.; Giranda, V.L.; Luo, Y. Small interfering RNA–mediated Polo-like kinase 1 depletion preferentially reduces the survival of p53-defective, oncogenic transformed cells and inhibits tumor growth in animals. *Caner Res.* **2005**, *65*, 2698–2704. [CrossRef] [PubMed]
12. Elez, R.; Piiper, A.; Kronenberger, B.; Kock, M.; Brendel, M.; Hermann, E.; Pliquett, U.; Neumann, E.; Zeuzem, S. Tumor regression by combination antisense therapy against Plk1 and Bcl-2. *Oncogene* **2003**, *22*, 69–80. [CrossRef] [PubMed]
13. Kneisel, L.; Strebhardt, K.; Bernd, A.; Wolter, M.; Binder, A.; Kaufmann, R.J. Expression of polo-like kinase (PLK1) in thin melanomas: A novel marker of metastatic disease. *J. Cutan. Pathol.* **2002**, *29*, 354–358. [CrossRef] [PubMed]
14. Huang, X.; Xie, Z.; Liao, C. Developing polo-like kinase 1 inhibitors. *Future Med. Chem.* **2020**, *12*, 869–871. [CrossRef]
15. Rudolph, D.; Steegmaier, M.; Hoffmann, M.; Grauert, M.; Baum, A.; Quant, J.; Haslinger, C.; Garin-Chesa, P.; Adolf, G.R. BI 6727, a Polo-like kinase inhibitor with improved pharmacokinetic profile and broad antitumor activity. *Clin. Cancer Res.* **2009**, *15*, 3094–3102. [CrossRef]
16. Rudolph, D.; Impagnatiello, M.A.; Blaukopf, C.; Sommer, C.; Gerlich, D.W.; Roth, M.; Tontsch-Grunt, U.; Wernitznig, A.; Savarese, F.; Hofmann, M.H.; et al. Efficacy and mechanism of action of volasertib, a potent and selective inhibitor of Polo-like kinases, in preclinical models of acute myeloid leukemia. *J. Pharmacol. Exp. Ther.* **2015**, *352*, 579–589. [CrossRef]
17. Sero, V.; Tavanti, E.; Vella, S.; Hattinger, C.M.; Fanelli, M.; Michelacci, F.; Versteeg, R.; Valsasina, B.; Gudeman, B.; Picci, P.J. Targeting polo-like kinase 1 by NMS-P937 in osteosarcoma cell lines inhibits tumor cell growth and partially overcomes drug resistance. *Investig. New Drugs* **2014**, *32*, 1167–1180. [CrossRef]
18. Eisenberg, D.; Schwarz, E.; Komaromy, M.; Wall, R.J. Analysis of membrane and surface protein sequences with the hydrophobic moment plot. *J. Mol. Biol.* **1984**, *179*, 125–142. [CrossRef]
19. Golbraikh, A.; Tropsha, A. Predictive QSAR modeling based on diversity sampling of experimental datasets for the training and test set selection. *J. Comput. Aided Mol. Des.* **2000**, *5*, 231–243.
20. Chirico, N.; Gramatica, P.J. Real external predictivity of QSAR models. Part 2. New intercomparable thresholds for different validation criteria and the need for scatter plot inspection. *J. Chem. Inf. Model.* **2012**, *52*, 2044–2058. [CrossRef]
21. Roy, K.; Chakraborty, P.; Mitra, I.; Ojha, P.K.; Kar, S.; Das, R.N. Some case studies on application of "rm2" metrics for judging quality of quantitative structure–activity relationship predictions: Emphasis on scaling of response data. *J. Comp. Chem.* **2013**, *34*, 1071–1082. [CrossRef] [PubMed]
22. Thibaut, U.; Folkers, G.; Klebe, G.; Kubinyi, H.; Merz, A.; Rognan, D. Recommendations for CoMFA studies and 3D QSAR publications. *Quant. Struct.-Act. Relat.* **1994**, *13*, 1–3. [CrossRef]
23. Poso, A.; von Wright, A.; Gynther, J. An empirical and theoretical study on mechanisms of mutagenic activity of hydrazine compounds. *Mutat. Res.* **1995**, *332*, 63–71. [CrossRef]
24. Dagan-Wiener, A.; Nissim, I.; Ben Abu, N.; Borgonovo, G.; Bassoli, A.; Niv, M.Y. Bitter or not? BitterPredict, a tool for predicting taste from chemical structure. *Sci. Rep.* **2017**, *7*, 1–13.
25. Cousins, K.R. Computer review of ChemDraw ultra 12.0. *J. Am. Chem. Soc.* **2011**, *133*, 8388. [CrossRef]
26. Brooks, W.H.; Daniel, K.G.; Sung, S.-S.; Guida, W.C. Computational validation of the importance of absolute stereochemistry in virtual screening. *J. Chem. Inf. Model.* **2008**, *48*, 639–645. [CrossRef]

27. Van den Driessche, G.; Fourches, D. Adverse drug reactions triggered by the common HLA-B* 57: 01 variant: A molecular docking study. *J. Cheminform.* **2017**, *9*, 1–17. [CrossRef]
28. Farid, R.; Day, T.; Friesner, R.A.; Pearlstein, R.A. New insights about HERG blockade obtained from protein modeling, potential energy mapping, and docking studies. *Bioorg. Med. Chem.* **2006**, *14*, 3160–3173. [CrossRef]
29. Giardina, S.F.; Werner, D.S.; Pingle, M.; Feinberg, P.B.; Foreman, K.W.; Bergstrom, D.E.; Arnold, L.D.; Barany, F. Novel, self-assembling dimeric inhibitors of human β tryptase. *J. Med. Chem.* **2020**, *63*, 3004–3027. [CrossRef]
30. Friesner, R.A.; Murphy, R.B.; Repasky, M.P.; Frye, L.L.; Greenwood, J.R.; Halgren, T.A.; Sanschagrin, P.C.; Mainz, D.T. Extra precision glide: Docking and scoring incorporating a model of hydrophobic enclosure for protein−ligand complexes. *J. Med. Chem.* **2006**, *49*, 6177–6196. [CrossRef]
31. Ghosh, S.; Keretsu, S.; Cho, S.J. Designing of the N-ethyl-4-(pyridin-4-yl) benzamide based potent ROCK1 inhibitors using docking, molecular dynamics, and 3D-QSAR. *PeerJ* **2021**, *9*, e11951. [CrossRef] [PubMed]
32. Cramer, R.D.; Patterson, D.E.; Bunce, J.D. Comparative molecular field analysis (CoMFA). 1. Effect of shape on binding of steroids to carrier proteins. *J. Am. Chem. Soc.* **1988**, *110*, 5959–5967. [CrossRef] [PubMed]
33. Klebe, G.; Abraham, U.; Mietzner, T. Molecular similarity indices in a comparative analysis (CoMSIA) of drug molecules to correlate and predict their biological activity. *J. Med. Chem.* **1994**, *37*, 4130–4146. [CrossRef] [PubMed]
34. Jayatilleke, P.R.; Nair, A.C.; Zauhar, R.; Welsh, W. Computational Studies on HIV-1 Protease Inhibitors: Influence of Calculated Inhibitor−Enzyme Binding Affinities on the Statistical Quality of 3D-QSAR CoMFA Models. *J. Med. Chem.* **2000**, *43*, 4446–4451. [CrossRef] [PubMed]
35. Gadhe, C.G.; Kothandan, G.; Cho, S.J. Large variation in electrostatic contours upon addition of steric parameters and the effect of charge calculation schemes in CoMFA on mutagenicity of MX analogues. *Mol. Sim.* **2012**, *38*, 861–871. [CrossRef]

pharmaceuticals

MDPI

Article

Ligand Growing Experiments Suggested 4-amino and 4-ureido pyridazin-3(2*H*)-one as Novel Scaffold for FABP4 Inhibition

Letizia Crocetti [1,*], Giuseppe Floresta [2], Chiara Zagni [2], Divya Merugu [3], Francesca Mazzacuva [4], Renan Rodrigues de Oliveira Silva [3], Claudia Vergelli [1], Maria Paola Giovannoni [1] and Agostino Cilibrizzi [3,5,*]

[1] Dipartimento NEUROFARBA—Pharmaceutical and Nutraceutical Section, via Ugo Schiff 6, Sesto Fiorentino, 50019 Florence, Italy

[2] Dipartimento di Scienze del Farmaco e della Salute, Università di Catania, Viale A. Doria 6, 95125 Catania, Italy

[3] Institute of Pharmaceutical Science, King's College London, Stamford Street, London SE1 9NH, UK

[4] School of Health, Sport and Bioscience, University of East London, London E15 4LZ, UK

[5] Medicines Development, Centre for Therapeutic Innovation, University of Bath, Bath BA2 7AY, UK

* Correspondence: letizia.crocetti@unifi.it (L.C.); agostino.cilibrizzi@kcl.ac.uk (A.C.); Tel.: +39-0554573683 (L.C.); +44-(0)-20-7848-9532 (A.C.)

Abstract: Fatty acid binding protein (FABP4) inhibitors are of synthetic and therapeutic interest and ongoing clinical studies indicate that they may be a promise for the treatment of cancer, as well as other diseases. As part of a broader research effort to develop more effective FABP4 inhibitors, we sought to identify new structures through a two-step computing assisted molecular design based on the established scaffold of a co-crystallized ligand. Novel and potent FABP4 inhibitors have been developed using this approach and herein we report the synthesis, biological evaluation and molecular docking of the 4-amino and 4-ureido pyridazinone-based series.

Keywords: fatty acid binding protein; FABP4; FABP4is; FABP4 inhibitors; pyridazinone; computing assisted molecular design

Citation: Crocetti, L.; Floresta, G.; Zagni, C.; Merugu, D.; Mazzacuva, F.; de Oliveira Silva, R.R.; Vergelli, C.; Giovannoni, M.P.; Cilibrizzi, A. Ligand Growing Experiments Suggested 4-amino and 4-ureido pyridazin-3(2*H*)-one as Novel Scaffold for FABP4 Inhibition. *Pharmaceuticals* **2022**, *15*, 1335. https://doi.org/10.3390/ph15111335

Academic Editors: Marialuigia Fantacuzzi and Mariangela Agamennone

Received: 6 October 2022
Accepted: 21 October 2022
Published: 28 October 2022

1. Introduction

Fatty acids (FAs) are long carbon chain organic carboxylic acids responsible for different actions in the human organism [1,2]. Their chronic high concentration in circulation leads to various disorders [3,4], including atherosclerosis [5], diabetes [6] and obesity [7]. Considering that their chemical structure is characterized by high lipophilicity, FAs are insoluble in water, and their trafficking into the body requires specific carriers such as the fatty acid-binding proteins (FABPs). [8]. Since their discovery, FABPs have been classified into different families based on their localization in the human body, such as A-FABP (adipocyte), B-FABP (brain), E-FABP (epidermal), H-FABP (muscle and heart), I-FABP (intestinal), Il-FABP (ileal), L-FABP (liver), M-FABP (myelin), and T-FABP (testis). FABP4 (aP2 or A-FABP) is the subtype expressed in adipocytes [9], and the research into small molecule inhibitors for such protein initially started when it was reported that knockout animal models of FABP4 produced protective effects against the development of insulin resistance [10], as well as several pathological events linked to the metabolic syndrome and atherosclerosis [11–13]. Interestingly, pharmacological approaches with small molecules that inhibit the normal function of the protein are also valid in this regard, demonstrating similar results as the genetic procedures by mimicking the phenotype of FABP4-deficient mice [14]. This family of transporter proteins also has a role in cancer progression [15], and it was discovered that non-physiological expressions of FABPs are present in some of the most common cancers such as renal cell carcinoma, bladder and prostate, as well as other types of cancer cells [16–18]. It was recently discovered that FABP4 promotes the metastasis and invasion of colon cancer and that the treatment with a classical small molecule inhibitor (BMS309403) weakened the migration and invasion of colon cancer cells [19]. FABP4 leads also to abnormal metastasis patterns in ovarian cancer, and recent findings demonstrate that the protein is responsible for the disease's aggressivity, contributing to poor prognosis in this tumor [20]. Moreover, the transporter has also been shown to play a role in accelerating glioblastoma cell growth [21]. All these recent findings related to cancer research proved that FABP4 targeting may represent an effective and promising therapeutic strategy against oncological conditions, in addition to the established effects on metabolic and cardiovascular diseases.

Recently, a variety of effective FABP4 inhibitors (FABP4i) have been developed, but unfortunately, none of them is currently in the clinical research phases [14,22]. Computer-aided drug design represents a promising and effective tool for the identification of molecular hits as FABP4i [23–27]. In line with our recent interest in the development of new antitumor compounds and the identification of novel bioactive heterocycles [28–32], herein we report the design, synthesis and in vitro characterization of 4-amino and 4-ureido pyridazinone-based series of FABP4i inspired by the scaffold hopping of an established ligand co-crystallized within the protein.

2. Results and Discussion

2.1. Heterocyclic Small-Molecule Design

To generate a novel series of FABP4 inhibitors we have exploited a two-step computing assisted molecular design. As shown in Figure 1, in the first step of the drug-design process we focused on the search for bioisosteric-replacements/scaffold hopping of the pyrimidine scaffold of the co-crystallyzed ligand (2-[(2-oxo-2-piperidin-1 -ylethyl)sulfanyl]-6-(trifluoromethyl)pyrimidin-4-ol; pdbID: 1TOU). Our bioisosteric replacement analysis led to the selection of three nitrogen-containing heterocyclic frameworks, i.e., pyridazinones, pyridines and benzo[*d*]thiazole (see Supplementary Materials). Considering the synthetic accessibility of pyridazinone-based molecules and that pyridazinone was not investigated earlier as a scaffold to access FABP4 inhibitors, we envisaged to use this heterocycle to carry out automated ligand growing experiments inside the FABP4 cavity, as described in the Section 3, leading to 52 target molecules. The compounds were then synthesized and screened against FABP4 and the chemical structures are reported in Tables 1 and 2. Both the scaffold hopping and the ligand growing experiments were conducted using Spark (https://www.cresset-group.com/products/spark/ accessed on 15 June 2022) [33].

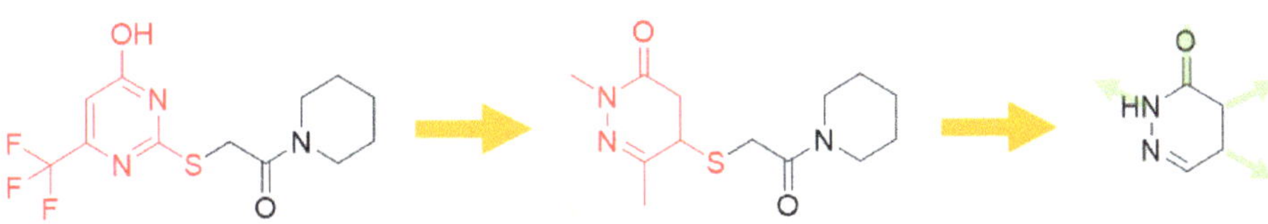

Figure 1. Schematic representation of the computer assisted design of the 4-amino and 4-ureido pyridazinones.

Table 1. 4-NH_2-pyridazinones synthesized and screened against FABP4.

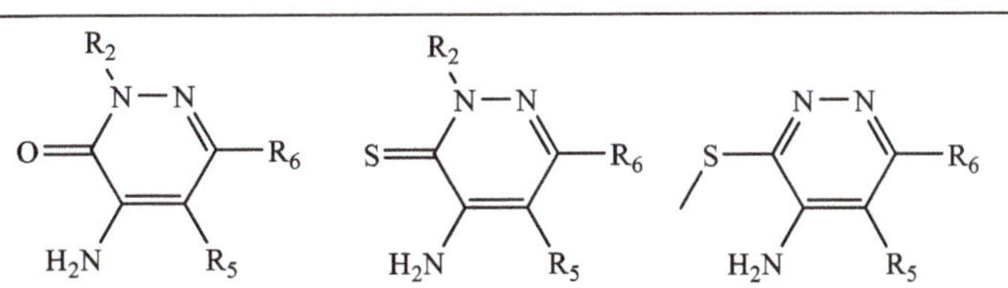

32 27

Table 1. *Cont.*

Comp.	R_2	R_5	R_6
4a	H	CONHPh	Ph
4b	H	CONHnC_3H_7	Ph
5a	C_2H_5	CONHPh	Ph
5b	C_2H_5	CONHnC_3H_7	Ph
6	H	$CONH_2$	Ph
7	H	CN	Ph
16	Ph	$CONH_2$	CH_3
17	Ph	CN	CH_3
18	Ph	$COCH_3$	H
21	cC_6H_{11}	$CONH_2$	Ph
22	cC_6H_{11}	CN	Ph
24d	iC_3H_7	H	Ph
24e	nC_3H_7	H	Ph
24f	nC_4H_9	H	Ph
27	-	H	Ph
32	CH_3	H	Ph
37a	H	H	3-thienyl
37c	H	H	cC_6H_{11}
37d	H	H	iC_3H_7
38a	CH_3	H	3-thienyl
38b	CH_3	H	cC_6H_{11}
38c	CH_3	H	iC_3H_7
38d	CH_3	H	CH_2-Ph
42a	CH_3	H	2-(OH)-Ph
42b	CH_3	H	4-(NH_2)-Ph
44	CH_3	H	4-($NHCOCH_3$)-Ph
48	CH_3	H	2-pyridinyl
54	Ph	H	CH_3
57	Ph	pyrazole	CH_3

Table 2. 4-Amino and 4-ureido pyridazinones synthesized and screened against FABP4.

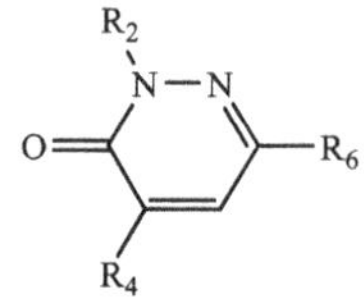

Comp.	R_2	R_4	R_6
25a	CH_3	$NHCONH_2$	Ph
25b	cC_6H_{11}	$NHCONH_2$	Ph
25c	C_2H_5	$NHCONH_2$	Ph
25d	iC_3H_7	$NHCONH_2$	Ph
25e	nC_3H_7	$NHCONH_2$	Ph
25f	nC_4H_9	$NHCONH_2$	Ph
28	H	$NHCONH_2$	Ph
29a	CH_3	$NHCOCH_3$	Ph
29b	CH_3	$NHCOC_2H_5$	Ph
29c	CH_3	NHCOiC_3H_7	Ph
29d	CH_3	NHCOnC_3H_7	Ph
30a	CH_3	NH-(3-CN)-Ph	Ph
30b	CH_3	NH-(2-CN)-Ph	Ph
31a	CH_3	NH-(3-$CONH_2$)-Ph	Ph
31b	CH_3	NH-(2-$CONH_2$)-Ph	Ph

Table 2. *Cont.*

Comp.	R_2	R_4	R_6
35	CH_3	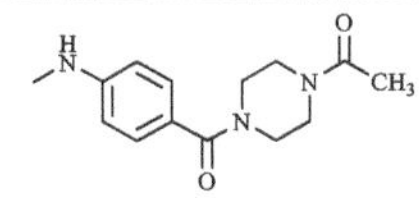	Ph
39a	CH_3	$NHCONH_2$	cC_6H_{11}
39b	CH_3	$NHCONH_2$	iC_3H_7
40	H	$NHCONH_2$	2-(OH)-Ph
43	CH_3	$NHCONH_2$	2-(OH)-Ph
49	CH_3	$NHCONH_2$	2-pyridinyl
51	CH_3	$CONH_2$	Ph
55	Ph	$NHCONH_2$	CH_3

2.2. Chemistry

The synthetic procedures carried out to obtain the target compounds containing the pyridazinone scaffold are reported in Schemes 1–9. The structures were confirmed on the basis of analytical and spectral data. Scheme 1 shows the synthetic pathway affording the final compounds **4a,b**, **5a,b**, **6** and **7**. Intermediate **2** [34] was obtained starting from isoxazole-pyridazinone **1**, synthesized by adopting previously reported protocols [29–32] and using methanol and triethylamine for opening the isoxazole *nucleus*. The subsequent hydrolysis (acid **3** [35]) and acylation with thionyl chloride, triethylamine and appropriate amine led to final compounds **4a,b**. Products **5a,b** were obtained from alkylation reaction of **4a,b** with ethyl bromide in standard conditions. The opening of isoxazole *core* of the starting material **1** with 33% NH_4OH afforded to amide **6** [28] which, by subsequent dehydration with $POCl_3$, led to compound **7**. The synthesis of final compounds **16–18** is reported in Scheme 2. The reaction between sodium salt of diketone **8** with the commercially available ethyl chloro(hydroximino)acetate **9** in ethanol led to a mixture of isomers **10** and **11 [36]** that were cyclized to isoxazole-pyridazinone **12** and **13** using phenylhydrazine and PPA. After chromatographic separation, the latter were subjected to a series of reactions to obtain the compounds **16–18**. The treatment of intermediate **13** with ammonium formate and Pd/C provided compound **18**, while the treatment of **12** with methanol and triethylamine led to pyridazinone **14**. Intermediate **14** was first hydrolyzed to acid (**15**), then converted to amide (**16**) and finally treated with $POCl_3$ to obtain the cyano derivative **17**. The final compounds **21** and **22** were obtained through a procedure similar to that shown in Scheme 1 for amide derivative **6** and cyano derivative **7**, using intermediate **20** as the starting material, which was obtained by reaction of cyclohexyl hydrazine and PPA with isoxazole **19 [34]** (see Scheme 3). Scheme 4 reports the synthesis of the pyridazinone-based derivatives of type **24** and **25** (unsubstituted at position 5), compound **28** and the thio-derivative **27**. Intermediate **23** [37] was reacted with the appropriate brominated alkylating agent in presence of potassium carbonate and dry DMF to afford **24a–f** derivatives (**24a**, [38]; **24c**, [34]). The formation of urea derivatives of type **25** was carried out using sodium acetate and triphosgene in dry THF at reflux, and then treated with ammonia. The urea **28** was directly obtained from intermediate **23** using the same conditions used for compound type **25**. The transformation of the carbonyl (C=O) in thiocarbonyl group (C=S) was carried out using the Lawesson's reagent in toluene (**26**) and the subsequently alkylation with methyl iodide in standard condition led to the thio derivative **27**. In Schemes 5 and 6 are reported the synthetic procedures of other un-substituted pyridazinones at position 5, but bearing different groups/functions at position 4 and 6. In particular, Scheme 5 depicts the synthetic pathways for compounds with a phenyl ring at position 6 and a methyl group at N-2, while different substituents are introduced at position 4. Starting from compound **24a [38]** (Scheme 4), the amino group at position 4 was acylated using the suitable anhydride in pyridine in a sealed/pressure vessel to obtain the final compounds **29a–d**. Moreover, the same amino group was also subjected to a coupling reaction using the appropriate

R-phenylboronic acid in presence of copper (II) acetate and triethylamine to furnish the derivatives **30a,b** and **33**. The substituent R on the phenyl at position 4 was further elaborated. The *m*/*o*-CN group of compounds **30a,b** was converted into *m*/*o*-$CONH_2$ (compounds **31a,b**, respectively) with 80% sulfuric acid under reflux. The 4-carbethoxy function in product **33** was firstly hydrolyzed to acid **34**, converted into the corresponding acid chloride with thionyl chloride and then acylated with 1-acetylpiperazine (compound **35**). Lastly, the carbonyl group of intermediate **24a** was converted in thiocarbonyl (**32**) using the same procedure discussed in Scheme 4. In Scheme 6 are depicted pyridazinone-based derivatives with a methyl group or a hydrogen at N-2, an amino group or urea functionality at position 4, but bearing different groups e/o functions (e.g., R-phenyl, alkyl, cycloalkyl) at position 6. Starting from commercially available intermediates **36a–f**, the introduction of an amino group at position 4 with hydrazine hydrate at high temperature led to compounds **37a–e** (**37e**, [38]) and the subsequent alkylation with methyl iodide provided products **38a–d**. The derivatives **39a,b** and **40** were obtained from reaction with triphosgene and ammonia in the same conditions reported in Scheme 4, starting from **38b,c** and **37e**, respectively. The direct alkylation of the intermediates **36e** and **36f** afforded the corresponding *N*-methyl derivatives **41a,b** (**41a**, [39]), which were subsequently converted into compounds **42a,b** through the same reaction used to obtain **37a–e**. In particular, the reaction conditions used to introduce an amino group at position 4 led also to the reduction in the nitro group in compound **42b**. The latter was subjected to acylation reaction with acetyl chloride to obtain product **44**. Instead, intermediate **42a** was subjected to triphosgene treatment to obtain the urea derivative **43**. Scheme 7 reports the synthesis of final compounds **48** and **49**. Intermediate **47** was obtained starting from isoxazole **45**, previously synthesized by us [34] by cyclization reaction with methyl hydrazine (**46**) and subsequent opening of the isoxazole ring with ammonium formate and palladium on carbon. The deacetylation (on **47**) with 48% bromic acid at high temperature led to compound **48**, which was subsequently treated with triphosgene and ammonia to obtain the urea derivative **49**. Compound **51** was obtained through alkylation reaction using standard conditions [40], but starting from product **50** [41] (Scheme 8). Lastly, the final compound **55** (Scheme 9) was obtained starting from intermediate **52** [42], through the same reactions of isoxazole nucleus opening (**53 [42]**), deacetylation (**54 [43]**) and formation of the urea function. In Scheme 9 is also illustrated the treatment of **52** with dimethylformamide dimethyl acetal to generate intermediate **56**, which was subsequently converted into compound **57** using hydrazine hydrate.

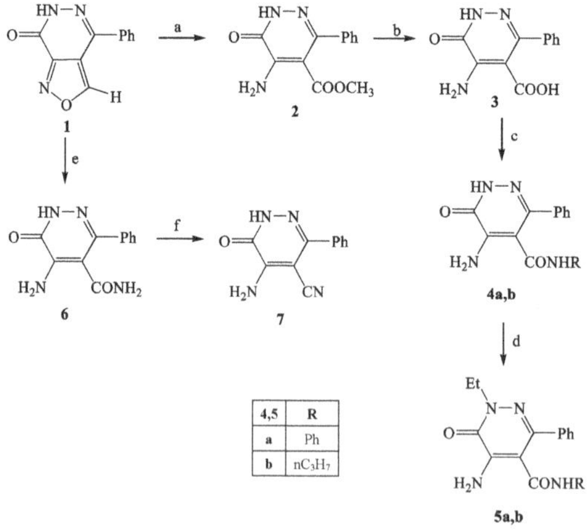

Scheme 1. Reagents and conditions: (**a**) Et_3N, CH_3OH, 60 °C, 2 h; (**b**) 6N NaOH, EtOH, reflux, 30 min; (**c**) (i) $SOCl_2$, Et_3N, r.t., 30 min; (ii) R-NH_2, anhydrous THF, r.t., 2 h; (**d**) CH_3CH_2Br, K_2CO_3, anhydrous DMF, reflux, 30–90 min; (**e**) 33% NH_4OH, $C_5H_{11}N$, 60 °C, 90 min; (**f**) $POCl_3$, 60 °C, 2h.

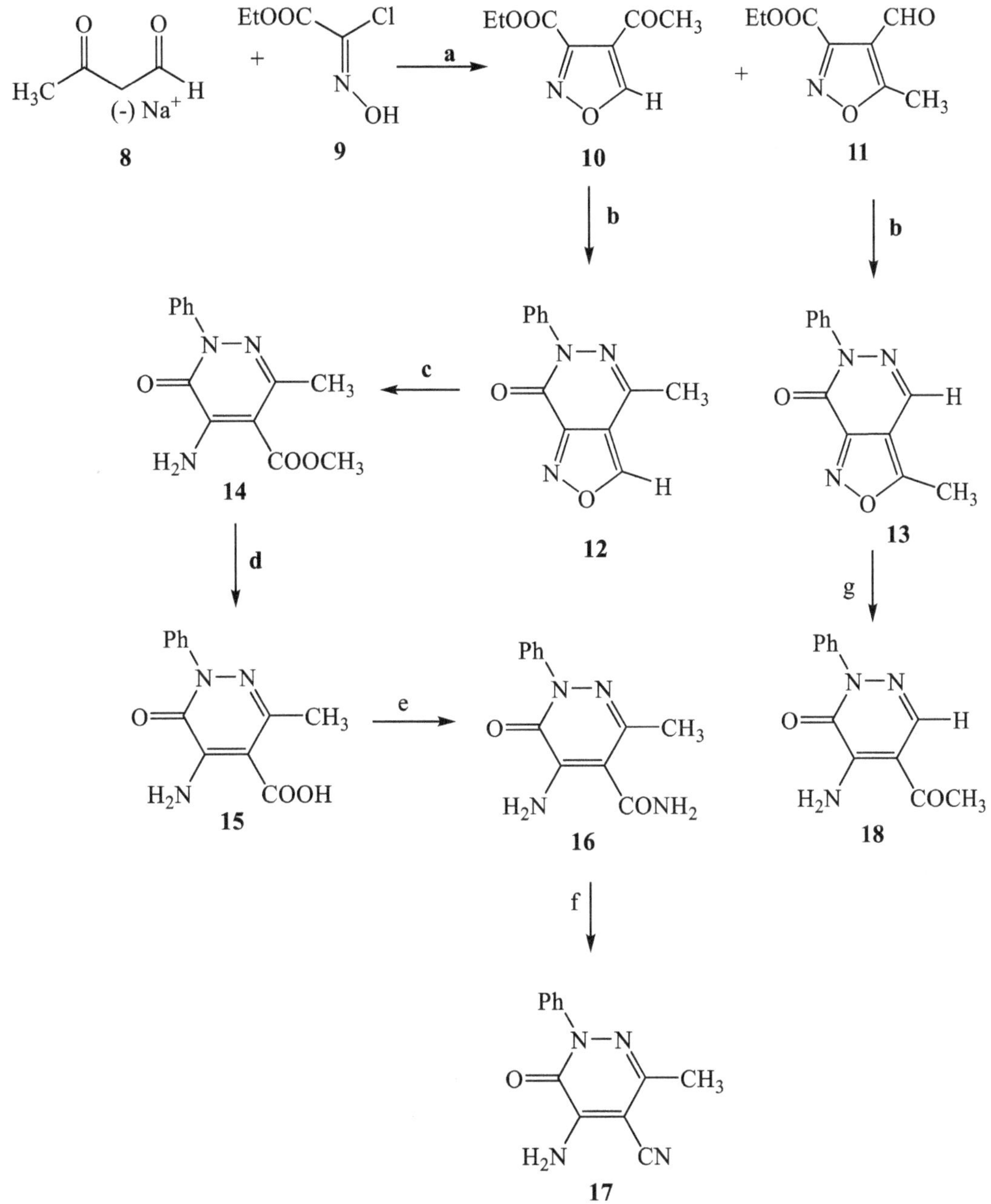

Scheme 2. Reagents and conditions: (**a**) anhydrous EtOH, 0 °C, 1h; (**b**) phenylhydrazine, PPA, EtOH, 70 °C, 30 min; (**c**) CH_3OH, Et_3N, 60 °C, 2h; (**d**) NaOH, EtOH, reflux, 30 min; (**e**) (i) $SOCl_2$, Et_3N, reflux, 30 min.; (ii) 33% NH_4OH, anhydrous THF, r.t., 15 min; (**f**) $POCl_3$, 60 °C, 2h; (**g**) $HCOONH_4$, 10% Pd/C, EtOH, reflux, 2 h.

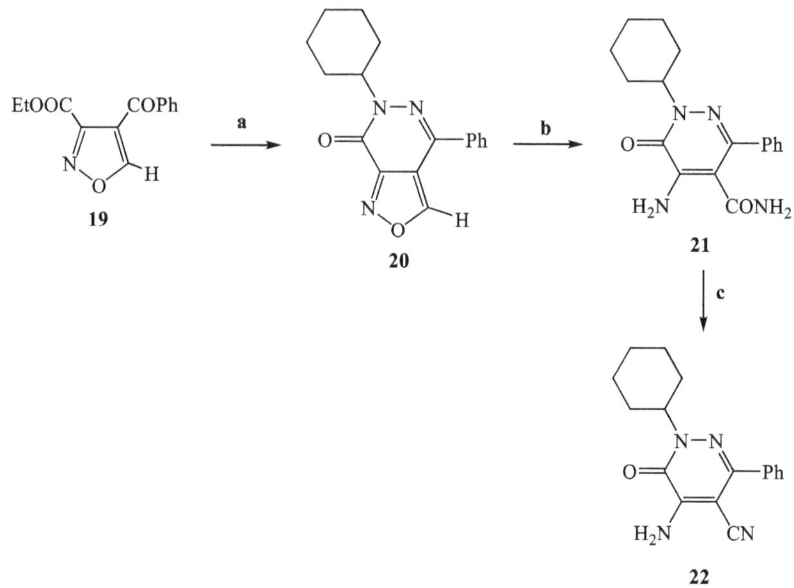

Scheme 3. Reagents and conditions: (**a**) cyclohexylhydrazine, PPA, EtOH, 70 °C, 30 min; (**b**) 33% NH_3, piperidine, 60 °C, 90 min; (**c**) $POCl_3$, 60 °C, 2h.

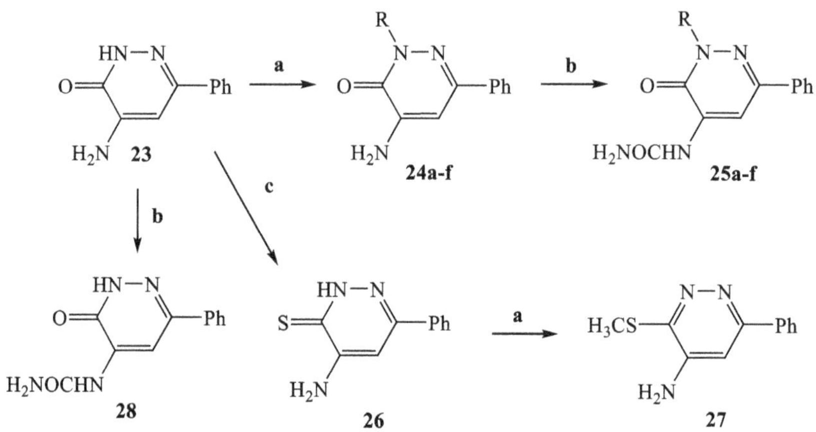

24,25	R
a	CH_3
b	cC_6H_{11}
c	C_2H_5
d	iC_3H_7
e	nC_3H_7
f	C_4H_9

Scheme 4. Reagents and conditions: (**a**) suitable R-Br, K_2CO_3, anhydrous DMF, reflux, 1–4 h; (**b**) (i) dry THF, CH_3COONa, 0 °C then triphosgene, reflux, 2 h; (ii) NH_3 33%, 0 °C, 1 h; (**c**) Lawesson's reagent, anhydrous toluene, reflux, 5 h.

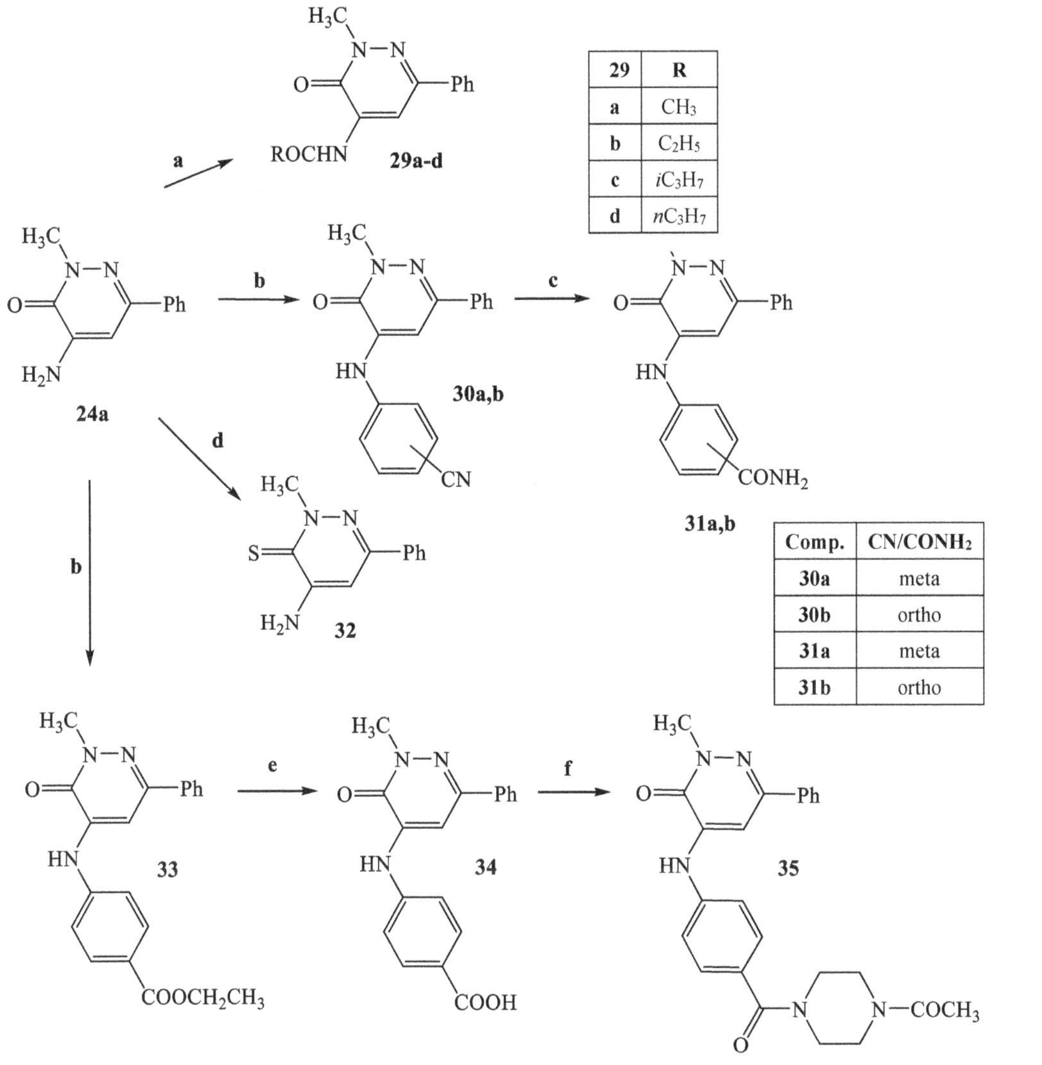

29	R
a	CH_3
b	C_2H_5
c	iC_3H_7
d	nC_3H_7

Comp.	CN/CONH2
30a	meta
30b	ortho
31a	meta
31b	ortho

Scheme 5. Reagents and conditions: (**a**) suitable $(R\text{-}CO_2)O$, anhydrous C_6H_5N, closed tube, 140 °C, 5 h; (**b**) 2/3-cyanophenylboronic acid (for **30a,b**) or 4-ethoxycarbonylphenylboronic acid (for **33**), $Cu(Ac)_2$, Et_3N, dry CH_2Cl_2, r.t., 12 h; (**c**) H_2SO_4 80%, 80 °C, 4 h; (**d**) Lawesson's reagent, anhydrous toluene, reflux, 10 h; (**e**) NaOH 6N, EtOH 96%, reflux, 1 h; (**f**) (i) $SOCl_2$, Et_3N (catalytic), reflux, 1 h; (ii) anhydrous THF, 1-acetylpiperazine, 0 °C then r.t., 1 h.

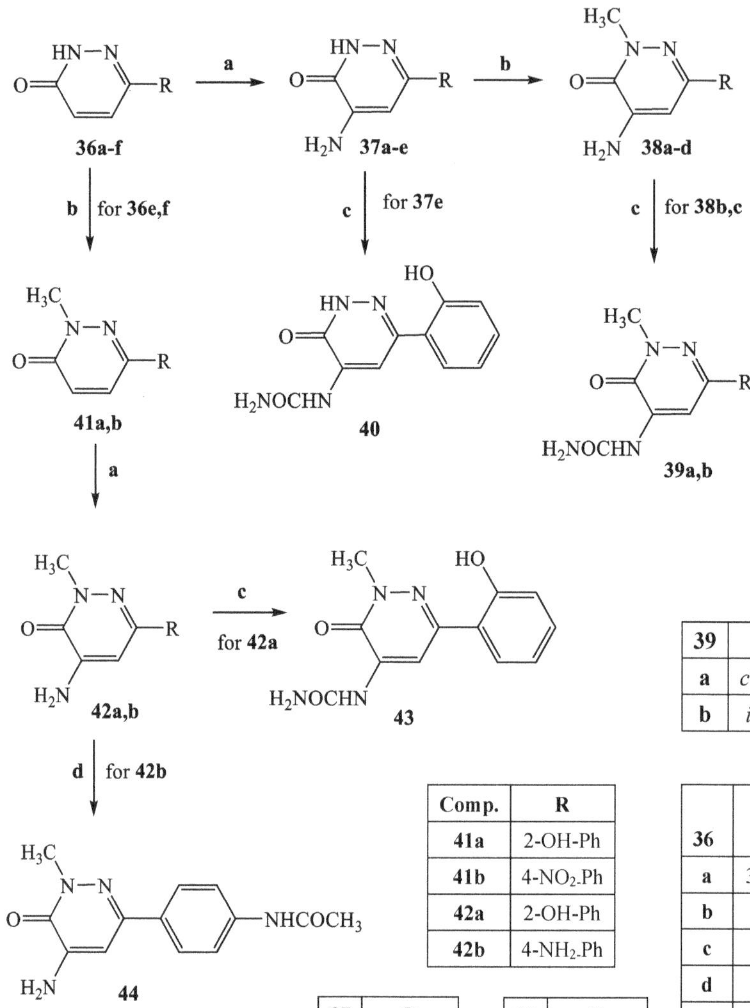

39	R
a	cC_6H_{11}
b	iC_3H_7

Comp.	R
41a	2-OH-Ph
41b	$4\text{-}NO_2\text{-}Ph$
42a	2-OH-Ph
42b	$4\text{-}NH_2\text{-}Ph$

36	R
a	3-thienyl
b	cC_6H_{11}
c	iC_3H_7
d	CH_2Ph
e	2-OH-Ph
f	$4\text{-}NO_2\text{-}Ph$

37	R
a	3-thienyl
b	cC_6H_{11}
c	iC_3H_7
d	CH_2Ph
e	2-OH-Ph

38	R
a	3-thienyl
b	cC_6H_{11}
c	iC_3H_7
d	CH_2Ph

Scheme 6. Reagents and conditions: (**a**) $NH_2NH_4 \cdot H_2O$, sealed/pressure vessel, 180 °C, 12 h; (**b**) CH_3I, K_2CO_3, anhydrous DMF, 80 °C, 2–4 h; (**c**) (i) anhydrous THF, CH_3COONa, 0 °C then triphosgene, reflux, 2 h; (ii) NH_4OH 33%, 0 °C, 1 h; (**d**) $ClCOCH_3$, anhydrous THF, 0 °C, then r.t., 20 min.

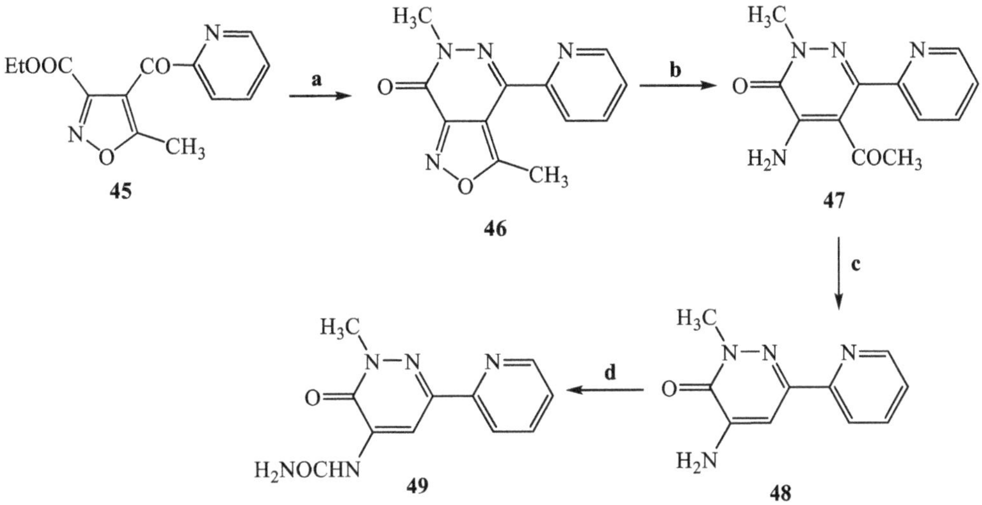

Scheme 7. Reagents and conditions: (a) $CH_3(NH)NH_2$, EtOH 96%, r.t., 2 h; (b) $HCOONH_4$, Pd/C, EtOH 96%, reflux, 2h; (c) HBr 48%, sealed/pressure vessel, 130 °C, 3 h; (d) (i) anhydrous THF, CH_3COONa, 0 °C then triphosgene, reflux, 2 h; (ii) 33% NH_4OH, 0 °C, 1 h.

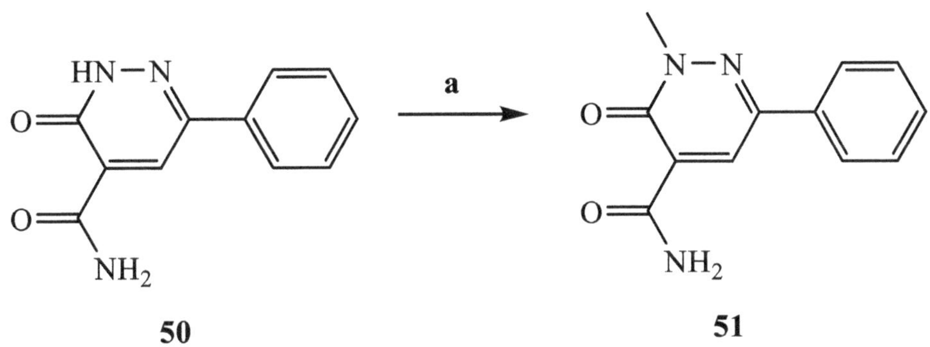

Scheme 8. Reagents and conditions: (a) CH_3I, K_2CO_3, anhydrous DMF, 80 °C, 2 h.

Based on the analytical and spectral data (proton and carbon NMR) and mass spectrometry (MS), all the new compounds confirmed the predicted chemical structures, as well as satisfactory results in terms of formulation and purity (See Section 3; in Supporting Information are reported representative examples of analytical characterization data of the compounds processed to FABP4 inhibition assay in vitro). Reversed phase liquid chromatography was used to perform a qualitative analysis of the dataset's purity. The formation of the products was monitored by UV absorbance at wavelengths of 281 nm and 254 nm. The retention times range was from 6 to 17 min (See Section 3 and Supporting Information). The overall feature of mass spectra (LC-MS) of this series of pyridazinone-derivatives is the presence of a predominant peak corresponding to the molecular ion $[M + H]^+$ (See Section 3 and Supporting Information).

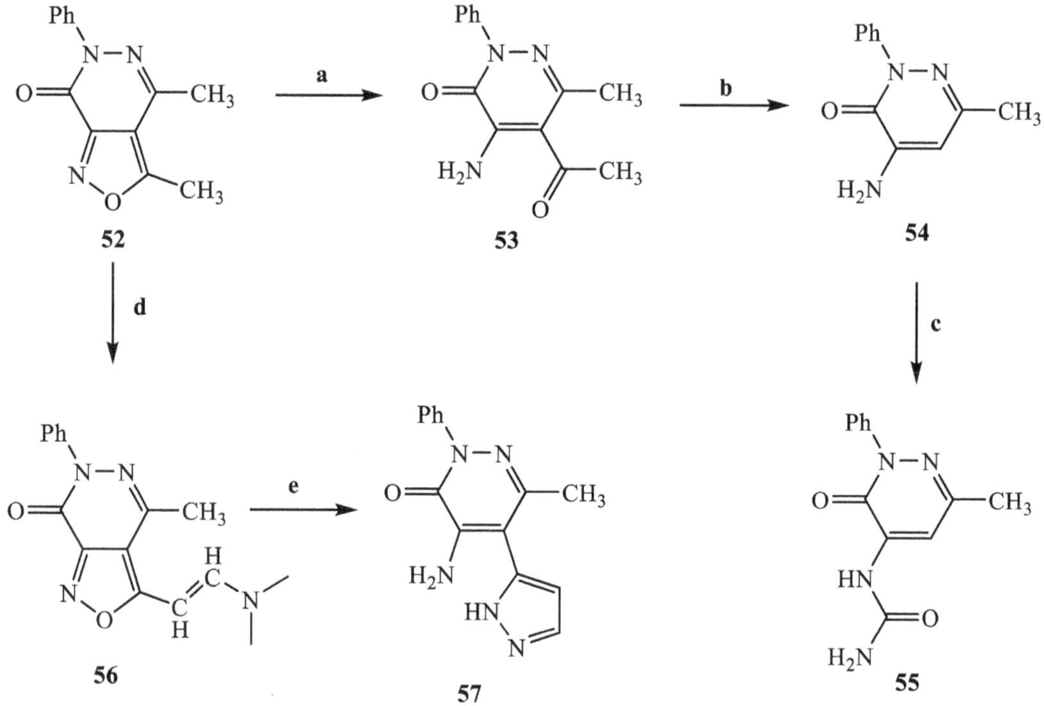

Scheme 9. Reagents and conditions: (**a**) $HCOONH_4$, Pd/C 10%, EtOH 96%, reflux, 2 h; (**b**) HBr 48%, sealed/pressure vessel, 130 °C, 3 h; (**c**) (i) anhydrous THF, CH_3COONa, 0 °C then triphosgene, reflux, 2 h; (ii) NH_4OH 33%, 0 °C, 1 h; (**d**) DMF-DMA, 90 °C, 1 h; (**e**) $NH_2NH_4{\cdot}H_2O$, anhydrous EtOH, 70 °C, 10 h.

2.3. *FABP4 Inhibition Evaluation*

FABP4 inhibitory activity was assessed by measuring the decrease in fluorescent signal of a detection reagent (DR) when displaced by a strong FABP4 ligand. Specifically, the DR exhibits an increased fluorescence intensity when bound to FABP4. Therefore, any effective ligand of the protein, which binds to the same binding pocket and can displace the DR, determines a reduction in the fluorescence read-out. The new molecular series was screened in a two-step procedure. Firstly, a single concentration of 5 μM was used to gain an estimation of the overall inhibitory effect of all the molecules. Subsequently, only the compounds that were able to reduce the fluorescence reading of at least 95% were further evaluated by measuring the IC_{50} values (μM), which were lastly compared with the activity of the arachidonic acid (i.e., FABP4 established ligand). The single point displacement results are reported in Figure 2. Based on the data of the first screening, 10 molecules were selected as most effective compounds—i.e., able to reduce the fluorescence of the DR to at least 95%, for which the IC_{50} (μM) was calculated. Arachidonic acid was used as a positive control, resulting with an IC_{50} of 3.42 μM. The IC_{50} values of our set of compounds are reported in Table 3. Compound **25a** demonstrated a potent inhibitory activity, with an IC_{50} value (i.e., 2.97 μM) lower than the reference arachidonic acid.

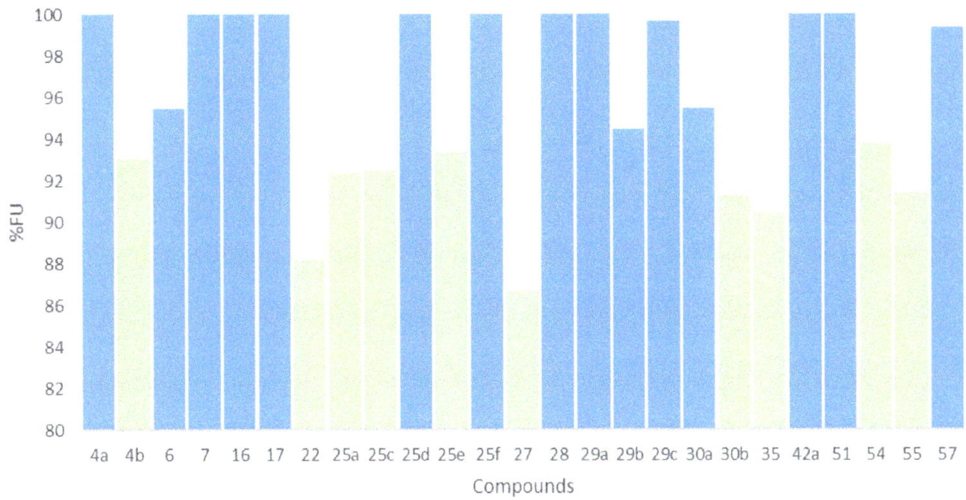

Figure 2. Single point displacement experiment for selected compounds.

Table 3. Measured IC_{50} values for selected compounds.

Compounds	IC_{50} (μM)
Arachidonic acid	3.42 ± 0.54
4b	8.27 ± 0.20
25a	2.97 ± 0.26
30b	23.18 ± 0.52
22	15.23 ± 0.76
25c	>50
35	>50
25e	>50
54	>50
55	>50
27	>50

2.4. Molecular Modelling Studies

Since the first apo-FABP crystal structure was published in 1992, many other holo-FABP structures with a variety of ligands have been solved. The hydrophobic pocket side chains engage a hydrogen bond to the carboxylate of FAs toward several amino acids. Moreover, a network of water molecules may be involved in mediating these interactions. The docking experiments of the molecular series compounds were conducted on the most active compounds **4b**, **25a**, **30b**, and **22**. Figure 3 shows the 2D binding interactions for the molecules, while Figure 4 displays the predicted poses inside the binding pocket of FABP4. All the compounds are able to engage several interactions with relevant residues in the binding pocket, such as R126 and Y128, as well as R106. R126 can interact with both the carbonyls of the most potent compound **25a**, that also interacts directly with Y128 and, through the network of water molecules, with S53. The **4b** is well allocated inside the binding pocket and is engaging a strong H-bond interaction with R126. Differently, compound **22** is not suitably allocated inside the pocket to generate appropriate binding with R126 and Y128 and most of the stabilizing interactions are due to pi–pi stacking with A75, F16 and M20. Lastly, the -CN group of **30b** results responsible of the stabilizing interaction with R126 and Y128, that are likely to account for the lower activity of the compound, as determined by the lower binding interaction for this group with the residues.

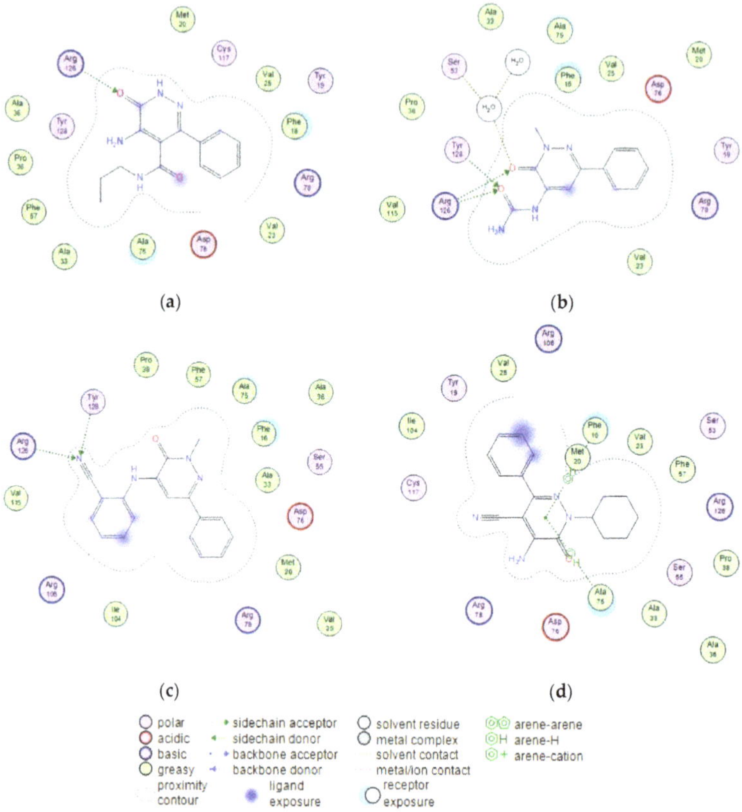

Figure 3. (**a**) 2D interaction between **4b** and FABP4. (**b**) 2D interaction between **25a** and FABP4. (**c**) 2D interaction between **30b** and FABP4. (**d**) 2D interaction between **22** and FABP4.

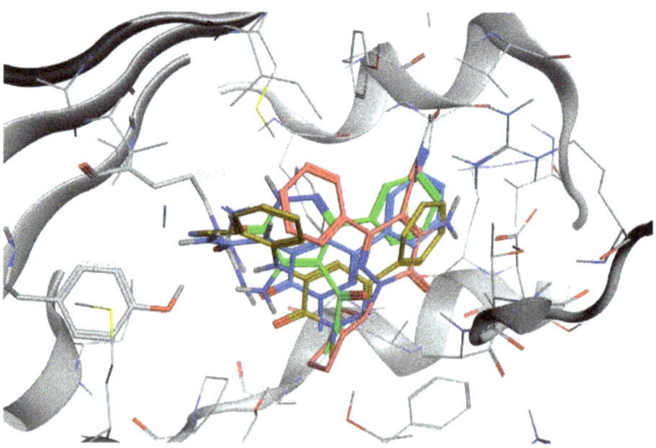

Figure 4. Docked poses inside FABP4 of molecules **4b** (green), **25a** (blue), **30b** (dark yellow) and **22** (light red).

3. Experimental Section

3.1. General Remarks

All the chemical reagents were purchased from Merk and Sigma Aldrich of reagent grade and were used without any further purification. Extracts were dried over Na_2SO_4 and the solvents were removed under reduced pressure. All reactions were monitored by thin-layer chromatography (TLC) using commercial plates (Merck) pre-coated with silica gel 60 F-254. Visualization was performed by UV fluorescence (λmax = 254 nm) or by staining with iodine or potassium permanganate. Chromatographic separations were performed on silica gel columns by gravity (Kieselgel 40, 0.063–0.200 mm; Merck) or flash chromatography (Kieselgel 40, 0.040–0.063 mm; Merck). Yields refer to chromatographically and spectroscopically pure compounds, unless otherwise stated. When reactions were performed in anhydrous conditions, the mixtures were maintained under nitrogen atmosphere. Compounds were named following IUPAC rules as applied by Beilstein-Institut AutoNom 2000 (4.01.305) or CA Index Name. All melting points were determined on a microscope hot stage Büchi apparatus and are uncorrected. ^{1}H-NMR and ^{13}C-NMR spectra were obtained on a Bruker AVANCE 400 spectrometer at 400 MHz and 100 MHz, respectively, using 5 mm i.d. glass tubes. Chemical shifts (δ) values are expressed as parts per million (ppm) using DMSO (d_6) (2.50 for proton and 39.52 for carbon), methanol (d_4) (3.31 for proton and 49.00 for carbon) or $CDCl_3$ (7.26 for proton and 77.16 for carbon) as solvents. The coupling constants (J) are reported in Hz. The following splitting patterns are identified: s, singlet; d, doublet; t, triplet; m, multiplet; or any combination of these e.g., dd, dt, etc. Analytical reversed-phase high performance liquid chromatography (reversed-phase HPLC) was conducted out on HP 1050 instrument (Agilent Technologies, Waldbronn, Germany) to ascertain the chromatographic purity of compounds. The system includes a quaternary pump, an autosampler, and a Kontron DEG 104 degasser (Kontron, Tokyo, Japan). A C_{18} column, Zorbax,80 Å, 3.5 μm, 2.1 × 100 mm was used with a total run time of 30 min. The mobile phase is composed of 0.1% Trifluoro acetic acid (TFA) in Milli-Q H_2O and Acetonitrile (can) at a flow rate of 0.3 mL/min with an injection volume of 10–30 μL [44]. The compounds were detected at 281 nm and 254 nm UV wavelengths. The values of the retention times (t_R) are given in minutes. Mass spectrometry (LC-MS) experiments were performed on all the samples. The stock solutions (1 mg/mL in MeOH) where diluted with 0.1% HCOOH in MeOH/H_2O (50:50) to a final concentration of 50 μg/mL prior to analysis. The instrument used consisted of a Thermo Accela LC system interfaced to a Thermo TSQ Access triple quadrupole mass spectrometer with a HESI source. The data were processed with Xcalibur software (version 2.0). An amount of 10 μL of sample was analyzed in flow injection, with a flow rate of 0.2 mL/min of mobile phase 0.1% HCOOC in MeOH/H_2O (50:50). Parameters used for the analysis in positive ion mode were: spray voltage 3500 V; vaporizer temperature 300 °C; sheath gas pressure 50 au; capillary temperature 350 °C; capillary offset 35.

3.2. Chemistry

3.2.1. General Procedure for Compounds **4a,b**

A mixture of **3** (0.35 mmol) [35], a catalytic amount of Et_3N (0.1 mL) and $SOCl_2$ (9.35 mmol) was stirred at room temperature for 30 min. Then the excess of $SOCl_2$ was removed in vacuo and the residue oil was dissolved in cold anhydrous THF (1 mL). To this suspension, the appropriate amine (0.75 mmol) was added and the mixture was stirred at room temperature for 2 h. After cooling, cold water was added (2–5 mL) and the suspension was extracted with CH_2Cl_2 (3 × 15 mL); the solvent was evaporated under vacuum to afford the desired final compounds, which were purified by flash column chromatography using cyclohexane/ethyl acetate 1:2 as eluent (4a), or by crystallization from ethanol (4b).

5-Amino-6-oxo-3-phenyl-1,6-dihydropyridazine-4-carboxylic acid phenylamide (**4a**)

Yield = 40%; mp = 228–229 °C (EtOH). Light brown solid, ^{1}H NMR (400 MHz, DMSO-d_6) δ 6.57 (s, 2H, NH_2), 7.02 (t, 1H, J = 7.4 Hz, ArCONH), 7.23 (t, 2H, J = 7.8 Hz, Ar), 7.32 (d,

2H, J = 7.3 Hz, Ar), 7.39 (d, 2H, J = 8.0 Hz, Ar), 7.47–7.49 (m, 2H, Ar), 10.04 (s, 1H, $CONH_2$), 12.88 (s, 1H, ArNH). ^{13}C NMR (100 MHz, DMSO-d_6) δ163.75, 156.35, 155.43, 145.54, 141.87, 138.61, 128.43, 128.24, 127.93, 123.86, 119.95, 109.88. MS-ESI for $C_{17}H_{14}N_4O_2$ (Calcd, 306.11), $[M + H]^+$ at *m/z* 306.96, t_R = 11.825. Anal. Calcd for $C_{17}H_{14}N_4O_2$: C, 66.66; H, 4.61; N, 18.29. Found C, 66.92; H, 4.63; N, 18.36.

5-Amino-6-oxo-3-phenyl-1,6-dihydropyridazine-4-carboxylic acid propylamide (**4b**)

Yield = 35%; mp = 228–230 °C (EtOH). Yellow coloured solid, ^1H NMR (400 MHz, DMSO-d_6) δ 0.57–0.61 (m, 2H, CH_3), 1.18 (dp, 2H, J = 14.2, 7.2 Hz, CH_2), 2.94 (q, 2H, J = 6.8 Hz, $NHCH_2$), 6.31 (s, 2H, NH_2), 7.36 (dt, 2H, J = 4.5, 1.6 Hz, Ar), 7.44 (dq, 2H, J = 6.4, 1.7 Hz, Ar), 7.98 (t, 1H, J = 5.8 Hz, Ar), 12.79 (s, 1H, ArNH). ^{13}C NMR (100 MHz, DMSO-d_6) δ 164.94, 156.30, 145.52, 141.61, 137.07, 128.17, 127.90, 127.80, 110.26, 40.55, 21.51, 11.24. MS-ESI for $C_{14}H_{16}N_4O_2$ (Calcd, 272.13), $[M + H]^+$ at *m/z* 273.02, t_R = 9.970. Anal. Calcd for $C_{14}H_{16}N_4O_2$: C, 61.75; H, 5.92; N, 20.58. Found C, 61.99; H, 5.94; N, 20.66.

3.2.2. General Procedure for Compounds **5a,b**

A mixture of **4a,b** (0.43 mmol), K_2CO_3 (0.86 mmol) and 0.50 mmol of ethyl bromide in anhydrous DMF (2 mL) was refluxed for 30–90 min. After cooling, the mixture was diluted with cold water (15 mL) and compound **5a** was recovered by filtration under vacuum. For compound **5b** the suspension was extracted with CH_2Cl_2 (3 × 15 mL) and the solvent was evaporated in vacuo. The crude products were purified by crystallization from ethanol.

5-Amino-1-ethyl-6-oxo-3-phenyl-1,6-dihydropyridazine-4-carboxylic acid phenylamide (**5a**)

Yield = 90%; mp = 172–173 °C (EtOH). ^1H NMR (400 MHz, $CDCl_3$) δ 1.42 (t, 3H, CH_3CH_2, J = 7.2 Hz), 4.26 (q, 2H, CH_3CH_2, J = 7.2 Hz), 6.70 (exch br s, 1H, CONH), 6.93 (d, 2H, Ar, J = 8.0 Hz), 7.04 (t, 1H, Ar, J = 8.0 Hz), 7.20 (t, 2H, Ar, J = 8.0 Hz), 7.49–7.54 (m, 3H, Ar), 7.55–7.60 (m, 2H, Ar). Anal. Calcd for $C_{19}H_{18}N_4O_2$: C, 68.25; H, 5.43; N, 16.76. Found C, 68.41; H, 5.44; N, 16.72.

5-Amino-1-ethyl-6-oxo-3-phenyl-1,6-dihydro-pyridazine-4-carboxylic acid propylamide (**5b**)

Yield = 80%; mp = 141–143 °C (EtOH). ^1H NMR (400 MHz, $CDCl_3$) δ 0.64 (t, 3H, NH-$CH_2CH_2CH_3$, J = 7.2 Hz), 1.15 (sex, 2H, NH-$CH_3CH_2CH_2$, J = 7.6 Hz), 1.43 (t, 3H, N-CH_2CH_3, J = 7.2 Hz), 3.05 (q, 2H, NH-$CH_2CH_2CH_3$, J = 7.2 Hz), 4.25 (q, 2H, N-CH_2CH_3, J = 7.2 Hz), 5.02 (exch br s, 1H, $CONHCH_2$), 6.95 (exch br s, 2H, NH_2), 7.45–7.51 (m, 5H, Ar). Anal. Calcd for $C_{16}H_{20}N_4O_2$: C, 63.98; H, 6.71; N, 18.65. Found C, 63.83; H, 6.70; N, 18.70.

3.2.3. 5-Amino-6-oxo-3-phenyl-1,6-dihydropyridazine-4-carboxylic acid amide (**6**)

A mixture of isoxazolopyridazinone **1** (0.94 mmol) [29], 2 mL of 33% NH_3 and a catalytic amount of piperidine was stirred at 60 °C for 90 min in a sealed/pressure vessel. After cooling the precipitate was recovered by suction and recrystallized with diethyl ether. Yield = 46%; mp > 300 °C (Et_2O). Light brown solid, ^1H NMR (400 MHz, DMSO-d_6) δ 6.38 (s, 2H, NH_2), 7.38 (dd, 3H, Ar, J = 5.0, 2.1 Hz), 7.47–7.49 (m, 2H, Ar), 12.79 (s, 1H, ArNH). ^{13}C NMR (100 MHz, DMSO-d_6) δ 167.25, 156.28, 145.36, 141.66, 137.17, 128.17, 127.97, 127.83, 109.75. MS-ESI for $C_{11}H_{10}N_4O_2$ (Calcd, 230.08), $[M + H]^+$ at *m/z* 230.95, t_R = 6.091. Anal. Calcd for $C_{11}H_{10}N_4O_2$: C, 57.39; H, 4.38; N, 24.34. Found C, 57.16; H, 4.36; N, 24.24.

3.2.4. 5-Amino-6-oxo-3-phenyl-1,6-dihydropyridazine-4-carbonitrile (**7**)

A suspension of **6** (0.40 mmol) in $POCl_3$ (8 mmol) was stirred at 60 °C for 1–2 h. After cooling, the reaction mixture was treated with cold water (15 mL) and the suspension was extracted with CH_2Cl_2 (3 × 15 mL). The organic solvent was evaporated to afford the desired final compound which was purified by crystallized from diethyl ether. Yield = 48%; mp = 287–289 °C (Et_2O). Yellow coloured solid, ^1H NMR (400 MHz, DMSO-d_6) δ 6.78 (s,

1H, NH_2), 7.48 (tt, 3H, J = 3.9, 2.4 Hz, Ar), 7.57–7.61 (m, 2H, Ar), 12.98 (s, 1H, ArNH). ^{13}C NMR (100 MHz, DMSO-d_6) δ154.47, 154.06, 149.95, 145.72, 135.21, 129.28, 128.29, 128.13, 115.68, 113.42. MS-ESI for $C_{11}H_8N_4O$ (Calcd, 212.07), $[M + H]^+$ at *m/z* 212.89, t_R = 10.234. Anal. Calcd for $C_{11}H_8N_4O$: C, 62.26; H, 3.80; N, 26.40. Found C, 62.01; H, 3.78; N, 26.29.

3.2.5. 4-Acetyl-isoxazole-3-carboxylic acid ethyl ester (**10**)

To a cooled (−5 °C) and stirred suspension of **8** (9.9 mmol) in anhydrous ethanol, a solution of ethyl chloro(hydroximino)acetate **9** (6.6 mmol) in the same solvent (11 mL) was added dropwise. The solvent was evaporated in vacuo, cold water was added (10 mL) and the suspension was extracted with CH_2Cl_2 (3 × 15 mL). A mixture of isoxazoles **10** and **11** [36] was obtained and they were separated by flash column chromatography using cyclohexane/ethyl acetate 2:1 as eluent. Yield = 15%; oil. ^1H NMR (400 MHz, $CDCl_3$) δ 1.45 (t, 3H, CH_2CH_3, J = 7.2 Hz), 2.55 (s, 3H, $COCH_3$), 4.51 (q, 2H, CH_2CH_3, J = 7.2 Hz), 8.95 (s, 1H, Ar). Anal. Calcd for $C_8H_9NO_4$: C, 52.46; H, 4.95; N, 7.65. Found C, 52.33; H, 4.94; N, 7.67.

3.2.6. General Procedure for Compounds **12** and **13**

To a cooled and stirred mixture of isoxazoles **10** or **11** (6.56 mmol) and 2.5 g of PPA (25 mmol) in 2 mL of anhydrous EtOH, 7.87 mmol of phenylhydrazine were added. The reaction was carried out at 70 °C for 30 min. After cooling the solvent was evaporated under vacuum, cold water was added (10 mL) and the suspension was extracted with CH_2Cl_2 (3 × 15 mL). Evaporation of the solvent afforded the desired compounds.

4-Methyl-6-phenyl-6H-isoxazolo [3,4-d]pyridazin-7-one (**12**)

Yield = 90%; mp = 200–201 °C (Cyclohexane). ^1H NMR (400 MHz, $CDCl_3$) δ 2.54 (s, 3H, CH_3), 7.38 (t, 1H, Ar, J = 7.6 Hz), 7.45–7.50 (m, 2H, Ar), 7.55–7.60 (m, 2H, Ar), 9.22 (s, 1H, C=CH). Anal. Calcd for $C_{12}H_9N_3O_2$: C, 63.43; H, 3.99; N, 18.49. Found C, 63.58; H, 4.00; N, 18.44.

3-Methyl-6-phenyl-6H-isoxazolo [3,4-d]pyridazin-7-one (**13**)

Yield = 80%; mp = 188–190 °C (Cyclohexane). ^1H NMR (400 MHz, $CDCl_3$) δ 2.88 (s, 3H, CH_3), 7.42 (t, 1H, Ar, J = 7.4 Hz), 7.51 (t, 2H, Ar, J = 7.8 Hz), 7.60 (d, 2H, Ar, J = 7.6 Hz), 8.16 (s, 1H, N=CH). Anal. Calcd for $C_{12}H_9N_3O_2$: C, 63.43; H, 3.99; N, 18.49. Found C, 63.55; H, 3.99; N, 18.46.

5-Amino-3-methyl-6-oxo-1-phenyl-1,6-dihydropyridazine-4-carboxylic acid methyl ester (**14**)

A mixture of **12** (6.21 mmol) and Et_3N (0.8 mL) in 2 mL of CH_3OH was heated at 60 °C for 2 h. After cooling, ice water (20 mL) was added and the suspension was extracted with CH_2Cl_2 (3 × 15 mL). Then the solvent was evaporated in vacuo to afford compound **14** which was purified by flash column chromatography using cyclohexane/ethyl acetate 1:1 as eluent. Yield = 80%; mp = 91–93 °C (Cyclohexane). ^1H NMR (400 MHz, $CDCl_3$) δ 2.53 (s, 3H, CH_3), 3.95 (s, 3H, $COOCH_3$), 7.39 (t, 1H, Ar, J = 7.4 Hz), 7.49 (t, 2H, Ar, J = 8.4 Hz), 7.65 (d, 2H, Ar, J = 8.4 Hz), 8.16 (exch br s, 2H, NH_2). Anal. Calcd for $C_{13}H_{13}N_3O_3$: C, 60.22; H, 5.05; N, 16.21. Found C, 60.08; H, 5.06; N, 16.26.

3.2.7. 5-Amino-3-methyl-6-oxo-1-phenyl-1,6-dihydropyridazine-4-carboxylic acid (**15**)

A mixture of **14** (2.12 mmol), ethanol (3 mL) and 6N NaOH (2 mL) was stirred at reflux for 30 min. After cooling, the solvent was evaporated under vacuum, cold water was added (2–3 mL) and the mixture was acidified with 6N HCl. The precipitate was recovered by vacuum filtration and crystallized from cyclohexane. Yield = 90%; mp = 214–216 °C (Cyclohexane). ^1H NMR (400 MHz, DMSO-d_6) δ 2.37 (s, 3H, CH_3), 7.39 (t, 1H, Ar, J = 7.2 Hz), 7.46 (t, 2H, Ar, J = 8.0 Hz), 7.54 (d, 2H, Ar, J = 8.0 Hz), 8.25 (exch br s, 2H, NH_2). Anal. Calcd for $C_{12}H_{11}N_3O_3$: C, 58.77; H, 4.52; N, 17.13. Found C, 58.61; H, 4.51; N, 17.16.

3.2.8. 5-Amino-3-methyl-6-oxo-1-phenyl-1,6-dihydropyridazine-4-carboxylic acid amide (**16**)

A mixture of **15** (1.88 mmol), a catalytic amount of Et_3N (0.1 mL) and $SOCl_2$ (51 mmol) was refluxed for 30 min. After cooling, the excess of $SOCl_2$ was removed in vacuo and the residue oil was dissolved in cold dry THF (1 mL). To this suspension a solution of 33% NH_3 (2 mL) in 1.5 mL of dry THF was added and the mixture was stirred at room temperature for 15 min. After evaporation of the solvent, the mixture was diluted with cold water (20 mL) and the precipitate obtained was filtered and crystallized from ethanol. Yield = 80%; mp = 247–249 °C (EtOH). White coloured solid, ^1H NMR (400 MHz, DMSO-d_6) δ 2.21 (s, 3H, CH_3), 7.35–7.40 (m, 1H, Ar), 7.45–7.51 (m, 5H, Ar), 7.65 (s, 1H, NH_2), 7.88 (s, 1H, $CONH_2$). ^{13}C NMR (100 MHz, DMSO-d_6) δ 167.39, 155.53, 143.12, 141.99, 141.63, 128.82, 127.99, 125.85, 110.42, 20.24. MS-ESI for $C_{12}H_{12}N_4O_2$ (Calcd, 244.10), $[M + H]^+$ at *m/z* 244.95, t_R = 7.921. Anal. Calcd for $C_{12}H_{12}N_4O_2$: C, 59.01; H, 4.95; N, 22.94. Found C, 59.24; H, 4.97; N, 23.03.

3.2.9. 5-Amino-3-methyl-6-oxo-1-phenyl-1,6-dihydropyridazine-4-carbonitrile (**17**)

Compound **17** was obtained starting from compound **16**, through the same procedure described for **7**. After dilution with cold water, the precipitate was recovered by filtration under *vacuum* and the solid obtained was purified by flash column chromatography using cyclohexane/ethyl acetate 1:1 as eluent. Yield = 20%; mp = 201–203 °C (EtOH). ^1H NMR (400 MHz, DMSO-d_6) δ 2.26 (s, 3H, CH_3), 7.38 (ddd, 1H, J = 7.7, 5.5, 3.6 Hz, Ar), 7.42–7.51 (m, 4H, Ar). ^{13}C NMR (100 MHz, DMSO-d_6) δ 153.37, 149.50, 144.17, 141.60, 130.03, 127.68, 125.90, 115.14, 100.83, 20.47. MS-ESI for $C_{12}H_{10}N_4O$ (Calcd, 226.08), 226.96 *m/z* $[M + H]^+$, 435.11 *m/z* $[2M+H-H_2O]^+$, 451.07 *m/z* $[2M-H_2+H]^+$. t_R =11.385. Anal. Calcd for $C_{12}H_{10}N_4O$: C, 63.71; H, 4.46; N, 24.76. Found C, 63.96; H, 4.48; N, 24.85.

3.2.10. 5-Acetyl-4-amino-2-phenylpyridazin-3(2H)-one (**18**)

Intermediate **13** (1.01 mmol) was suspended in 3.5 mL of EtOH, then 6.08 mmol of $HCOONH_4$ and 40 mg of 10% Pd/C were added. The mixture was refluxed for 2 h and after cooling, CH_2Cl_2 (5 mL) was added. The solution was stirred for 5 min, then the catalyst was filtered off and the solvent was evaporated in vacuo to furnish desiderd compound **18**. Yield = 98%; mp = 181–183 °C (Cyclohexane). ^1H NMR (400 MHz, $CDCl_3$) δ 2.60 (s, 3H, $COCH_3$), 6.95 (exch br s, 1H, NH_2), 7.42 (t, 1H, Ar, J = 7.6 Hz), 7.51 (t, 2H, Ar, J = 7.6 Hz), 7.64 (d, 2H, Ar, J = 7.6 Hz), 8.13 (s, 1H, C6-H), 9.15 (exch br s, 1H, NH_2). Anal. Calcd for $C_{12}H_{11}N_3O_2$: C, 62.87; H, 4.84; N, 18.33. Found C, 62.69; H, 4.83; N, 18.28.

3.2.11. 6-Cyclohexyl-4-phenyl-6H-isoxazolo [3,4-d]pyridazin-7-one (**20**)

Compound **20** was obtained starting from **19** [34] adopting the general procedure described for compounds **12** and **13**, but using cyclohexyl hydrazine as reagent. The mixture was heated at 70 °C for 5 h. After dilution with ice-water, the precipitate was recovered by filtration under vacuum and crystallized from ethanol. Yield = 45%; mp = 211–213 °C (EtOH). ^1H NMR (400 MHz, $CDCl_3$) δ 1.27–1.33 (m, 1H, C_6H_{11}), 1.45–1.51 (m, 3H, C_6H_{11}), 1.75–1.80 (m, 1H, C_6H_{11}), 1.85–1.95 (m, 5H, C_6H_{11}), 5.05–5.10 (m, 1H, C_6H_{11}), 7.50–7.60 (m, 3H, Ar), 7.85 (d, 2H, Ar, J = 7.2 Hz), 9.30 (s, 1H, isoxazole). Anal. Calcd for $C_{17}H_{17}N_3O_2$: C, 69.14; H, 5.80; N, 14.23. Found C, 69.33; H, 4.82; N, 18.28.

3.2.12. 5-Amino-1-cyclohexyl-6-oxo-3-phenyl-1,6-dihydropyridazine-4-carboxylic acid amide (**21**)

A mixture of **20** (0.64 mmol) and 33% NH_3 was stirred at 120 °C for 3 h in a sealed/pressure vessel. After cooling, ice-water was added and the suspension was extracted with CH_2Cl_2 (3 × 15 mL). Evaporation of the solvent afforded the desired final compound. Yield = 20%; mp = 125–128 °C (Cyclohexane). ^1H NMR (400 MHz, $CDCl_3$) δ 1.20–1.25 (m, 1H, C_6H_{11}), 1.35–1.48 (m, 2H, C_6H_{11}), 1.60–1.65 (m, 2H, C_6H_{11}), 1.70–1.88 (m, 5H, C_6H_{11}), 4.77–4.82 (m, 1H, C_6H_{11}), 6.55 (exch br s, 2H, NH_2), 7.25–7.31 (m, 3H, Ar), 7.44 (d, 2H, Ar,

J = 7.6 Hz), 8.50 (exch br s, 2H, $CONH_2$). Anal. Calcd for $C_{17}H_{20}N_4O_2$: C, 65.37; H, 6.45; N, 17.94. Found C, 65.52; H, 6.46; N, 17.99.

3.2.13. 5-Amino-1-cyclohexyl-6-oxo-3-phenyl-1,6-dihydropyridazine-4-carbonitrile (**22**)

Compound **22** was obtained starting from compound **21**, through the same procedure described for **7** and **17**. After dilution with cold water, the precipitate was recovered by suction and the solid was purified by crystallization from etanol. Yield = 90%; mp = 170–172 °C (EtOH). Yellow coloured solid, ^{1}H NMR (400 MHz, $CDCl_3$) δ 1.23 (qt, 1H, J = 13.2, 3.3 Hz, CH_2-cyclohexane), 1.46 (ttt, 2H, J = 13.8, 7.8, 3.0 Hz, CH_2-cyclohexane), 1.71 (dt, 1H, J = 13.3, 3.4 Hz, CH_2-cyclohexane), 1.84–1.91 (m, 6H, J = 4.4, 3.5 Hz, CH_2-cyclohexane), 4.84–4.93 (m, 1H, CH_2-cyclohexane), 7.45–7.50 (m, 3H, Ar), 7.72–7.74 (m, 2H, Ar). ^{13}C NMR (100 MHz, $CDCl_3$) δ 152.65, 148.93, 144.34, 135.02, 129.91, 128.75, 128.20, 114.83, 85.25, 58.19, 30.99, 25.60. MS-ESI for $C_{17}H_{18}N_4O$ (Calcd, 294.15), $[M + H]^+$ at *m/z* 294.99, $[M + ACN + H]^+$ at *m/z* 336.01, t_R = 17.509. Anal. Calcd for $C_{17}H_{18}N_4O$: C, 69.37; H, 6.16; N, 19.03. Found C, 69.09; H, 6.13; N, 18.95.

3.2.14. General Procedure for **24b, 24d-f**

A mixture of **23** [37] (0.80 mmol), K_2CO_3 (1.60 mmol) and 0.96–1.44 mmol of the appropriate alkyl or cycloalkyl bromide in anhydrous DMF (1 mL) was refluxed for 2–4 h. After cooling, the mixture was diluted with cold water (20 mL) and extracted with CH_2Cl_2 (3 × 15 mL). Evaporation of the solvent afforded the desired final compounds which were purified by flash column chromatography using cyclohexane/ethyl acetate 1:1 (for **24b,e,f**) or 1:2 (for **24d**) as eluent.

4-Amino-2-cyclohexyl-6-phenylpyridazin-3(2H)-one (**24b**)

Yield = 21%; mp = 120–124 °C (EtOH). ^{1}H NMR (400 MHz, $CDCl_3$) δ 1.20–1.35 (m, 1H, C_6H_{11}), 1.40–1.50 (m, 2H, C_6H_{11}), 1.70–1.80 (m, 1H, C_6H_{11}), 1.90–2.05 (m, 6H, C_6H_{11}), 4.85–5.10 (m, 3H, 1H C_6H_{11} + 2H NH_2), 6.75 (s, 1H, -CH pyridaz.), 7.35–7.50 (m, 3H, Ar), 7.70 (d, 2H, Ar, J = 7.6 Hz). Anal. Calcd for $C_{16}H_{19}N_3O$: C, 71.35; H, 7.11; N, 15.60. Found C, 71.52; H, 7.10; N, 15.56.

4-Amino-2-isopropyl-6-phenylpyridazin-3(2H)-one (**24d**)

Yield = 85%; mp = 122–124 °C (EtOH). ^{1}H NMR (400 MHz, $CDCl_3$) δ 1.45 (d, 6H, $CH(CH_3)_2$, J = 6.8 Hz), 4.97 (exch br s, 2H, NH_2), 5.41 (quin, 1H, $CH(CH_3)_2$, J = 6.8 Hz), 6.75 (s, 1H, -CH pyridaz.), 7.40–7.50 (m, 3H, Ar), 7.81 (d, 2H, Ar, J = 7.6 Hz). Anal. Calcd for $C_{13}H_{15}N_3O$: C, 68.10; H, 6.59; N, 18.33 Found C, 68.31; H, 6.60; N, 18.29.

4-Amino-6-phenyl-2-propylpyridazin-3(2H)-one (**24e**)

Yield = 83%; mp = 79–81 °C (EtOH). ^{1}H NMR (400 MHz, $CDCl_3$) δ 1.02 (t, 3H, $CH_2CH_2CH_3$, J = 7.2 Hz), 1.93 (sex, 2H, $CH_2CH_2CH_3$, J = 7.2 Hz), 4.23 (t, 2H, $CH_2CH_2CH_3$, J = 7.2 Hz), 4.99 (exch br s, 2H, NH_2), 6.73 (s, 1H, -CH pyridaz.), 7.38–7.50 (m, 3H, Ar), 7.78 (d, 2H, Ar, J = 8.0 Hz). Anal. Calcd for $C_{13}H_{15}N_3O$: C, 68.10; H, 6.59; N, 18.33 Found C, 68.28; H, 6.60; N, 18.31.

4-Amino-2-butyl-6-phenylpyridazin-3(2H)-one (**24f**)

Yield = 94%; mp = 67–69 °C (EtOH). ^{1}H NMR (400 MHz, $CDCl_3$) δ 1.00 (t, 3H, $CH_2CH_2CH_2CH_3$, J = 7.2 Hz), 1.45 (m, 2H, $CH_2CH_2CH_2CH_3$), 1.87 (m, 2H, $CH_2CH_2CH_2CH_3$), 4.26 (t, 2H, $CH_2CH_2CH_2CH_3$, J = 7.2 Hz), 4.99 (exch br s, 2H, NH_2), 6.75 (s, 1H, -CH pyridaz.), 7.38–7.50 (m, 3H, Ar), 7.76 (d, 2H, Ar, J = 8.0 Hz). Anal. Calcd for $C_{14}H_{17}N_3O$: C, 69.11; H, 7.04; N, 17.27 Found C, 69.29; H, 7.03; N, 17.32.

3.2.15. General Procedure for Compounds **25a–f**

To a cooled (0 °C) and stirred suspension of the appropriate pyridazinone **24a–f** (0.65 mmol) in anhydrous THF (1–3 mL), anhydrous sodium acetate (1.55 mmol) and triphosgene (2.26 mmol) were added. The mixture was stirred for 10 min at room temperature and refluxed for 2 h. Then, the suspension was cooled to 0 °C and 1 mL of 33% NH_3 was added and the mixture was stirred for 30–90 min at room temperature. After evaporation of the solvent, ice/cold water was added (15 mL) and the precipitate obtained was recovered by filtration under vacuum and purified by crystallization from ethanol to obtain the pure samples of **25a–f.**

(2-Methyl-3-oxo-6-phenyl-2,3-dihydro-pyridazin-4-yl)urea (**25a**)

Yield = 65%; mp > 300 °C (EtOH). ^{1}H NMR (400 MHz, DMSO-d_6) δ 3.77 (s, 3H, CH_3), 7.42–7.51 (m, 3H, Ar), 7.73–7.76 (m, 2H, Ar), 8.35 (s, 1H, Ar), 8.98 (s, 1H, $NHCONH_2$). ^{13}C NMR (100 MHz, DMSO-d_6) δ 155.69, 155.11, 145.12, 137.92, 135.75, 129.39, 129.09, 126.02, 106.62, 20.93. MS-ESI for $C_{12}H_{12}N_4O_2$ (Calcd, 244.10), $[M + H]^+$ at *m/z* 244.95, t_R = 11.531. Anal. Calcd for $C_{12}H_{12}N_4O_2$: C, 59.01; H, 4.95; N, 22.94 Found C, 59.24; H, 4.97; N, 23.03.

(2-Cyclohexyl-3-oxo-6-phenyl-2,3-dihydro-pyridazin-4-yl)urea (**25b**)

Yield = 35%; mp = 261–263 °C (EtOH). ^{1}H NMR (400 MHz, $CDCl_3$) δ 1.18–1.31 (m, 1H, C_6H_{11}), 1.40–1.51 (m, 2H, C_6H_{11}), 1.64–1.72 (m, 1H, C_6H_{11}), 1.70–1.90 (m, 6H, C_6H_{11}), 4.87 (m, 1H, C_6H_{11}), 6.80 (exch br s, 2H, NH_2), 7.45–7.55 (m, 3H, Ar), 7.79 (d, 2H, Ar, J = 7.6 Hz), 8.37 (s, 1H, -CH pyridaz.), 8.96 (exch br s, 1H, $NHCONH_2$). Anal. Calcd for $C_{17}H_{20}N_4O_2$: C, 65.37; H, 6.45; N, 17.94. Found C, 65.18; H, 6.46; N, 17.91.

(2-Ethyl-3-oxo-6-phenyl-2,3-dihydro-pyridazin-4-yl)urea (**25c**)

Yield = 25%; mp = 270–271 °C (EtOH). White coloured solid, ^{1}H NMR (400 MHz, DMSO-d_6) δ 1.33 (t, 3H, J = 7.1 Hz, NCH_2CH_3), 4.20 (q, 2H, J = 7.2 Hz, NCH_2), 6.70 (exch br s, 2H, $NHCONH_2$) 7.47 (dt, 3H, J = 13.1, 7.1 Hz, Ar), 7.73 (dd, 2H, J = 23.5, 7.6 Hz, Ar), 8.34 (s, 1H, Ar), 8.98 (s, 1H, $NHCONH_2$). ^{13}C NMR (100 MHz, DMSO-d_6) δ 155.36, 154.51, 150.89, 145.06, 137.97, 135.75, 128.88, 125.84, 106.29, 47.00, 13.39. MS-ESI for $C_{13}H_{14}N_4O_2$ (Calcd, 258.11), $[M + H]^+$ at *m/z* 259.02, 215.90 *m/z* $[M\text{-}CONH_2 + H]^+$. t_R = 12.090. Anal. Calcd for $C_{13}H_{14}N_4O_2$: C, 60.45; H, 5.46; N, 21.69. Found C, 60.21; H, 5.44; N, 21.60.

(2-Isopropyl-3-oxo-6-phenyl-2,3-dihydro-pyridazin-4-yl)urea (**25d**)

Yield = 68%; mp = 260–263 °C (EtOH). White coloured solid, ^{1}H NMR (400 MHz, DMSO-d_6) δ 1.36 (d, 6H, J = 6.7 Hz, $NCH(CH_3)_2$), 5.24 (q, 1H, J = 6.6 Hz, NCH), 7.47 (dt, 2H, J = 15.9, 7.2 Hz, Ar), 7.67–7.79 (m, 3H, Ar), 8.34 (s, 1H, Ar), 8.95 (s, 1H, $NHCONH_2$). ^{13}C NMR (100 MHz, DMSO-d_6) δ 155.69, 154.31, 144.75, 137.69, 136.13, 132.04, 129.30, 129.08, 127.97, 125.91, 105.93, 49.71, 20.96. MS-ESI for $C_{14}H_{16}N_4O_2$ (Calcd, 272.13), $[M + H]^+$ at *m/z* 272.95, t_R = 14.248. Anal. Calcd for $C_{14}H_{16}N_4O_2$: C, 61.75; H, 5.92; N, 20.58. Found C, 61.99; H, 5.94; N, 20.66.

(3-Oxo-6-phenyl-2-propyl-2,3-dihydro-pyridazin-4-yl)urea (**25e**)

Yield = 60%; mp = 273–275 °C (EtOH). White coloured solid, ^{1}H NMR (400 MHz, DMSO-d_6) δ 0.89 (td, 3H, J = 7.4, 2.7 Hz, $NCH_2CH_2CH_3$), 1.79 (q, 2H, J = 7.3 Hz, NCH_2CH_2), 4.13 (t, 2H, J = 7.1 Hz, NCH_2), 7.42–7.51 (m, 2H, Ar), 7.69 (d, 1H, J = 5.6 Hz, Ar), 7.73–7.76 (m, 2H, Ar), 8.33 (s, 1H, Ar), 8.96 (s, 1H, $NHCONH_2$). ^{13}C NMR (100 MHz, DMSO-d_6) δ 155.68, 154.88, 145.06, 137.94, 135.87, 132.03, 129.36, 129.07, 128.06, 126.03, 106.37, 53.19, 21.39, 11.13. MS-ESI for $C_{14}H_{16}N_4O_2$ (Calcd, 272.13), $[M + H]^+$ at *m/z* 272.95, 229.90 *m/z* $[M\text{-}CONH_2 + H]^+$. t_R = 14.037. Anal. Calcd for $C_{14}H_{16}N_4O_2$: C, 61.75; H, 5.92; N, 20.58. Found C, 61.99; H, 5.94; N, 20.66.

(2-Butyl-3-oxo-6-phenyl-2,3-dihydro-pyridazin-4-yl)urea (**25f**)

Yield = 85%; mp = 265–267 °C (EtOH). White solid, ^{1}H NMR (400 MHz, DMSO-d_6 + D_2O) δ 0.88 (t, 3H, J = 7.4 Hz, CH_3), 1.27 (q, 2H, J = 7.3 Hz, CH_2CH_3), 1.73 (q, 2H, J = 7.2 Hz, $CH_2CH_2CH_3$), 4.13–4.16 (m, 2H, CH_2ArN), 7.41–7.51 (m, 3H, Ar), 7.73–7.75 (m, 2H, Ar), 8.08 (s, 1H, $NHCONH_2$), 8.36 (s, 1H, Ar), 9.10 (s, 2H, $CONH_2$). ^{13}C NMR (100 MHz, DMSO-

d_6) δ 154.80, 154.60, 137.56, 135.69, 129.20, 128.94, 127.78, 125.79, 118.03, 106.44, 51.11, 29.92, 19.24, 13.53. MS-ESI for $C_{15}H_{18}N_4O_2$ (Calcd, 286.14), $[M + H]^+$ at *m/z* 286.94, t_R = 31.162. Anal. Calcd for $C_{15}H_{18}N_4O_2$: C, 62.92; H, 6.34; N, 19.57. Found C, 62.66; H, 6.31; N, 19.49.

3.2.16. 4-Amino-6-phenylpyridazine-3(2H)-thione (**26**)

A mixture of **23** [37] (0.86 mmol) and Lawesson's reagent (1.71 mmol) in anhydrous toluene (2–3 mL) was heated at 90 °C for 5 h. After cooling the solvent was evaporated under vacuum, cold water was added (10 mL) and the mixture was extracted with CH_2Cl_2 (3 × 15 mL). Evaporation of the solvent afforded **26** which was purified by flash column chromatography using CH_2Cl_2/CH_3OH 10:1 as eluent. Yield = 63%; mp = 175–178 °C (Cyclohexane). ^1H NMR (400 MHz, $CDCl_3$) δ 5.77 (exch br s, 2H, NH_2), 6.80 (s, 1H, -CH pyridaz.), 7.45–7.55 (m, 3H, Ar), 7.75–7.81 (m, 2H, Ar). Anal. Calcd for $C_{10}H_9N_3S$: C, 59.09; H, 4.46; N, 20.67. Found C, 59.23; H, 4.45; N, 20.62.

3.2.17. 3-Methylsulfanyl-6-phenyl-pyridazin-4-ylamine (**27**)

Compound **27** was obtained, starting from compound **26**, through the general procedure described for **24b** and **24d–f**. After dilution with cold water, the precipitate was recovered by suction and purified by crystallization. Yield = 40%; mp = 168–170 °C (Cyclohexane). Greenish colour solid, ^1H NMR (400 MHz, DMSO-d_6) δ 2.64 (s, 3H, SCH_3), 6.27 (exch br s, 2H, NH_2), 7.00 (s, 1H, Ar), 7.46 (dt, 3H, ArH, J = 12.6, 6.9 Hz), 7.90–7.93 (m, 2H, Ar). ^{13}C NMR (100 MHz, DMSO-d_6) δ155.31, 147.71, 144.50, 137.20, 129.38, 129.02, 126.51, 102.80, 12.72. MS-ESI for $C_{11}H_{11}N_3S$ (Calcd, 217.07), $[M + H]^+$ at *m/z* 217.86, t_R = 9.922. Anal. Calcd for $C_{11}H_{11}N_3S$: C, 60.80; H, 5.10; N, 19.34. Found C, 60.56; H, 5.08; N, 19.26.

3.2.18. (3-Oxo-6-phenyl-2,3-dihydro-pyridazin-4-yl)urea (**28**)

Compound **28** was obtained, starting from **23** [37], through the same procedure described for **25a–f**. Yield = 85%; mp >300 °C (EtOH). ^1H NMR (400 MHz, DMSO-d_6) δ 7.41–7.50 (m, 3H, Ar), 7.72–7.75 (m, 2H, Ar), 8.33 (s, 1H, Ar), 8.94 (s, 1H, $NHCONH_2$), 13.21 (s, 1H, ArNH). ^{13}C NMR (100 MHz, DMSO-d_6) δ156.00, 155.33, 145.54, 139.57, 138.27, 135.77, 134.50, 128.87, 125.70, 106.97. MS-ESI for $C_{11}H_{10}N_4O_2$ (Calcd, 230.08), $[M + H]^+$ at *m/z* 230.88, t = 10.042. Anal. Calcd for $C_{11}H_{10}N_4O_2$: C, 57.39; H, 4.38; N, 24.34. Found C, 57.62; H, 4.39; N, 24.44.

3.2.19. General Procedure for Compounds **29a–d**

A mixture of **24a** [38] (0.39 mmol) and the appropriate R-anhydride (13.1 mmol) in 1 mL of pyridine was heated at 140 °C for 5 h in a sealed/pressure vessel. After cooling, ice/cold water was added (50 mL), the precipitate was recovered by filtration under vacuum and purified by crystallization from ethanol to obtain the desired compounds.

N-(2-Methyl-3-oxo-6-phenyl-2,3-dihydro-pyridazin-4-yl)acetamide (**29a**)

Yield = 90%; mp = 211–212 °C (EtOH). Brownish black coloured solid, ^1H NMR (400 MHz, $CDCl_3$) δ 2.28 (s, 3H, CH_3CONH), 3.91 (s, 3H, CH_3ArN), 7.42–7.48 (m, 3H, Ar), 7.80–7.83 (m, 2H, Ar), 8.61 (s, 1H, ArH). ^{13}C NMR (100 MHz, $CDCl_3$) δ 196.96, 155.66, 146.61, 135.61, 135.59, 129.59, 128.97, 126.46, 110.82, 40.91, 24.98. MS-ESI for $C_{13}H_{13}N_3O_2$ (Calcd, 243.10), $[M + H]^+$ at *m/z* 243.90, t_R = 13.311. Anal. Calcd for $C_{13}H_{13}N_3O_2$: C, 64.19; H, 5.39; N, 17.27. Found C, 64.45; H, 5.41; N, 17.34.

N-(2-Methyl-3-oxo-6-phenyl-2,3-dihydro-pyridazin-4-yl)propionamide (**29b**)

Yield = 93%; mp = 210–211 °C (EtOH). Ash coloured solid, ^1H NMR (400 MHz, $CDCl_3$) δ 1.26 (td, 3H, J = 7.5, 1.1 Hz, CH_3CH_2CONH), 2.49–2.55 (m, 2H, CH_2CONH), 3.92 (s, 3H, CH_3ArN), 7.41–7.47 (m, 3H, Ar), 7.81–7.84 (m, 2H, Ar), 8.65 (s, 1H, Ar). ^{13}C NMR (100 MHz, $CDCl_3$) δ 173.77, 155.73, 146.62, 135.63, 135.60, 129.58, 128.95, 126.43, 110.76, 40.90, 31.00, 9.24. MS-ESI for $C_{14}H_{15}N_3O_2$ (Calcd, 257.12), $[M + H]^+$ at *m/z* 257.90, t_R = 14.604. Anal. Calcd for $C_{14}H_{15}N_3O_2$: C, 65.36; H, 5.88; N, 16.33. Found C, 65.10; H, 5.90; N, 16.39.

N-(2-Methyl-3-oxo-6-phenyl-2,3-dihydro-pyridazin-4-yl)isobutyramide (**29c**)

Yield = 95%; mp = 146–148 °C (EtOH). Brown coloured solid, ^{1}H NMR (400 MHz, $CDCl_3$) δ 1.28 (dd, 6H, J = 6.9, 1.2 Hz, $(CH_3)_2CHCONH$), 2.64–2.71 (m, 1H, CHCONH), 3.92 (s, 3H, CH_3ArN), 7.41–7.46 (m, 3H, Ar), 7.81–7.85 (m, 2H, Ar), 8.66 (s, 1H, Ar). ^{13}C NMR (100 MHz, $CDCl_3$) δ 177.10, 155.82, 146.61, 135.71, 135.59, 129.57, 128.94, 126.42, 110.83, 40.87, 36.95, 19.45. MS-ESI for $C_{15}H_{17}N_3O_2$ (Calcd, 271.13), $[M + H]^+$ at *m/z* 272.04, t_R = 16.829. Anal. Calcd for $C_{15}H_{17}N_3O_2$: C, 66.40; H, 6.32; N, 15.49. Found C, 66.66; H, 6.34; N, 15.55.

N-(2-Methyl-3-oxo-6-phenyl-2,3-dihydro-pyridazin-4-yl)butyramide (**29d**)

Yield = 92%; mp = 187–189 °C (EtOH). ^{1}H NMR (400 MHz, $CDCl_3$) δ 1.05 (t, 3H, $CH_2CH_2CH_3$, J = 7.2 Hz), 1.80 (sex, 2H, $CH_2CH_2CH_3$, J = 7.2 Hz), 2.49 (q, 2H, $CH_2CH_2CH_3$, J = 7.2 Hz), 3.94 (s, 3H, N-CH_3), 7.45–7.50 (m, 3H, Ar), 7.85 (d, 2H, Ar, J = 7.6 Hz), 8.63 (exch br s, 1H, NH), 8.68 (s, 1H, -CH pyridaz.). Anal. Calcd for $C_{15}H_{17}N_3O_2$: C, 66.40; H, 6.32; N, 15.49. Found C, 66.25; H, 6.31; N, 15.44.

3.2.20. General procedure for compounds **30a,b** and **33**

A mixture of compound **24a** [38] (0.79 mmol), the appropriate R-phenylboronic acid (0.79 mmol), copper acetate (1.19 mmol) and triethylamine (1.59 mmol) in CH_2Cl_2 (5 mL) was stirred at room temperature for 3–12 h. After evaporation of the solvent, ethyl acetate was added (15–20 mL) and the solution was extracted first with 33% NH_3 (3 × 5 mL) and then with water (2 × 5 mL). The organic layer was evaporated under vacuum and the residue was purified by crystallization from ethanol.

3.2.21. 3-(2-Methyl-3-oxo-6-phenyl-2,3-dihydro-pyridazin-4-ylamino)benzonitrile (**30a**)

Yield = 60%; mp = 234–235 °C (EtOH). White coloured solid, ^{1}H NMR (400 MHz, DMSO-d_6) δ 3.80 (s, 3H, CH_3), 7.23 (s, 1H, Ar), 7.42–7.48 (m, 3H, Ar), 7.55–7.59 (m, 2H, Ar), 7.80 (dd, 3H, J = 8.0, 1.8 Hz, ArCN), 7.86 (d, 1H, J = 1.8 Hz, ArCN), 9.03 (exch br s, 1H, NH). ^{13}C NMR (100 MHz, DMSO-d_6) δ 193.78, 157.05, 151.81, 140.85, 136.62, 129.18, 126.83, 111.09, 100.21, 23.94. MS-ESI for $C_{18}H_{14}N_4O$ (Calcd, 302.12), $[M + H]^+$ at *m/z* 302.90, $[M + ACN + H]^+$ at *m/z* 343.92, t_R = 16.247. Anal. Calcd for $C_{18}H_{14}N_4O$: C, 71.51; H, 4.67; N, 18.53. Found C, 71.22; H, 4.65; N, 18.45.

3.2.22. 2-(2-Methyl-3-oxo-6-phenyl-2,3-dihydro-pyridazin-4-ylamino)benzonitrile (**30b**)

Yield = 32%; mp = 178–180 °C (EtOH). White coloured solid, ^{1}H NMR (400 MHz, $CDCl_3$) δ 3.95 (s, 3H, CH_3), 7.12 (s, 1H, Ar), 7.41–7.47 (m, 3H, ArCN), 7.55 (d, 1H, J = 8.3 Hz, ArCN), 7.65 (td, 1H, J = 7.8, 1.6 Hz, Ar), 7.72 (ddd, 3H, J = 7.6, 3.6, 1.7 Hz, Ar), 7.96 (exch br s, 1H, NH). ^{13}C NMR (100 MHz, $CDCl_3$) δ 147.91, 138.56, 13.22, 129.42, 128.97, 126.44, 124.63, 121.19, 114.68, 100.94, 40.59. MS-ESI for $C_{18}H_{14}N_4O$ (Calcd, 302.12), $[M + H]^+$ at *m/z* 302.97, $[M + ACN + H]^+$ at *m/z* 344.06, t_R = 16.180. Anal. Calcd for $C_{18}H_{14}N_4O$: C, 71.51; H, 4.67; N, 18.53. Found C, 71.22; H, 4.65; N, 18.45.

3.2.23. 4-(2-Methyl-3-oxo-6-phenyl-2,3-dihydro-pyridazin-4-ylamino)benzoic acid ethyl ester (**33**)

Yield = 80%; mp = 171–172 °C (Cyclohexane). ^{1}H NMR (400 MHz, $CDCl_3$) 1.42 (t, 3H, CH_2CH_3, J = 7.2 Hz), 3.96 (s, 3H, CH_3), 4.41 (q, 2H, CH_2CH_3, J = 7.2 Hz), 7.30–7.40 (m, 3H, 2H Ar + CH pyridaz.), 7.45–7.50 (m, 3H, Ar), 7.77 (d, 2H, Ar, J = 8.8 Hz), 7.90 (exch br s, 1H, NH), 8.12 (d, 2H, Ar, J = 8.8 Hz). Anal. Calcd for $C_{20}H_{19}N_3O_3$: C, 68.75; H, 5.48; N, 12.03. Found C, 68.58; H, 5.47; N, 12.06.

3.2.24. General Procedure for Compounds **31a,b**

A mixture of appropriate pyridazin-benzonitrile **30a** or **30b** (0.165 mmol) and 80% H2SO4 (2 mL) was stirred at 80 °C for 4 h. After cooling, ice/cold water (2–3 mL) was slowly added, the precipitate obtained was recovered by filtration under vacuum and purified by crystallization.

3.2.25. 3-(2-Methyl-3-oxo-6-phenyl-2,3-dihydro-pyridazin-4-ylamino)benzamide (**31a**)

Yield = 93%; mp = 214–216 °C (Cyclohexane). ^{1}H NMR (400 MHz, DMSO-d_6) δ 3.81 (s, 3H, CH_3), 7.13 (s, 1H, -CH pyridaz.), 7.40–7.50 (m, 5H, 4H Ar + 1H $CONH_2$), 7.60 (d, 1H, Ar, J = 9.2 Hz), 7.64 (d, 1H, Ar, J = 7.2 Hz), 7.76 (d, 2H, Ar, J = 8.0 Hz), 7.91 (s, 1H, Ar), 8.02 (exch br s, 1H, $CONH_2$), 8.93 (exch br s, 1H, NH). Anal. Calcd for $C_{18}H_{16}N_4O_2$: C, 67.49; H, 5.03; N, 17.49. Found C, 67.36; H, 5.04; N, 17.53.

3.2.26. 2-(2-Methyl-3-oxo-6-phenyl-2,3-dihydro-pyridazin-4-ylamino)benzamide (**31b**)

Yield = 95%; mp = 140–142 °C (Cyclohexane). ^{1}H NMR (400 MHz, DMSO-d_6) δ 3.80 (s, 3H, CH_3), 7.15 (t, 1H, Ar, J = 7.6 Hz), 7.38 (s, 1H, -CH pyridaz.), 7.43–7.50 (m, 3H, Ar), 7.57 (t, 1H, Ar, J = 7.6 Hz), 7.66 (exch br s, 1H, $CONH_2$), 7.77 (t, 2H, Ar, J = 9.2 Hz), 7.84 (d, 2H, Ar, J = 6.8 Hz), 8.17 (exch br s, 1H, $CONH_2$), 10.68 (exch br s, 1H, NH). Anal. Calcd for $C_{18}H_{16}N_4O_2$: C, 67.49; H, 5.03; N, 17.49. Found C, 67.36; H, 5.04; N, 17.53.

3.2.27. 4-Amino-2-methyl-6-phenylpyridazine-3(2H)-thione (**32**)

Compound **32** was obtained, starting from compound **24a** [38], through the same procedure described for 26. In this case, the mixture was refluxed for 10 h. After cooling, ice/cold water was added. The precipitate was recovered by suction and purified by flash column chromatography using cyclohexane/ethyl acetate 1:1 as eluent. Yield = 85%; mp = 134–135 °C (EtOH). ^{1}H NMR (400 MHz, $CDCl_3$) δ 4.38 (s, 3H, CH_3), 5.90 (exch br s, 2H, NH_2), 6.78 (s, 1H, -CH pyridaz.), 7.45–7.50 (m, 3H, Ar), 7.80–7.85 (m, 2H, Ar). Anal. Calcd for $C_{11}H_{11}N_3S$: C, 60.80; H, 5.10; N, 19.34. Found C, 60.97; H, 5.11; N, 19.30.

3.2.28. 4-(2-Methyl-3-oxo-6-phenyl-2,3-dihydro-pyridazin-4-ylamino)benzoic acid (**34**)

Compound **34** was obtained through the general procedure described for 15. After cooling, the mixture was acidified with 6N HCl and the final product was filtered off to obtain the desired compound. Yield = 90%; mp = 280–281 °C (Diethyl ether). ^{1}H NMR (400 MHz, DMSO-d_6) δ 3.82 (s, 3H, CH_3), 7.37 (s, 1H, -CH pyridaz.), 7.40–7.50 (m, 3H, Ar), 7.58 (d, 2H, Ar, J = 8.8 Hz), 7.84 (d, 2H, Ar, J = 8.4 Hz), 7.95 (d, 2H, Ar, J = 8.4 Hz), 9.16 (exch br s, 1H, NH), 12.78 (exch br s, 1H, OH). Anal. Calcd for $C_{18}H_{15}N_3O_2$: C, 67.28; H, 4.71; N, 13.08. Found C, 67.44; H, 4.71; N, 13.05.

3.2.29. 4-[4-(4-Acetyl-piperazine-1-carbonyl)-phenylamino]-2-methyl-6-phenylpyridazin-3(2H)-one (**35**)

Compound **35** was obtained starting from 34 through the same procedure described for **4a,b**. In this case the mixture was stirred at room temperature for 40 min. After cooling, THF was removed in vacuo and cold water was added (10 mL). The crude precipitate was recovered by filtration under vacuum and purified by crystallization. Yield = 94%; mp = 213–215 °C (Cyclohexane). Ligrownown solid, ^{1}H NMR (400 MHz, $CDCl_3$) δ 2.14 (s, 3H, CH_3CONH), 3.94 (s, 3H, CH_3), 3.60 (d, 8H, J = 46.0 Hz, 2 × NCH_2CH_2N), 7.22 (s, 1H, Ar), 7.33 (d, 2H, J = 7.9 Hz, Ar), 7.47 (dd, 6H, J = 24.3, 7.9 Hz, Ar), 7.71–7.80 (m, 2H, NH + Ar). ^{13}C NMR (100 MHz, $CDCl_3$) δ 170.15, 169.37, 156.24, 146.00, 140.96, 139.58, 136.38, 130.72, 129.39, 129.32, 128.96, 126.42, 120.74, 99.58, 40.75, 21.55. MS-ESI for $C_{24}H_{25}N_5O_3$ (Calcd, 431.20), $[M + H]^+$ at *m/z* 432.10, $[M + Na]^+$ at *m/z* 454.08, t_R = 13.347. Anal. Calcd for $C_{24}H_{25}N_5O_3$: C, 66.81; H, 5.84; N, 16.23. Found C, 66.54; H, 5.82; N, 16.16.

3.2.30. General procedure for compounds **37a–d** and **42a,b**

A suspension of appropriate pyridazinone **36a–d** (1.29 mmol), commercially available, and hydrazine hydrate (48 mmol) was stirred in a sealed/pressure vessel at 180–200 °C for 6–12 h. After cooling, ice-cold water was added (15 mL) and the precipitate obtained was recovered by filtration under vacuum to obtain the desired compounds **37a–d.** To obtain compounds **42a,b** we adopted the same procedure, using **41a,b** (**41a**, [39]) as starting materials.

3.2.31. 4-Amino-6-thiophen-3-yl-pyridazin-3(2H)-one (**37a**)

Yield = 52%; mp > 300 °C (EtOH). ^{1}H NMR (400 MHz, DMSO-d$_6$) δ 6.40 (exch br s, 2H, NH_2), 6.68 (s, 1H, -CH pyridaz.), 7.45 (dd, 1H, thiophene, J_1 = 1.2 Hz and J_2 = 4.8 Hz), 7.60 (dd, 1H, thiophene, J_1 = 2.8 Hz and J_2 = 4.8 Hz), 7.81 (ds, 1H, thiophene, J = 1.2 Hz), 12.54 (exch br s, 1H, NH). Anal. Calcd for $C_8H_7N_3OS$: C, 49.73; H, 3.65; N, 21.75. Found C, 49.61; H, 3.65; N, 21.69.

3.2.32. 4-Amino-6-cyclohexylpyridazin-3(2H)-one (**37b**)

Yield = 46%; mp = 284–287 °C (EtOH). ^{1}H NMR (400 MHz, DMSO-d$_6$) δ 1.20–1.39 (m, 5H, C_6H_{11}), 1.72–1.83 (m, 5H, C_6H_{11}), 2.30–2.35 (m, 1H, C_6H_{11}), 6.14 (s, 1H, CH pyridaz.), 6.18 (exch br s, 2H, NH_2), 12.24 (exch br s, 1H, NH). Anal. Calcd for $C_{10}H_{15}N_3O$: C, 62.15; H, 7.82; N, 21.74. Found C, 62.29; H, 7.80; N, 21.79.

3.2.33. 4-Amino-6-isopropylpyridazin-3(2H)-one **(37c)**

Yield = 40%; mp = 246–248 °C (EtOH). ^{1}H NMR (400 MHz, DMSO-d$_6$) δ 1.11 (d, 6H, $CH(CH_3)_2$, J = 7.2 Hz), 2.67 (m, 1H, $CH(CH_3)_2$), 6.16 (s, 1H, -CH pyridaz.), 6.18 (exch br s, 2H, NH_2), 12.23 (exch br s, 1H, NH). Anal. Calcd for $C_7H_{11}N_3O$: C, 54.89; H, 7.24; N, 27.43. Found C, 54.76; H, 7.23; N, 27.51.

3.2.34. 4-Amino-6-benzylpyridazin-3(2H)-one (**37d**)

Yield = 42%; mp = 247–250 °C (EtOH). ^{1}H NMR (400 MHz, DMSO-d$_6$) δ 3.70 (s, 2H, CH_2Ph), 6.04 (s, 1H, -CH pyridaz.), 6.22 (exch br s, 2H, NH_2), 7.20–7.40 (m, 5H, Ar), 12.31 (exch br s, 1H, NH). Anal. Calcd for $C_{11}H_{11}N_3O$: C, 65.66; H, 5.51; N, 20.88. Found C, 65.84; H, 5.50; N, 20.83.

3.2.35. 4-Amino-6-(2-hydroxyphenyl)-2-methylpyridazin-3(2H)-one (**42a**)

Yield = 58%; mp = 212–213 °C (EtOH). White coloured solid, ^{1}H NMR (400 MHz, $CDCl_3$) δ 3.85 (s, 3H, CH_3), 6.84 (s, 1H, Ar), 6.92 (td, 1H, J = 7.7, 1.2 Hz, Ar), 7.00–7.05 (m, 1H, Ar), 7.29 (td, 1H, J = 8.3, 7.8, 1.6 Hz, Ar), 7.58–7.50 (m, 1H, Ar). ^{13}C NMR (100 MHz, $CDCl_3$) δ 157.81, 131.24, 126.35, 119.47, 118.30, 98.93, 29.87. MS-ESI for $C_{11}H_{11}N_3O_2$ (Calcd, 217.08), $[M + H]^+$ at *m/z* 217.93, t_R = 11.693. Anal. Calcd for $C_{11}H_{11}N_3O_2$: C, 60.82; H, 5.10; N, 19.34. Found C, 60.57; H, 5.08; N, 19.26.

3.2.36. 4-Amino-6-(4-aminophenyl)-2-methylpyridazin-3(2H)-one (**42b)**

Yield = 55%; mp = 208–209 °C (Cyclohexane). ^{1}H NMR (400 MHz, DMSO-d$_6$) δ 3.65 (s, 3H, N-CH_3), 5.35 (exch br s, 2H, NH_2), 6.33 (exch br s, 2H, Ph-NH_2), 6.59 (d, 2H, Ar, J = 8.0 Hz), 6.63 (s, 1H, -CH pyridaz.), 7.42 (d, 2H, Ar, J = 8.0). Anal. Calcd for $C_{11}H_{11}N_4O$: C, 61.10; H, 5.59; N, 25.91. Found C, C, 61.27; H, 5.58; N, 25.85.

3.2.37. General Procedure for Compounds **38a–d**

A mixture of the appropriate pyridazinone **37a–d** (0.67 mmol), K_2CO_3 (1.34 mmol) and CH_3I (1.01 mmol) in anhydrous DMF (1.5 mL) was stirred at 80 °C for 1–4 h. After cooling, the mixture was diluted with cold water (15 mL) and compound **38a** was recovered by suction and crystallized from ethanol. For compounds **38b–d** the suspension was extracted with CH_2Cl_2 (3 × 15 mL) and the solvent was evaporated in vacuo. The final compounds were purified by flash column chromatography using cyclohexane/ethyl acetate 1:2 (for **38b,d**) or CH_2Cl_2/CH_3OH 9.5:0.5 (for **38c**) as eluents.

3.2.38. 4-Amino-2-methyl-6-thiophen-3-yl-pyridazin-3(2H)-one (**38a**)

Yield = 62%; mp = 178–179 °C (EtOH). ^{1}H NMR (400 MHz, DMSO-d$_6$) δ 3.67 (s, 3H, N-CH_3), 6.50 (exch br s, 2H, NH_2), 6.68 (s, 1H, -CH pyridaz.), 7.47 (d, 1H, thiophene,

J = 4.8 Hz), 7.61 (m, 1H, thiophene), 7.84 (s, 1H, thiophene). Anal. Calcd for $C_9H_9N_3OS$: C, 52.16; H, 4.38; N, 20.27. Found C, 52.05; H, 4.37; N, 20.22.

3.2.39. 4-Amino-6-cyclohexyl-2-methylpyridazin-3(2H)-one (**38b**)

Yield = 58%; oil. ^{1}H NMR (400 MHz, $CDCl_3$) δ 1.30–1.43 (m, 5H, C_6H_{11}), 1.68–1.92 (m, 5H, C_6H_{11}), 2.40 (m, 1H, C_6H_{11}), 3.75 (s, 3H, N-CH_3), 4.91 (exch br s, 2H, NH_2), 6.21 (s, 1H, -CH pyridaz.). Anal. Calcd for $C_{11}H_{17}N_3O$: C, 63.74; H, 8.27; N, 20.27. Found C, 63.87; H, 8.29; N, 20.23.

3.2.40. 4-Amino-6-isopropyl-2-methyl-2H-pyridazin-3-one (**38c**)

Yield = 49%; oil. ^{1}H NMR (400 MHz, $CDCl_3$) δ 1.18 (d, 6H, $CH(CH_3)_2$, J = 7.2 Hz), 2.75 (m, 1H, $CH(CH_3)_2$), 3.74 (s, 3H, N-CH_3), 4.96 (exch br s, 2H, NH_2), 6.21 (s, 1H, -CH pyridaz.). Anal. Calcd for $C_8H_{13}N_3O$: C, 57.46; H, 7.84; N, 25.13. Found C, 57.58; H, 7.82; N, 25.07.

3.2.41. 4-Amino-6-benzyl-2-methylpyridazin-3(2H)-one (**38d**)

Yield = 48%; mp = 104–108 °C (EtOH). ^{1}H NMR (400 MHz, $CDCl_3$) δ 3.80 (s, 3H, N-CH_3), 3.82 (s, 2H, CH_2Ph), 4.81 (exch br s, 2H, NH_2), 6.07 (s, 1H, -CH pyridaz.), 7.22–7.35 (m, 5H, Ar). Anal. Calcd for $C_{12}H_{13}N_3O$: C, 66.96; H, 6.09; N, 19.52. Found C, 66.83; H, 5.50; N, 20.83.

3.2.42. General Procedure for Compounds **39a,b, 40** and **43**

Compounds **39a,b, 40** and **43** were obtained starting from **38b,c, 37e** and **42a**, respectively, through the same procedure described for compound **25a–f**.

3.2.43. (6-Cyclohexyl-2-methyl-3-oxo-2,3-dihydropyridazin-4-yl)urea (**39a**)

Yield = 66%; mp = 251–254 °C (EtOH). ^{1}H NMR (400 MHz, DMSO-d_6) δ 1.30–1.40 (m, 5H, C_6H_{11}), 1.70–1.85 (m, 5H, C_6H_{11}), 2.45 (m, 1H, C_6H_{11}), 3.64 (s, 3H, N-CH_3), 6.74 (exch br s, 2H, $CONH_2$), 7.79 (s, 1H, -CH pyridaz.), 8.84 (exch br s, 1H, NHCO). Anal. Calcd for $C_{12}H_{18}N_4O_2$: C, 57.58; H, 7.25; N, 22.38. Found C, 57.41; H, 7.23; N, 22.43.

3.2.44. (6-Isopropyl-2-methyl-3-oxo-2,3-dihydropyridazin-4-yl)urea (**39b**)

Yield = 60%; mp = 248–251 °C (EtOH). ^{1}H NMR (400 MHz, DMSO-d_6) δ 1.15 (d, 6H, $CH(CH_3)_2$, J = 6.8 Hz), 2.80 (m, 1H, $CH(CH_3)_2$), 3.65 (s, 3H, N-CH_3), 6.74 (exch br s, 2H, $CONH_2$), 7.81 (s, 1H, -CH pyridaz.), 8.85 (exch br s, 1H, CONH). Anal. Calcd for $C_9H_{14}N_4O_2$: C, 51.42; H, 6.71; N, 26.65. Found C, 51.31; H, 6,70; N, 26.61.

3.2.45. [6-(2-Hydroxy-phenyl)-3-oxo-2,3-dihydro-pyridazin-4-yl]-urea (**40**)

Yield = 85%; mp > 300 °C (EtOH). ^{1}H NMR (400 MHz, DMSO-d_6) δ 6.75 (exch br s, 2H, $CONH_2$), 6.90–6.95 (m, 2H, Ar), 7.27 (t, 1H, Ar, J = 8.4 Hz), 7.45 (dd, 1H, Ar, J_1 = 1.2 Hz and J_2 = 8.0 Hz), 8.39 (s, 1H, -CH pyridaz.), 8.92 (exch br s, 1H, NHCO), 10.43 (exch br s, 1H, OH), 13.18 (exch br s, 1H, NH). Anal. Calcd for $C_{11}H_{10}N_4O_3$: C, 53.66; H, 4.09; N, 22.75. Found C, 53.51; H, 4,08; N, 22.81.

3.2.46. [6-(2-Hydroxyphenyl)-2-methyl-3-oxo-2,3-dihydropyridazin-4-yl]-urea (**43**)

Yield = 95%; mp = 278–280 °C (EtOH). ^{1}H-NMR (400 MHz, DMSO-d_6) δ 3.76 (s, 3H, N-CH_3), 6.79 (exch br s, 2H, NH_2), 6.88–6.95 (m, 2H, Ar), 7.27 (t, 1H, Ar, J = 7.2 Hz), 7.44 (d, 1H, Ar, J = 6.8 Hz), 8.37 (s, 1H, -CH pyridaz.), 8.93 (exch br s, 1H, NHCO), 10.17 (exch br s, 1H, OH). Anal. Calcd for $C_{12}H_{12}N_4O_3$: C, 55.38; H, 4.65; N, 21.53. Found C, 55.49; H, 4,65; N, 21,49.

3.2.47. N-[4-(5-Amino-1-methyl-6-oxo-1,6-dihydropyridazin-3-yl)-phenyl]-acetamide (**44**)

To a cooled (0 °C) and stirred solution of **42b** (0.93 mmol) in anhydrous THF (2–3 mL), 1.02 mmol of acetyl chloride was added and the mixture was stirred at room temperature for 20 min. After dilution with cold water (20–30 mL), the precipitate was recovered by filtration under vacuum and purified by crystallization. Yield = 92%; mp = 270–272 °C (Cyclohexane). ^{1}H NMR (400 MHz, DMSO-d$_6$) δ 2.06 (s, 3H, COCH$_3$), 3.69 (s, 3H, N-CH$_3$), 6.48 (exch br s, 2H, NH$_2$), 6.71 (s, 1H, -CH pyridaz.), 7.60–7.70 (m, 4H, Ar), 8.80 (exch br s, 1H, NHCO). Anal. Calcd for $C_{13}H_{14}N_4O_2$: C, 60.45; H, 5.46; N, 21.69. Found C, 60.58; H, 5.45; N, 21.63.

3.2.48. 3,6-Dimethyl-4-pyridin-2-yl-isoxazolo [3,4-d]pyridazin-7(6H)-one (**46**)

To a cooled (0–4 °C) solution of **45** [34] (0.38 mmol) in EtOH (2–3 mL), methylhydrazine (1.30 mmol) was added and the mixture was stirred at room temperature for 90 min. The precipitate was recovered by filtration under vacuum to obtain the desired compound. Yield = 84%; mp = 154–155 °C (EtOH). ^{1}H NMR (400 MHz, DMSO-d$_6$) δ 2.94 (s, 3H, C3-CH$_3$), 3.74 (s, 3H, N-CH$_3$), 7.57 (t, 1H, Ar, J = 5.2 Hz), 7.95–8.05 (m, 2H, Ar), 8.76 (d, 1H, Ar, J = 5.2 Hz). Anal. Calcd for $C_{12}H_{10}N_4O_2$: C, 59.50; H, 4.16; N, 23.13. Found C, 59.66; H, 4.16; N, 23.20.

3.2.49. 5-Acetyl-4-amino-2-methyl-6-pyridin-2-yl-pyridazin-3(2H)-one (**47**)

A mixture of **46** (0.82 mmol), 10% Pd/C (20 mg) and ammonium formate (4.9 mmol) in EtOH (5 mL), was refluxed for 2 h. After addition of CH_2Cl_2 (4–5 mL) and filtration of charcoal, evaporation of the solvent afforded the product **47**. Yield = 65%; mp = 201–203 °C (EtOH). ^{1}H NMR (400 MHz, DMSO-d$_6$) δ 1.89 (s, 3H, COCH$_3$), 3.73 (s, 3H, N-CH$_3$), 7.23 (exch br s, 2H, NH$_2$), 7.44–7.50 (m, 1H, Ar), 7.89 (d, 1H, Ar, J = 7.2 Hz), 7.96 (t, 1H, Ar, J = 7.2 Hz), 8.57 (d, 1H, Ar, J = 4.4 Hz). Anal. Calcd for $C_{12}H_{12}N_4O_2$: C, 59.01; H, 4.95; N, 22.94. Found C, 59.18; H, 4.96; N, 22.99.

3.2.50. 4-Amino-2-methyl-6-pyridin-2-yl-pyridazin-3(2H)-one (**48**)

A suspension of **47** (0.53 mmol) in 1 mL of 48% HBr was stirred in a sealed/pressure vessel at 130 °C for 3 h. After cooling ice-cold water was added and the precipitate was recovered by filtration under vacuum to obtain the desired product 48. Yield = 65%; mp = 294–295 °C (EtOH). ^{1}H NMR (400 MHz, DMSO-d$_6$) δ 3.75 (s, 3H, N-CH$_3$), 7.24 (s, 1H, -CH pyridaz.), 7.51 (m, 1H, Ar), 8.00 (t, 1H, Ar, J = 7.2 Hz), 8.12 (d, 1H, Ar, J = 8.0 Hz), 8.66 (d, 1H, Ar, J = 4.4 Hz). Anal. Calcd for $C_{10}H_{10}N_4O$: C, 59.40; H, 4.98; N, 27.71. Found C, 59.51; H, 4.99; N, 27.75.

3.2.51. (2-Methyl-3-oxo-6-pyridin-2-yl-2,3-dihydropyridazin-4-yl)-urea (**49**)

Compound **49** was obtained starting from **48**, through the same procedure described for compounds **25a–f**, **39a,b**, **40** and **43.** Yield = 15%; mp > 300 °C (EtOH). ^{1}H NMR (400 MHz, DMSO-d$_6$) δ 3.82 (s, 3H, N-CH$_3$), 6.80 (exch br s, 2H, NH$_2$), 7.46 (m, 1H, Ar), 7.92 (t, 1H, Ar, J = 7.6 Hz), 8.10 (d, 1H, Ar, J = 7.6 Hz), 8.67 (d, 1H, Ar, J = 4.8 Hz), 8.83 (s, 1H, -CH pyridaz.), 8.97 (exch br s, 1H, NHCO). Anal. Calcd for $C_{11}H_{11}N_5O_2$: C, 53.87; H, 4.52; N, 28.56. Found C, 53.78; H, 5.00; N, 27.71.

3.2.52. 2-Methyl-3-oxo-6-phenyl-2,3-dihydropyridazine-4-carboxamide (**51**)

Compound **51** was obtained starting from **50** [41], through the same procedure described for compounds **38a–d**. The compound was purified by crystallization from diethyl ether. Yield = 55%; mp = 215–217 °C (Et$_2$O). Light brown solid, ^{1}H NMR (400 MHz, CDCl$_3$) δ 3.99 (s, 3H, CH$_3$), 5.98 (exch br s, 1H, CONH$_2$), 7.43–7.51 (m, 3H, Ar), 7.84–7.88 (m, 2H, Ar), 8.72 (s, 1H, Ar), 9.41 (exch br s, 1H, CONH$_2$). ^{13}C NMR (100 MHz, CDCl$_3$) δ 163.58, 160.21, 145.33, 134.16, 132.64, 130.02, 129.23, 129.04, 126.15, 41.53. MS-ESI for $C_{12}H_{11}N_3O_2$ (Calcd, 229.08), $[M + H]^+$ at *m/z* 229.90, t_R = 12.205. Anal. Calcd for $C_{12}H_{11}N_3O_2$: C, 62.87; H, 4.84; N, 18.33. Found C, 62.62; H, 4.82; N, 18.26.

3.2.53. 1-(6-Methyl-3-oxo-2-phenyl-2,3-dihydropyridazin-4-yl)urea (**55**)

Compound **55** was obtained strating from **54** [43], through the same procedure for the formation of urea described for compounds **25a–f, 39a,b, 40** and **43**. Yield = 95%; mp = 288–290 °C (EtOH). White coloured solid, ^{1}H NMR (400 MHz, DMSO-d_6) δ 2.25 (s, 3H, CH_3), 7.40–7.42 (m, 1H, Ar), 7.47 (d, 2H, J = 8.2 Hz, Ar), 7.50–7.54 (m, 2H, Ar), 7.78 (s, 1H, Ar), 8.91 (s, 1H, $NHCONH_2$). ^{13}C NMR (100 MHz, DMSO-d_6) δ 155.56, 154.89, 146.52, 141.98, 138.04, 128.86, 128.12, 125.97, 109.55, 21.48. MS-ESI for $C_{12}H_{12}N_4O_2$ (Calcd, 244.10), $[M + H]^+$ at *m/z* 244.88, t_R = 10.100. Anal. Calcd for $C_{12}H_{12}N_4O_2$: C, 59.01; H, 4.95; N, 22.94. Found C, 59.25; H, 4.97; N, 23.03.

3.2.54. (E)-3-(2-(Dimethylamino)vinyl)-4-methyl-6-phenylisoxazolo [3,4-d]pyridazin-7(6H)-one (**56**)

A mixture of **52** (1.04 mmol) [42] in 2.5 mL of DMF-DMA was hetaed at 90–100 °C for 1 h. After cooling, ice/cold water was added (15 mL) and the precipitate obtained was recovered by filtration under vacuum to obtained the pure desired compound. Yield = 90%; mp = 224–226 °C dec. (Cyclohexane). ^{1}H NMR (400 MHz, $CDCl_3$) δ 2.50 (s, 3H, CH_3), 3.00–3.20 (m, 6H, $N(CH_3)_2$), 5.25 (d, 1H, CH=CH-N, J = 10.0 Hz), 7.30–7.35 (m, 1H, Ar), 7.45–7.50 (m, 2H, Ar), 7.59–7.64 (m, 3H, 1H CH=CH-N + 2H Ar). Anal. Calcd for $C_{16}H_{16}N_4O_2$: C, 64.85; H, 5.44; N, 18.91. Found C, 65.10; H, 5.46; N, 18.98.

3.2.55. 4-Amino-6-methyl-2-phenyl-5-(1H-pyrazol-5-yl)pyridazin-3(2H)-one (**57**)

A mixture of intermediate **56** (0.81 mmol) and 1 mL of hydrazine hydrate (excess) in 2 mL of abs. EtOH was hetaed at 70 °C for 10 h. After cooling, ice/cold water was added (15 mL). The precipitate obtained was recovered by filtration under vacum and purified by crystallization from ethanol. Yield = 65%; mp = 119–121 °C. (Cyclohexane). Yellow coloured solid, ^{1}H NMR (400 MHz, Methanol-d_4) δ 2.31 (s, 3H, CH_3), 6.58 (exch br s, 2H, NH_2), 7.43 (t, 1H, J = 7.3 Hz, Ar), 7.52 (t, 3H, J = 7.6 Hz, Ar), 7.58 (d, 2H, J = 7.6 Hz, Ar), 7.83 (s, 1H, NH). ^{13}C NMR (100 MHz, Methanol-d_4) δ 174.64, 147.92, 143.30, 129.88, 129.32, 127.14, 106.95, 24.30. MS-ESI for $C_{14}H_{13}N_5O$ (Calcd, 267.11), $[M + H]^+$ at *m/z* 267.98. t_R = 10.882. Anal. Calcd for $C_{14}H_{13}N_5O$: C, 62.91; H, 4.90; N, 26.20. Found C, 62.66; H, 4.88; N, 26.09.

3.3. *Molecular Modeling and Biological Data*

The 2D chemical structures were built using Marvin Sketch and all the structures were subjected to molecular mechanics energy minimization using the MMFF94 force field present in the same software [45]. The 3D geometry of all compounds was then optimized using the PM3 Hamiltonian [46], as implemented in MOPAC 2016 package assuming a pH of 7.0 [47]. Once built and optimized, all structures were used in the bioisostere replacement tool Spark 10.4.0. Five hundred compounds were generated for the substitution (50 best compounds reported in the Supplementary Materials). The isosteric replacement was performed using the same 178,558 fragments for each part; in particular, the fragments derive from ChEMBL and Zinc databases with a protocol already reported and validated [27,48,49]. Ligand growing experiments were performed in the selected pyridazinone structure using an already reported protocol [50]. Docking calculations were made using AutoDock with the default docking parameters and a validated protocol [51,52]. The setup was done with YASARA [47]. The Lamarckian genetic algorithm implemented in AutoDock was used for the calculations. The ligand-centered maps were generated by AutoGrid with a spacing of 0.375 Å and dimensions that encompass all atoms extending 5 Å from the surface of the ligand. All of the parameters were inserted at their default settings. The X-ray crystal structures of the co-crystal FABP4/(2-[(2-oxo-2-piperidin-1-ylethyl)sulfanyl]-6-(trifluoromethyl)pyrimidin-4-ol) (PDBid: 1TOU) was downloaded from the Protein Data Bank (www.rcsb.org accessed on 15 June 2022).

3.4. FABP Inhibitory Activity Assays

To analyze the inhibitory activity of FABP4 ligands, a displacement assay was utilized as described by the Cayman's instruction, FABP4 Inhibitor/Ligand Screening Assay Kit, Item 10,010,231 (see Supplementary Materials for additional details). The samples of compounds for activity determination were prepared as a stock solution (1 mM) in DMSO. On the day of activity assay, the compounds were all diluted in phosphate buffer solution (PBS, pH 7.4) to different concentrations (100, 50, 10, 5, 2, 1, and 0 μM). Appropriate concentrations of DMSO in PBS were used as control. The detection reagent (FABP Assay Detection Reagent, Item 10010376) was used as provided by the Cayman's kit. The diluted Detection Reagent probe was mixed with FABP4 protein present in the kit and incubated for 10 min at room temperature. Compounds were then added and equilibrated for another 10 min. Lastly, the fluorescence signal was recorded at 470 nm (i.e., emission, with the excitation fixed at 370 nm) with a CytoFluor® Series 4000 Fluorescence Multi-Well Plate Reader. The IC_{50} was calculated as indicated in the kit booklet of FABP4 Inhibitor/Ligand Screening Assay Kit (Item No. 10010231) Cayman chemicals, as follows: 1) calculate the average fluorescence of each sample; 2) calculate the background corrected fluorescence (BCF) by subtracting the blank; 3) divide the BCF of each sample by the maximum BCF and multiply by 100% (this is the value in percent fluorescence units, i.e., % FU); 4) plot the % FU values against the concentration of inhibitor/ligand used; 5) find the concentration of inhibitor/ligand that corresponds to 50% FU, to determine IC_{50} values.

4. Conclusions

We have identified novel 4-amino and 4-ureido pyridazinone-based FABP4 inhibitors whose design was directed by computing assisted molecular design of bioisosteric-replacements/ scaffold hopping of the pyrimidine skeleton of the co-crystallyzed ligand 1TOU. Selected compounds have been synthesized and tested for their ability to inhibit FABP4. Among the new series, ten compounds were further evaluated on the basis of their inhibitory activity on FABP4 established via a single point displacement assay. In particular, **4b**, **25a**, **30b** and **22** exhibited high FABP4 inhibitory activity with IC_{50} in the low micromolar range. The results demonstrated that compound **25a** was the most potent analogue in terms of displacement of the arachidonic acid, with an IC_{50} value of 2.97 μM, which is lower than the IC_{50} of the positive control (3.42 μM). Docking experiments, conducted with the most active compounds **4b**, **25a**, **30b**, **22**, confirmed the ability of these molecules to interact with several amino acid residues present inside the FABP4 binding pocket, with the stronger interaction exhibited by compound **25a**. This result is in agreement with the higher activity recorded in vitro for **25a**, in comparison to the other 4-amino and 4-ureido pyridazinone-based analogues developed in this study.

Supplementary Materials: The following supporting information can be downloaded at: https://www.mdpi.com/article/10.3390/ph15111335/s1, ^{1}H NMRs of selected compounds; ^{13}C NMRs of selected compounds; Mass spectra of selected compounds; HPLC/UV chromatograms of selected compounds; 50 'best-fit' compounds generated with scaffold hopping replacement; Info on the FABP4 inhibitor assay kit; Averaged data as Background corrected fluorescence for IC_{50} measured compounds.

Author Contributions: Conceptualization, L.C., G.F. and A.C.; methodology, L.C., G.F., D.M., R.R.d.O.S., F.M. and C.V.; software, G.F. and C.Z.; formal analysis, L.C., G.F., D.M., R.R.d.O.S., F.M., C.V. and A.C.; resources, G.F, C.Z., M.P.G., A.C.; data curation, L.C., G.F., D.M., R.R.d.O.S., F.M.; writing—original draft preparation, L.C., G.F.,C.Z. and D.M.; writing—review and editing, L.C., G.F., A.C.; supervision, M.P.G. and A.C.; project administration, M.P.G. and A.C.; funding acquisition, M.P.G. and A.C. All authors have read and agreed to the published version of the manuscript.

Funding: This research has received funding for a scholarship to R.R.d.O.S from the Coordination for the Improvement of Higher Education Personnel—Brazil (CAPES-PRINT, funding number 88887.570120/2020-00).

Institutional Review Board Statement: Not applicable.

Informed Consent Statement: Not applicable.

Data Availability Statement: Data is contained within the article or supplementary material.

Conflicts of Interest: The authors declare no conflict of interest.

References

1. Das, U.N. Essential Fatty acids—A review. *Curr. Pharm. Biotechnol.* **2006**, *7*, 467–482. [CrossRef] [PubMed]
2. Furuhashi, M.; Hotamisligil, G.S. Fatty acid-binding proteins: Role in metabolic diseases and potential as drug targets. *Nat. Rev. Drug Discov.* **2008**, *7*, 489–503. [CrossRef] [PubMed]
3. Boden, G. Free fatty acids (FFA), a link between obesity and insulin resistance. *Front. Biosci.* **1998**, *3*, d169–d175. [CrossRef]
4. Hotamisligil, G.S.; Bernlohr, D.A. Metabolic functions of FABPs–mechanisms and therapeutic implications. *Nat. Rev. Endocrinol.* **2015**, *11*, 592–605. [CrossRef] [PubMed]
5. Boden, G. Obesity and free fatty acids. *Endocrinol. Metab. Clin. N. Am.* **2008**, *37*, 635–646. [CrossRef]
6. DeFronzo, R.A. Dysfunctional fat cells, lipotoxicity and type 2 diabetes. *Int. J. Clin. Pract. Suppl.* **2004**, *58*, 9–21. [CrossRef]
7. Sheth, S.G.; Gordon, F.D.; Chopra, S. Nonalcoholic steatohepatitis. *Ann. Intern. Med.* **1997**, *126*, 137–145. [CrossRef]
8. Storch, J.; Thumser, A.E. The fatty acid transport function of fatty acid-binding proteins. *Biochim. Biophys. Acta* **2000**, *1486*, 28–44. [CrossRef]
9. Queipo-Ortuno, M.I.; Escote, X.; Ceperuelo-Mallafre, V.; Garrido-Sanchez, L.; Miranda, M.; Clemente-Postigo, M.; Perez-Perez, R.; Peral, B.; Cardona, F.; Fernandez-Real, J.M.; et al. FABP4 dynamics in obesity: Discrepancies in adipose tissue and liver expression regarding circulating plasma levels. *PLoS ONE* **2012**, *7*, e48605. [CrossRef]
10. Syamsunarno, M.R.; Iso, T.; Hanaoka, H.; Yamaguchi, A.; Obokata, M.; Koitabashi, N.; Goto, K.; Hishiki, T.; Nagahata, Y.; Matsui, H.; et al. A critical role of fatty acid binding protein 4 and 5 (FABP4/5) in the systemic response to fasting. *PLoS ONE* **2013**, *8*, e79386. [CrossRef]
11. Thompson, B.R.; Mazurkiewicz-Munoz, A.M.; Suttles, J.; Carter-Su, C.; Bernlohr, D.A. Interaction of Adipocyte Fatty Acid-binding Protein (AFABP) and JAK2. *J. Biol. Chem.* **2009**, *284*, 13473–13480. [CrossRef]
12. Adida, A.; Spener, F. Adipocyte-type fatty acid-binding protein as inter-compartmental shuttle for peroxisome proliferator activated receptor gamma agonists in cultured cell. *BBA-Mol. Cell Biol.* **2006**, *1761*, 172–181. [CrossRef]
13. Fu, Y.; Luo, L.; Luo, N.; Garvey, W.T. Lipid metabolism mediated by adipocyte lipid binding protein (ALBP/aP2) gene expression in human THP-1 macrophages. *Atherosclerosis* **2006**, *188*, 102–111. [CrossRef]
14. Floresta, G.; Pistarà, V.; Amata, E.; Dichiara, M.; Marrazzo, A.; Prezzavento, O.; Rescifina, A. Adipocyte fatty acid binding protein 4 (FABP4) inhibitors. A comprehensive systematic review. *Eur. J. Med. Chem.* **2017**, *138*, 854–873. [CrossRef]
15. Nieman, K.M.; Kenny, H.A.; Penicka, C.V.; Ladanyi, A.; Buell-Gutbrod, R.; Zillhardt, M.R.; Romero, I.L.; Carey, M.S.; Mills, G.B.; Hotamisligil, G.S.; et al. Adipocytes promote ovarian cancer metastasis and provide energy for rapid tumor growth. *Nat. Med.* **2011**, *17*, 1498–1503. [CrossRef]
16. Tolle, A.; Suhail, S.; Jung, M.; Jung, K.; Stephan, C. Fatty acid binding proteins (FABPs) in prostate, bladder and kidney cancer cell lines and the use of IL-FABP as survival predictor in patients with renal cell carcinoma. *BMC Cancer* **2011**, *11*, 302. [CrossRef] [PubMed]
17. Uehara, H.; Takahashi, T.; Oha, M.; Ogawa, H.; Izumi, K. Exogenous fatty acid binding protein 4 promotes human prostate cancer cell progression. *Int. J. Cancer* **2014**, *135*, 2558–2568. [CrossRef]
18. Yang, A.; Zhang, H.; Sun, Y.; Wang, Y.; Yang, X.; Yang, X.; Zhang, H.; Guo, W.; Zhu, G.; Tian, J.; et al. Modulation of FABP4 hypomethylation by DNMT1 and its inverse interaction with miR-148a/152 in the placenta of preeclamptic rats and HTR-8 cells. *Placenta* **2016**, *46*, 49–62. [CrossRef]
19. Tian, W.; Zhang, W.; Zhang, Y.; Zhu, T.; Hua, Y.; Li, H.; Zhang, Q.; Xia, M. FABP4 promotes invasion and metastasis of colon cancer by regulating fatty acid transport. *Cancer Cell Int.* **2020**, *20*, 512. [CrossRef] [PubMed]
20. Gharpure, K.M.; Pradeep, S.; Sans, M.; Rupaimoole, R.; Ivan, C.; Wu, S.Y.; Bayraktar, E.; Nagaraja, A.S.; Mangala, L.S.; Zhang, X.N.; et al. FABP4 as a key determinant of metastatic potential of ovarian cancer. *Nat. Commun.* **2018**, *9*. [CrossRef]
21. Li, H.Y.; Lv, B.B.; Bi, Y.H. FABP4 accelerates glioblastoma cell growth and metastasis through Wnt10b signalling. *Eur. Rev. Med. Pharmacol. Sci.* **2018**, *22*, 7807–7818. [CrossRef] [PubMed]
22. Floresta, G.; Patamia, V.; Zagni, C.; Rescifina, A. Adipocyte fatty acid binding protein 4 (FABP4) inhibitors. An update from 2017 to early 2022. *Eur. J. Med. Chem.* **2022**, *240*, 114604. [CrossRef] [PubMed]
23. Wang, Y.; Law, W.K.; Hu, J.S.; Lin, H.Q.; Ip, T.M.; Wan, D.C. Discovery of FDA-approved drugs as inhibitors of fatty acid binding protein 4 using molecular docking screening. *J. Chem. Inf. Model* **2014**, *54*, 3046–3050. [CrossRef]
24. Zhou, Y.; Nie, T.; Zhang, Y.; Song, M.; Li, K.; Ding, M.; Ding, K.; Wu, D.; Xu, Y. The discovery of novel and selective fatty acid binding protein 4 inhibitors by virtual screening and biological evaluation. *Bioorg. Med. Chem.* **2016**, *24*, 4310–4317. [CrossRef]
25. Floresta, G.; Cilibrizzi, A.; Abbate, V.; Spampinato, A.; Zagni, C.; Rescifina, A. FABP4 inhibitors 3D-QSAR model and isosteric replacement of BMS309403 datasets. *Data Brief* **2019**, *22*, 471–483. [CrossRef]
26. Floresta, G.; Gentile, D.; Perrini, G.; Patamia, V.; Rescifina, A. Computational Tools in the Discovery of FABP4 Ligands: A Statistical and Molecular Modeling Approach. *Mar. Drugs* **2019**, *17*, 624. [CrossRef]

27. Floresta, G.; Cilibrizzi, A.; Abbate, V.; Spampinato, A.; Zagni, C.; Rescifina, A. 3D-QSAR assisted identification of FABP4 inhibitors: An effective scaffold hopping analysis/QSAR evaluation. *Bioorg. Chem.* **2019**, *84*, 276–284. [CrossRef]
28. Crocetti, L.; Floresta, G.; Nazir, S.; Vergelli, C.; Bhogal, A.; Biancalani, C.; Cesari, N.; Giovannoni, M.P.; Cilibrizzi, A. Synthesis and inverse virtual screening of new bi-cyclic structures towards cancer-relevant cellular targets. *Struct Chem.* **2022**, *33*, 769–793. [CrossRef]
29. Floresta, G.; Crocetti, L.; Giovannoni, M.P.; Biagini, P.; Cilibrizzi, A. Repurposing strategies on pyridazinone-based series by pharmacophore- and structure-driven screening. *J. Enzime Inhib. Med. Chem.* **2020**, *35*, 1137–1144. [CrossRef]
30. Giovannoni, M.P.; Vergelli, C.; Cilibrizzi, A.; Crocetti, L.; Biancalani, C.; Graziano, A.; Dal Piaz, V.; Loza, M.I.; Cadavid, M.I.; Díaz, J.L.; et al. Pyrazolo[1′,5′:1,6]pyrimido[4,5-d]pyridazin-4(3H)-ones as selective human A1 adenosine receptor ligands. *Bioorg. Med. Chem.* **2010**, *18*, 7890–7899. [CrossRef]
31. Biagini, P.; Biancalani, C.; Graziano, A.; Cesaria, N.; Giovannoni, M.P.; Cilibrizzi, A.; Dal Piaz, V.; Vergelli, C.; Crocetti, L.; Delcanale, M.; et al. Functionalized pyrazoles and pyrazolo[3,4-d]pyridazinones: Synthesis and evaluation of their phosphodiesterase 4 inhibitory activity. *Bioorg. Med. Chem.* **2010**, *18*, 3506–3517. [CrossRef] [PubMed]
32. Giovannoni, M.P.; Ciciani, G.; Cilibrizzi, A.; Crocetti, L.; Daniele, S.; Di Cesare Mannelli, L.; Ghelardini, C.; Giacomelli, C.; Guerrini, G.; Martini, C.; et al. Further studies on pyrazolo[1′,5′:1,6]pyrimido[4,5-d]pyridazin-4(3H)-ones as potent and selective human A1 adenosine receptor antagonists. *Eur. J. Med. Chem.* **2015**, *89*, 32–41. [CrossRef] [PubMed]
33. Cheeseright, T.; Mackey, M.; Rose, S.; Vinter, A. Molecular field extrema as descriptors of biological activity: Definition and validation. *J. Chem. Inf. Model* **2006**, *46*, 665–676. [CrossRef] [PubMed]
34. Dal Piaz, V.; Aguilar, I.N.; Buil Albero, M.A.; Garrido, R.Y.; Giovannoni, M.P.; Gracia, F.J.; Lumeras, A.W.; Vergelli, C. Preparation of pyridazin-3(2H)-ones and their use as PDE4 inhibitors. PCT WO2005049581A120050602, 2 June 2005.
35. Aguilar Izquierdo, N.; Carrascal Riera, M.; Dal Piaz, V.; Gracia Ferrer, J.; Lumeras Amador, W.; Masdeu Margalef, M.; Warrellow, G. Preparation of pyridazin-3(2H)-one derivatives as PDE4 inhibitors for the treatment of pathological diseases. PCT WO2005123693A120051229, 29 December 2005.
36. Biancalani, C.; Giovannoni, M.P.; Pieretti, S.; Cesari, N.; Graziano, A.; Vergelli, C.; Cilibrizzi, A.; Di Gianuario, A.; Colucci, M.; Mangano, G.; et al. Further Studies on Arylpiperazinyl Alkyl Pyridazinones: Discovery of an Exceptionally Potent, Orally Active, Antinociceptive Agent in Thermally Induced Pain. *J. Med. Chem.* **2009**, *52*, 7397–7409. [CrossRef]
37. Sircar, I. Synthesis of 4-amino-6-phenyl-3(2H)-pyridazinones: A general procedure. *J. Heterocycl. Chem.* **1983**, *20*, 1473–1476. [CrossRef]
38. Coates, W.J.; McKillop, A. Preparation of 4-amino-3(2H)-pyridazinones by direct amination of 3(2H)-pyridazinones with hydrazine. *Heterocycles* **1989**, *29*, 1077–1090. [CrossRef]
39. Hubbard, R.D.; Wang, L.; Park, C.H.; Sun, C.; Mcdaniel, K.F.; Pratt, J.K.; Soltwedel, T.N.; Wendt, M.D.; Holms, J.H.; Liu, D. Preparation of pyridinone and pyridazinone derivatives for the treatment of inflammatory diseases, diabetes, obesity, cancer, and AIDS. U.S. Patent US 20130331382A120131212, 12 December 2013.
40. Vergelli, C.; Schepetkin, I.A.; Ciciani, G.; Cilibrizzi, A.; Crocetti, L.; Giovannoni, M.P.; Guerrini, G.; Iacovone, A.; Kirpotina, L.N.; Khlebnikov, A.I.; et al. 2-Arylacetamido-4-phenylamino-5-substituted pyridazinones as formyl peptide receptors agonists. *Bioorg. Med. Chem.* **2016**, *24*, 2530–2543. [CrossRef] [PubMed]
41. Wermuth, C.G.; Schlewer, G.; Bourguignon, J.J.; Maghioros, G.; Bouchet, M.J.; Moire, C.; Kan, J.P.; Worms, P.; Biziere, K. 3-Aminopyridazine derivatives with atypical antidepressant, serotonergic and dopaminergic activities. *J. Med. Chem.* **1989**, *32*, 528–537. [CrossRef]
42. Vergelli, C.; Giovannoni, M.P.; Pieretti, S.; Giannuario, A.D.; Piaz, V.D.; Biagini, P.; Biancalani, C.; Graziano, A.; Cesari, N. 4-Amino-5-vinyl-3(2H)-pyridazinones and analogues as potent antinociceptive agents: Synthesis, SARs, and preliminary studies on the mechanism of action. *Bioorg. Med. Chem.* **2007**, *15*, 5563–5575. [CrossRef]
43. Cilibrizzi, A.; Quinn, M.T.; Kirpotina, L.N.; Schepetkin, I.A.; Holderness, J.; Ye, R.D.; Rabiet, M.J.; Biancalani, C.; Cesari, N.; Graziano, A.; et al. 6-methyl-2,4-disubstituted pyridazin-3(2H)-ones: A novel class of small-molecule agonists for formyl peptide receptors. *J. Med. Chem.* **2009**, *52*, 5044–5057. [CrossRef]
44. Memdouh, S.; Gavrilović, I.; Ng, K.; Cowan, D.; Abbate, V. Advances in the detection of growth hormone releasing hormone synthetic analogs. *Drug Test. Anal.* **2021**, *13*, 1871–1887. [CrossRef] [PubMed]
45. Cheng, A.; Best, S.A.; Merz, K.M., Jr.; Reynolds, C.H. GB/SA water model for the Merck molecular force field (MMFF). *J. Mol. Graph. Model* **2000**, *18*, 273–282. [CrossRef]
46. Stewart, J.J. Optimization of parameters for semiempirical methods IV: Extension of MNDO, AM1, and PM3 to more main group elements. *J. Mol. Model* **2004**, *10*, 155–164. [CrossRef] [PubMed]
47. Stewart, J.J. MOPAC: A semiempirical molecular orbital program. *J. Comput. Aided Mol. Des.* **1990**, *4*, 1–105. [CrossRef]
48. Floresta, G.; Pittalà, V.; Sorrenti, V.; Romeo, G.; Salerno, L.; Rescifina, A. Development of new HO-1 inhibitors by a thorough scaffold-hopping analysis. *Bioorg. Chem.* **2018**, *81*, 334–339. [CrossRef]
49. Floresta, G.; Amata, E.; Dichiara, M.; Marrazzo, A.; Salerno, L.; Romeo, G.; Prezzavento, O.; Pittalà, V.; Rescifina, A. Identification of Potentially Potent Heme Oxygenase 1 Inhibitors through 3D-QSAR Coupled to Scaffold-Hopping Analysis. *ChemMedChem* **2018**, *13*, 1336–1342. [CrossRef]
50. Floresta, G.; Fallica, A.N.; Salerno, L.; Sorrenti, V.; Pittalà, V.; Rescifina, A. Growing the molecular architecture of imidazole-like ligands in HO-1 complexes. *Bioorg. Chem.* **2021**, *117*, 105428. [CrossRef]

51. Varrica, M.G.; Zagni, C.; Mineo, P.G.; Floresta, G.; Monciino, G.; Pistarà, V.; Abbadessa, A.; Nicosia, A.; Castilho, R.M.; Amata, E.; et al. DNA intercalators based on (1,10-phenanthrolin-2-yl)isoxazolidin-5-yl core with better growth inhibition and selectivity than cisplatin upon head and neck squamous cells carcinoma. *Eur. J. Med. Chem.* **2018**, *143*, 583–590. [CrossRef]
52. Zagni, C.; Citarella, A.; Oussama, M.; Rescifina, A.; Maugeri, A.; Navarra, M.; Scala, A.; Piperno, A.; Micale, N. Hydroxamic Acid-Based Histone Deacetylase (HDAC) Inhibitors Bearing a Pyrazole Scaffold and a Cinnamoyl Linker. *Int. J. Mol. Sci.* **2019**, *20*, 945. [CrossRef]

pharmaceuticals

Article

Exploring the Natural Compounds in Flavonoids for Their Potential Inhibition of Cancer Therapeutic Target MEK1 Using Computational Methods

Wejdan M. AlZahrani [1], Shareefa A. AlGhamdi [1,2,*], Torki A. Zughaibi [2,3] and Mohd Rehan [2,3,*]

1 Department of Biochemistry, Faculty of Sciences, King Abdulaziz University, Jeddah 21589, Saudi Arabia; wmuhammadalzahrani@stu.kau.edu.sa

2 King Fahd Medical Research Center, King Abdulaziz University, Jeddah 22252, Saudi Arabia; taalzughaibi@kau.edu.sa

3 Department of Medical Laboratory Sciences, Faculty of Applied Medical Sciences, King Abdulaziz University, Jeddah 21589, Saudi Arabia

* Correspondence: saaalghamdi1@kau.edu.sa (S.A.A.); mrtahir@kau.edu.sa (M.R.)

Abstract: The Mitogen-Activated Protein Kinase (MAPK) signaling pathway plays an important role in cancer cell proliferation and survival. MAPKs' protein kinases MEK1/2 serve as important targets in drug designing against cancer. The natural compounds' flavonoids are known for their anticancer activity. This study aims to explore flavonoids for their inhibition ability, targeting MEK1 using virtual screening, molecular docking, ADMET prediction, and molecular dynamics (MD) simulations. Flavonoids (n = 1289) were virtually screened using molecular docking and have revealed possible inhibitors of MEK1. The top five scoring flavonoids based on binding affinity (highest score for MEK1 is −10.8 kcal/mol) have been selected for further protein–ligand interaction analysis. Lipinski's rule (drug-likeness) and absorption, distribution, metabolism, excretion, and toxicity predictions were followed to find a good balance of potency. The selected flavonoids of MEK1 have been refined with 30 (ns) molecular dynamics (MD) simulation. The five selected flavonoids are strongly suggested to be promising potent inhibitors for drug development as anticancer therapeutics of the therapeutic target MEK1.

Keywords: MEK1; flavonoids; virtual screening; molecular docking; ADMET; molecular dynamic (MD) simulation

check for updates

Citation: AlZahrani, W.M.; AlGhamdi, S.A.; Zughaibi, T.A.; Rehan, M. Exploring the Natural Compounds in Flavonoids for Their Potential Inhibition of Cancer Therapeutic Target MEK1 Using Computational Methods. *Pharmaceuticals* **2022**, *15*, 195. https://doi.org/10.3390/ph15020195

Academic Editors: Marialuigia Fantacuzzi and Mariangela Agamennone

Received: 8 December 2021
Accepted: 31 January 2022
Published: 3 February 2022

1. Introduction

Cancer is a leading cause of death in nearly every country in the world, as well as being the most significant barrier to extending life expectancy in the twenty-first century [1]. More than nearly 20 million new cases are predicted to be registered by 2025. Various implications are considered as hallmarks of cancer cells, including proliferative signaling, growth suppressor escape, cell death resistance, immortality, and angiogenesis induced by invasion–metastasis [2]. Since cancer is generally associated with several mutations that affect the main signaling pathways [3], targeted cancer therapy takes advantage of tumor-specific genetic vulnerabilities and mutations and designs therapeutics directly targeting cancer cells. The important therapeutic targets (mostly enzymes) are those whose suppression eventually kills cancer cells and that have minimal effect on normal tissues as they are either mutated or over-expressed in cancer cells [4,5]. One of the pathways that is commonly altered and activated, having a role in oncogenesis, the progression of the tumor, and drug resistance in cancer, is the Mitogen-Activated Protein Kinases (MAPK) signaling pathway [3,6]. The MAPK cascade, which is also known as RAS-regulated RAF-MEK1/2-ERK1/2 or ERK signaling pathway [7], is comprised of a number of kinases that carry out specific cell fate decisions in response to the input signal, explaining why numerous targeted therapies have been targeting this pathway [3,8]. By activating the epidermal growth factor receptor (EGFR) and the RAS small guanosine triphosphatases (GTPases) upstream, MAPK signaling promotes cell proliferation, survival, and metastasis. Mitogen-Activated Protein Kinase/Extracellular Regulated Kinase/Kinase 1 and 2 (MEK1/2) dual-specificity protein kinases are phosphorylated and activated by RAF kinases. MEK1/2 then phosphorylate and activate the Extracellular Regulated Kinase 1 and 2 (ERK1/2). Activated ERKs phosphorylate and regulate the activities of over 160 proteins that are estimated to be involved [9].

Protein kinases and phosphatases control the phosphorylation of proteins which makes them control nearly every aspect of the cell, therefore making them ideal candidates for evolutionary studies [10,11]. MEK1 and MEK2 are dual-specificity protein kinases; both are tyrosine (Y-) and serine/threonine (S/T-) protein kinases. A particular aspect of these protein kinases is that they are the core component of kinases for the MAPK/ERK signaling cascade as well as the gatekeepers of ERK1/2 activity, which is especially exciting since downstream ERK has multiple targets and transcription factors that control the critical processes in the cell. Thus, therapeutic targeting of MEK1/2 is relatively specific [12–14]. MEK1 and MEK2 are nearly identical, with a distinct pocket structure which is proximate to the ATP-binding site despite being distinct at the same time. Several conformational changes occur when an inhibitor binds to this region, locking unphosphorylated MEK1/2 into a catalytically inactive state. This occurs through the highly conserved DFG-out motif (Asp, Phe, and Gly) in the activation loop which exposes a site adjacent to the ATP binding site where inhibitors could bind to and lock the protein in an inactive state. Furthermore, since this ATP-noncompetitive process does not inhibit other protein kinases via the highly conserved ATP-binding pocket, unwanted side effects, such as inadvertent inhibition of other protein kinases and the challenge of competing with millimolar intracellular ATP concentrations, are largely avoided [15–17]. Unfortunately, most ATP-competitive kinase inhibitors interact with numerous members of the protein kinase family, making it challenging to build selective inhibitors for a single kinase target. Moreover, as a result of changes in the kinase domain, cancer cells can develop acquired drug resistance, making specific kinase inhibitors less effective. For these reasons, finding ATP-noncompetitive kinase inhibitors is becoming more desirable as a therapeutic development method [18]. Various popular pharmaceutical companies have shown great interest in MEK1 since it's highly selective of the MAPK/ERK pathway [19], and several MEK inhibitors are currently under clinical development [20].

Natural products can help treat a range of human disorders, including the world's second largest cause of death: cancer [21]. Lately, the interest in the natural products to be used as anticancer agents has increased, and phytochemicals have become valuable in anticancer drug development. More than 75% of the approved anticancer drugs between 1981 and 2007 are either natural products or have been developed based on them [22,23]. Flavonoids are among the proposed natural products for cancer prevention that have been increasingly found to have a major health impact [24]. Flavonoids are polyphenolic compounds obtained as small plant derivatives' secondary metabolites [25]. Since the discovery of the first flavonoid (rutin) in 1930, more than 6500 different flavonoids have been reported in various plant species, and their actual number is estimated to exceed 8000 [26,27]. A wide variety of biological and pharmacological effects have been found in flavonoids, such as their numerous antioxidant, antiproliferative, anti-inflammatory, antihypertensive, anti-carcinogenic, etc. effects, but the most notable activity is their potential role as anticancer agents [22,25,26,28–31]. Regarding the exciting feature of flavonoids is that they are by far the largest group of natural compounds that inhibit protein kinases and fit well into the ATP binding pocket and the neighboring area [32]. Others reported flavonoids' inhibition activity against Ser/The protein kinases 18 and receptor tyrosine kinases (RTKs) [33,34]. Various in silico, in vitro, and in vivo studies demonstrate the anticancer activities of flavonoids in several types of cancer [35–44]. Few studies have shown their anticancer relationship with the ERK and MAPK cascade in various cancer types [45–51]. Myricetin is one of the flavonoids which has been found to inhibit MEK1 in vivo and in vitro, in addition to being a potent ATP-noncompetitive inhibitor of MEK1 [52,53]. Another flavonoid is nobiletin, which has shown an in vitro antitumor effect against MEK in human fibrosarcoma HT−1080 cells [54]. Isorhamnetin is a flavonoid that inhibits MEK1 in the ATP-noncompetitive binding site [55]. A flavonoid that has shown potent and specific ATP-noncompetitive inhibition activity of MEK1 is PD 098059, which is the first MEK inhibitor to be discovered [56] and mostly used in laboratory experiments.

The current work involved the use of computational methods for the design of an MEK1 inhibitor. The computational methods, including molecular docking and binding simulation analysis, have increasingly been used for binding pose and designing of novel inhibitors or better derivatives of known existing inhibitors [57–68].

The current study proposed five novel MEK1 inhibitors and anticancer drug candidates by virtual screening of the natural compound flavonoids. A library of all known flavonoids was prepared and the top five screened compounds were proposed as potential MEK1 inhibitors and anticancer agents. The proposed compounds were further checked for key interacting residues, molecular interactions, binding energy, and dissociation constant using various methods. Finally, to show the stability of the protein–ligand, complexes were subjected to MD simulation analysis. This study will provide novel MEK1 inhibitors and anticancer agents with their action mechanism. The proposed flavonoids can further be tested experimentally for their potential use as novel anticancer agents.

2. Results and Discussion

2.1. Virtual Screening of Natural Compound Class Flavonoids for MEK1 Inhibition

The capacity of molecular docking studies to anticipate the right binding conformations of small molecules as ligands to the appropriate target binding site is an important feature of in silico drug discovery. In this view, the goal of our study was to do virtual screening of flavonoids for possible MEK1 inhibition, which is an essential protein target in the MAPK pathway in cancer cells. (Figure 1) shows the histogram of dock scores for the screened compounds (Supplementary File S1) where the selected top five ranked compounds are highlighted in the left. Most of the compounds are providing docking affinity (>−7.0 kcal/mol), but the highest scoring and best fit compounds were selected and the redundant compounds were omitted to give the best and most diverse structure compounds. The top five selected flavonoids (Figure 2) against MEK1 in (Table 1) show the highest docking score of (−10.8 kcal/mol) and the fifth rank with a score of (−10.4 kcal/mol) compared to the docking score of native inhibitor (−9.0 kcal/mol). The native inhibitor of the 3D structure is PD318008 (5-bromo-*N*-(2,3-dihydroxypropoxy)-3,4-dofluoro−2-[(2-fluoro-4-iodophenyl)amino]benzamide), an analog of PD184352 which is a highly selective, ATP-noncompetitive, and potent MEK1 and MEK2 inhibitor [16]. These results suggest that the selected flavonoids are probably better inhibitors of MEK1. Further, the molecular docking of trametinib, a specific and highly selective inhibitor of MEK1/2, was performed and gave the score of (−9.7 kcal/mol), which is also lower than those of the selected flavonoids, providing credence to the selected compounds [69].

Table 1. The selected flavonoids and the binding strength score against MEK1 (Autodock Vina docking energy, X-Score binding energy and pK_d). The higher the absolute values of the scores, the better the binding.

Rank	Flavonoids	Docking Affinity (kcal/mol)	Binding Energy (Kcal/mol)	pK_d or $-\log(K_d)$
1	129696793	−10.8	−10.25	7.52
2	10813589	−10.6	−10.96	8.03
3	10991656	−10.5	−9.49	6.95
4	10524567	−10.5	−10.83	7.94
5	10575055	−10.4	−10.11	7.41
Native		−9.0	−8.95	6.56

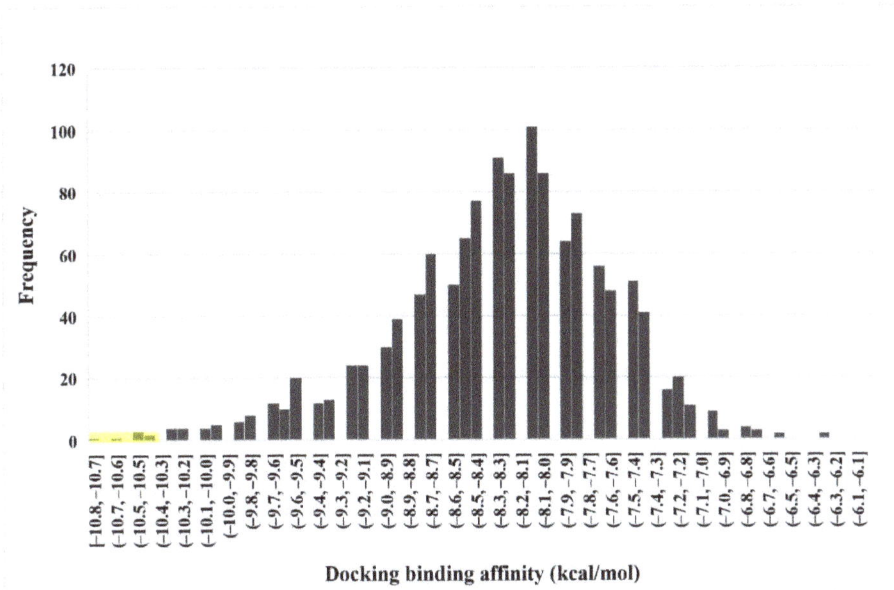

Figure 1. Histogram plot of dock binding affinity of all flavonoids. The dock scores of the selected flavonoids in the left part are highlighted in yellow.

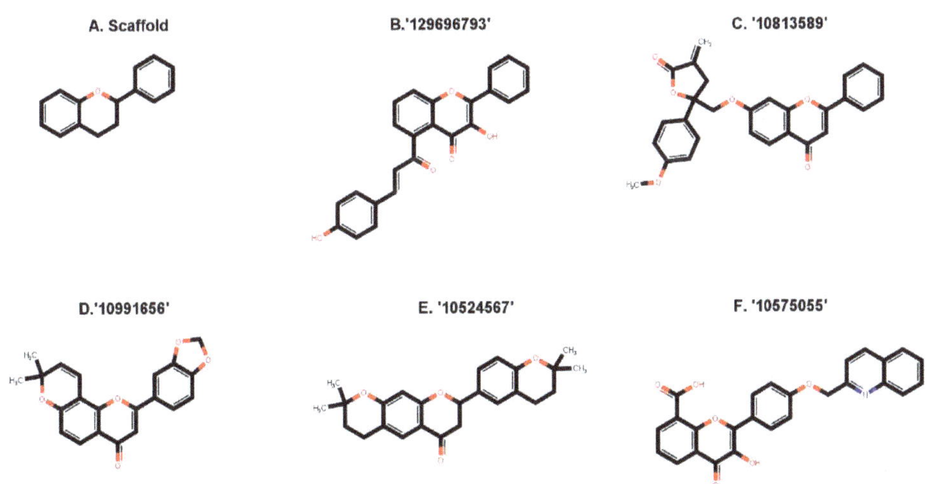

Figure 2. Two-dimensional sketch of flavonoid scaffold (**A**) and top selected flavonoids for MEK1 (**B–F**). The flavonoids are represented by their PubChem CID. Heteroatoms oxygen (O), and nitrogen (*N*)with their balancing hydrogens are shown as red, blue, and green, respectively.

The molecular docking results of the first rank flavonoid (CID: 129696793) against the MEK1 binding pocket showed that the pocket fit well in the catalytic site (Figure 3) and interacted with 16 amino acids: Leu-115, Leu-118, Val-127, Gly-128, Phe-129, Ile-141, Arg-189, Asp-208, Phe-209, Gly-210, Val-211, Ser-212, Leu-215, Ile-216, Met-219, and Arg-234 (Figure 4B). The binding strength scores' docking affinity (−10.8 Kcal/mol), binding energy (−10.25 Kcal/mol), and dissociation constant (pK_d, 7.52) showed quality binding as required for adequate inhibition (Table 1). The molecular interaction (Table 2) showed 2 hydrogen bonds and 62 (16) non-bonded contacts (hydrophobic interactions). The key interacting residues were Leu-118, Ile-141, Asp-208, Phe-209, and Ile-216 with 5, 6, 5, 9, and 9 non-bond contacts, respectively. The Met-219 turned out to have the maximum ΔASA (loss in Accessible Surface Area) (42.37 Å2) followed by Asp-208 (39.18 Å2). Hydrogen bonds were formed by Val-127 and Gly-128 measure 2.42 Å and 3.21 Å, respectively. Observing amino acid residues from the flavonoid binding site within the target protein seeks to predict the interactions that occur and that are thought to contribute to the flavonoid compound's pharmacological activities, such as MEK kinase inhibition. Inhibition of enzymatic activity of a protein by a compound is largely due to non-covalent bonds, including non-bonded contacts and hydrogen bonds [70]. The binding of native inhibitor with interacting residues and their interactions is also provided for comparison (Figure 4A). Comparing to the binding of the native inhibitor, there are six residues, Leu-118, Ile-141, Asp-208, Phe-209, Leu-215, and Met-219, common to the lists of interacting residues of this compound and the native inhibitor, and they include the key residues Asp-208 and Phe-209. The fact that the proposed compound binds to the similar group of residues in the catalytic site validates our prediction that it is comparable to the native inhibitor.

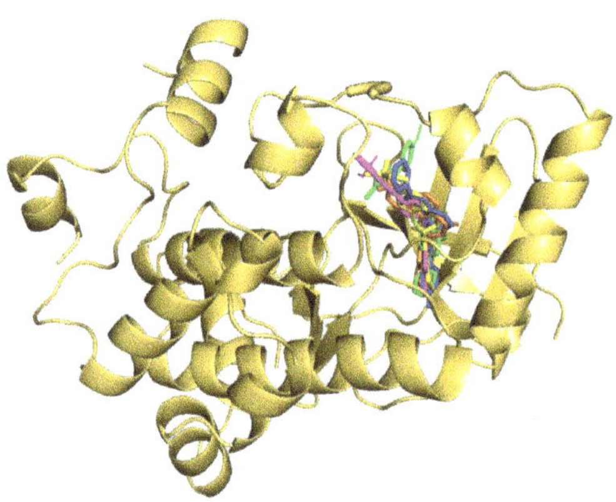

Figure 3. Molecular docking of top 5 selected flavonoids to MEK1. The protein is shown in cartoon representation colored yellow-orange, while compounds are shown in stick representations in various colors: binding of flavonoids '129696793' (blue), '10813589' (green), '10991656' (orange), '10524567' (yellow), and '10575055' (magenta).

The second highest ranked flavonoid (CID: 10813589) against the MEK1 binding pocket docked well (Figure 3) and interacted with ATP and 19 amino acids residues: Gly-79, Gly-80, Lys-97, Leu-98, Ile-99, His-100, Leu-115, Leu-118, Val-127, Gly-128, Phe-129, Ile-141, Asp-190, Asn-195, Asp-208, Phe-209, Val-211, Leu-215, and Met-219 (Figure 4C). The strength of binding scores' docking affinity of (−10.6 Kcal/mol), binding energy (−10.96 Kcal/mol) and dissociation constant (pK_d, 8.03) were reasonably high, as required for good inhibition (Table 1). The protein–ligand complex was stabilized by non-bonding interactions (Table 3) through 78 (20) non-bonded contacts and 2 hydrogen bonds. The key interacting residues were Ile-99, Asp-208, Lys-97, and Phe-209 with 15, 9, 8, and 6 non-bonding interactions of each, respectively. Met-219 turned out to have the maximum ΔASA with (43.97 Å^2), followed by Asp-208 with (39.24 Å^2) and Lys-97 with (31.09 Å^2). Hydrogen bonds in Gly-80 and Lys-97 measured 2.80 Å and 2.99 Å, respectively. Of the 19 interacting residues, 7 residues were common with those of the binding of the native inhibitor: Lys-97, Leu-118, Ile-141, Asp-208, Phe-209, Leu-215, and Met-219 (Figure 4A), including the key residues Lys-97, Asp-208, and Phe-209. The proposed compound bound to the same set of residues in the catalytic site, making it promising as the native inhibitor.

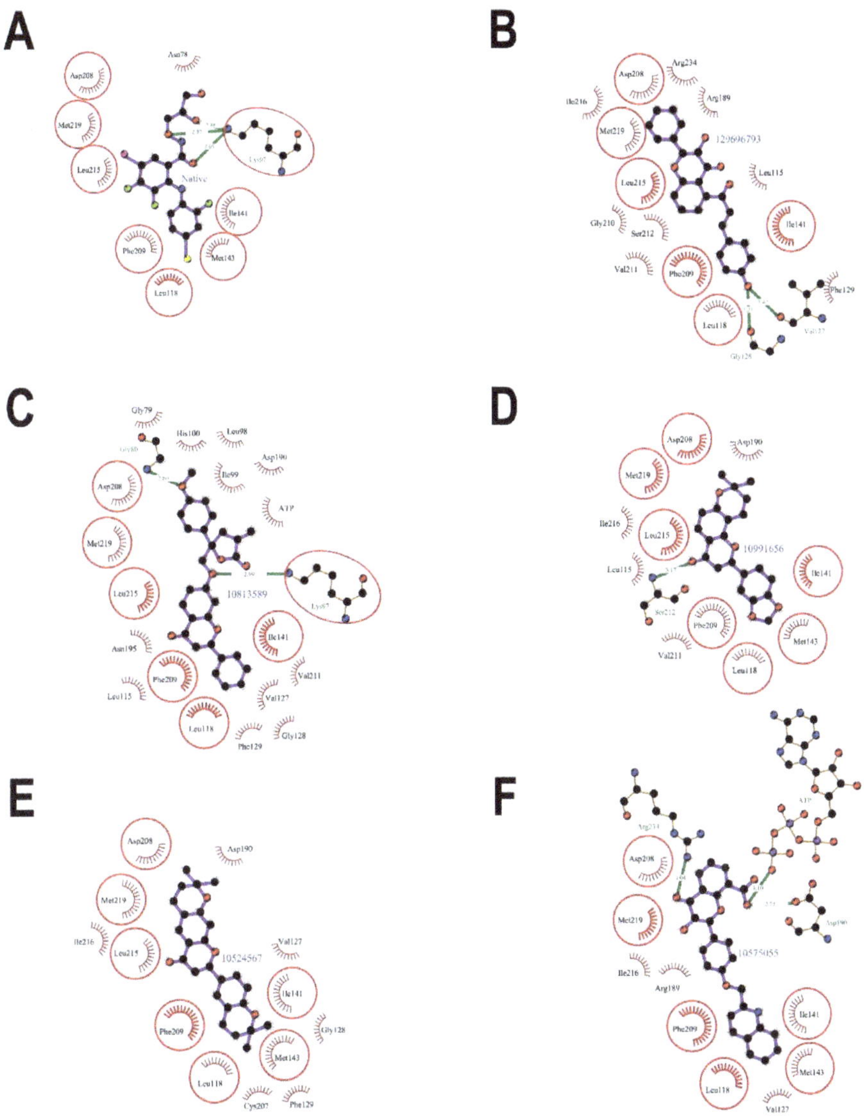

Figure 4. MEK1 protein–ligand interaction plots of native inhibitor (**A**) and selected flavonoids (**B**–**F**). The residues forming non-bonding interactions are shown as red bristles, while residues forming hydrogen bond and the bound ligand are shown as ball-and-stick representations. The carbon atoms are shown as black balls, nitrogen atoms as blue balls, oxygen atoms as red balls, fluorine atoms as green balls, bromine atom as pink balls, and iodine atom as a yellow ball. The interacting residues common with those of the native inhibitor are shown in circles. The hydrogen bonds are shown as green dashed lines labeled with bond length (in Å).

Table 2. The MEK1 residues interacting with selected flavonoid (CID: 129696793) are listed with the number of non-bonding interactions and ΔASA.

Interacting Residues	Hydrogen Bonds	Non-Bonding Interactions	ΔASA (In Å^2)
Leu-115	0	1	3.89
Leu-118	0	5	17.12
Val-127	1	2	10.27
Gly-128	1	1	0.78
Phe-129	0	2	6.45
Ile-141	0	6	21.79
Arg-189	0	3	11.93
Asp-208	0	5	39.18
Phe-209	0	9	25.06
Gly-210	0	1	8.22
Val-211	0	4	4.59
Ser-212	0	4	2.41
Leu-215	0	4	14.8
Ile-216	0	9	25.95
Met-219	0	4	42.37
Arg-234	0	2	16.78

Table 3. The MEK1 residues interacting with selected flavonoid (CID: 10813589) are listed with the number of non-bonding interactions and ΔASA.

Interacting Residues	Hydrogen Bonds	Non-Bonding Interactions	ΔASA (In Å^2)
Gly-79	0	3	17.63
Gly-80	1	4	15.78
Lys-97	1	8	31.09
Leu-98	0	1	0.55
Ile-99	0	15	27.12
His-100	0	2	16.96
Leu-115	0	2	3.89
Leu-118	0	3	16.47
Val-127	0	2	9.47
Gly-128	0	1	0.78
Phe-129	0	1	5.57
Ile-141	0	2	21.79
Asp-190	0	1	20.27
Asn-195	0	2	4.05
Asp-208	0	9	39.24
Phe-209	0	6	22.18
Val-211	0	3	4.59
Leu-215	0	1	12.45
Met-219	0	5	43.97
ATP	0	7	

The third rank flavonoid (CID: 10991656) bound to the MEK1 binding pocket (Figure 3) and showed interactions with 12 residues: Leu-115, Leu-118, Ile-141, Met-143, Asp-190, Asp-208, Phe-209, Val-211, Ser-212, Leu-215 Ile-216, and Met-219 (Figure 4D). The predicted binding strength scores for the protein–ligand complex were docking affinity of (−10.5 Kcal/mol), binding energy (−9.49 Kcal/mol), and dissociation constant (pK_d, 6.95) and showed quality binding as required for good inhibition (Table 1). The molecular interaction showed one hydrogen bond through Ser-212 measures 3.17 Å and 54 (12) non-bonded contacts (hydrophobic interactions). The key interacting residues (Table 4) are Asp-208 and Phe-209, with 13 and 10 non-bonding interactions of each, respectively. The maximum ΔASA is with Met-219 residue (43.97 Å^2) followed by Asp-208 with (40.1 Å^2).

The interacting residues and their interactions with the native inhibitor are also provided for comparison (Figure 4A). Out of the 12 interacting residues, 7 residues were common with those of the native inhibitor: Leu-118, Ile-141, Met-143, Asp-208, Phe-209, Leu-215, and Met-219, which included the key residues Asp-208 and Phe-209. This also suggested that the proposed compound was blocking the same set of residues as the native inhibitor, thereby inhibiting the protein's action.

Table 4. The MEK1 residues interacting with selected flavonoid (CID: 10991656) are listed with the number of non-bonding interactions and ΔASA.

Interacting Residues	Hydrogen Bonds	Non-Bonding Interactions	ΔASA (In Å^2)
Leu-115	0	2	3.89
Leu-118	0	1	15.6
Ile-141	0	2	21.79
Met-143	0	1	9.92
Asp-190	0	4	27.2
Asp-208	0	13	40.1
Phe-209	0	10	24.33
Val-211	0	3	4.59
Ser-212	1	3	2.41
Leu-215	0	5	14.8
Ile-216	0	5	13.79
Met-219	0	5	46.4

The dock results of the fourth ranked flavonoid (CID: 10524567) bound in the MEK1 binding pocket (Figure 3) showed interactions with 61 non-bonded contacts (hydrophobic interactions) with 13 amino acids: Leu-118, Val-127, Gly-128, Phy-129, Ile-141, Met-143, Asp-190, Cys-207, Asp-208, Phe-209, Leu-215 Ile-216, and Met-219 (Figure 4E). The binding strength scores' docking affinity (−10.5 Kcal/mol), binding energy (−10.83 Kcal/mol), and dissociation constant (pK_d, 7.94) showed as quality binding, as required for good inhibition (Table 1). The key interacting residues are Asp-208 and Phe-209 with 11 and 14 non-bonding interactions of each, respectively. Met-219 turned out to have the maximum ΔASA with (52.28 Å^2), then Asp-208 with (38.96 Å^2) and next Asp-190 with (30.23 Å^2) (Table 5). Compared to the native inhibitor (Figure 4A), there are seven common amino acids: Leu-118, Ile-141, Met-143, Asp-208, Phe-209, Leu-215, and Met-219, and they share the same key residue of Asp-208 and Phe-209 and thus, they may inhibit MEK1 kinase activity in the same way as the native inhibitor does.

Table 5. The MEK1 residues interacting with the selected flavonoid (CID: 10524567) are listed with the number of non-bonding interactions and ΔASA.

Interacting Residues	Hydrogen Bonds	Non-Bonding Interactions	ΔASA (In Å^2)
Leu-118	0	6	16.62
Val-127	0	2	9.35
Gly-128	0	1	0.78
Phy-129	0	1	5.48
Ile-141	0	5	21.79
Met-143	0	6	9.92
Asp-190	0	2	30.23
Cys-207	0	2	4.33
Asp-208	0	11	38.96
Phe-209	0	14	24.77
Leu-215	0	4	14.8
Ile-216	0	3	21.97
Met-219	0	4	52.28

This compound is a chiral compound and possesses R and S stereoisomeric configurations, as shown in Supplementary Figure S1. The above docking results of this compound are in R-stereoisomer configuration. In order to check the effect of S-stereoisomer of this compound on the binding to MEK1, the molecular docking was performed. The molecular docking of the (S) configuration of the chiral compound also showing high affinity, with (−10.9 Kcal/mol), binding energy (−10.85 Kcal/mol), and dissociation constant (pK_d, 7.96). The interaction showed 1 hydrogen bond and 58 non-bonded contacts with ATP and 16 amino acids: Lys-97, Ile-99, Leu-115, Leu-118, Val-127, Ile-141, Met-143, Asp-190, Leu-206, Cys-207, Asp-208, Phe-209, Gly-210, Val-211, Leu-215, and Met-219 (Table S2 in Supplementary file2). The hydrogen bond with Lys-97 measures 3.01 Å. The key interacting residues are: Phe-209 and Met-219 with 12 and 6 non-bonded contacts, respectively. Met-210 has the maximum with (48.6 Å^2), then Asp-208 with (44.44 Å^2). Comparing the interaction to the native inhibitor, the eight residues are common (Supplementary Figure S2): Lys-97, Leu-118, Ile-141, Met-143, Asp-208, Phe-209, Leu-215, and Met-219, and they share the same key residue, Phe-209. However, there is no literature upon the stereoisomerism of this compound. This can be a good chance to study the effect of stereoisomerism of this compound on the binding to the MEK1 kinase.

The dock results of the fifth ranked flavonoid (CID: 10575055) against the MEK1 binding pocket fit well within the catalytic site (Figure 3) and interacted with ATP and 11 amino acids residues, namely Leu-118, Val-127, Gly-128, Ile-141, Met-143, Arg-189, Asp-190, Asp-208, Phe-209, Met-219, and Arg-234 (Figure 4F). The results showed docking affinity of (−10.4 Kcal/mol), binding energy (−10.11 Kcal/mol), and dissociation constant (pK_d, 7.41) were also reasonably high as required for adequate MEK1 kinase inhibition (Table 1). The molecular interaction shows 3 hydrogen bonds and 44 (11) non-bonded contacts (hydrophobic interactions). The three hydrogen bonds with Asp-190, Arg-234, and ATP measure 2.73 Å, 3.04 Å, and 3.10 Å, respectively. The key interacting residues are Asp-208 and Phe-209, with 9 and 11 non-bonding interactions, respectively. Met-219 turned out to have the maximum ΔASA with (50.93 Å2), and then Asp-208 with (40.01 Å2) (Table 6). Of the 11 interacting residues, 6 residues were common among the interacting residues of the native inhibitor: Leu-118, Ile-141, Met-143, Asp-208, Phe-209, and Met-219 (Figure 4A), including the same key residues Asp-208 and Phe-209; thus, they might inhibit the MEK1 kinase activity similar to the native inhibitor.

Table 6. The MEK1 residues interacting with selected flavonoid (CID: 10575055) are listed with the number of non-bonding interactions and ΔASA.

Interacting Residues	Hydrogen Bonds	Non-Bonding Interactions	ΔASA (In Å^2)
Leu-118	0	4	16.73
Val-127	0	1	9.63
Ile-141	0	4	21.79
Met-143	0	3	9.92
Arg-189	0	3	14.72
Asp-190	1	4	41.14
Asp-208	0	9	40.01
Phe-209	0	11	25.06
Ile-216	0	1	18.14
Met-219	0	2	50.93
Arg-234	1	2	13.09
ATP	1	0	

2.2. Drug-Likeness and Pharmacokinetics Prediction

The strength of the ligand binding on the target protein is not the only factor in the discovery of novel medications. To assess the degree of effectiveness and therapeutic efficacy, it is also studied in terms of drug-likeness, pharmacokinetics, and toxicity. Lipinski's rule of five predicts the drug-likeness of the selected compounds. Moreover, pharmacokinetics including Absorption, Distribution, Metabolism, Excretion, and Toxicity (ADMET) play an important role in medicinal chemistry, which describes how drugs move through the body. The prediction of drug-likeness for the selected inhibitor for MEK1 is presented in (Table 7). Considering the desired values: molecular weight < 500, H-bond donors < 5, H-bond acceptors < 10, Rotatable bonds < 10, and lipophilicity (logP) < 5, all the selected compounds follow the desired values, except for (CID: 10813589) and (CID: 10524567), where the lipophilicity is slightly higher than 5. All the ADMET predicted properties of the top selected inhibitors for MEK1 are presented in (Table 8). The efficacy of the selected compounds as oral medicine was determined using two models for measuring absorption properties, including CaCO2 permeability and intestinal absorption. Where the desired CaCO2 permeability is >0.90 and intestinal absorption >30%, the prediction of the screened compounds shows the CaCO2 permeability values all in positive integers, with exception of (CID: 10575055). While intestinal absorption shows a high percentage, all are higher than 70%, which is considered as good absorption. Skin permeability is the next important factor in absorption. The ideal skin permeability is >−2.5 log Kp and all of the compounds under study have permeability values of less than −2.5 log Kp, indicating poor skin permeability. The ATP-binding cassette (ABC) transporter, which is necessary for efficient molecular transport across cell membranes, contains P-glycoprotein. P-glycoprotein substrates, and inhibitors of P-glycoprotein I and II which were examined in all of the substances that were screened. Except for (CID: 10813589) and (CID: 10524567), all of the compounds were found to be substrates, indicating that they can be transported through the cell membrane through the ABC transporter. The (CID: 10575055) was found to be ineffective as an inhibitor of P-glycoprotein I transporter, suggesting that they could be incapable of inhibiting these drug efflux pumps. The distribution of the substances in the body was determined using four distinct assays: volume of distribution (VDss), fraction unbound, BBB permeability, and central nervous system (CNS) permeability. To begin with, in the VDss assay, which is used to evaluate the total amount of drugs needed for uniform drug distribution in the bloodstream, readings less than −0.15 log are considered negative, while values greater than 0.45 log are considered good diffusion. Thus, (CID: 10991656) and (CID: 10524567) show average VDss values, while other compounds have low distribution volume. The potential of a drug to reach the brain is determined by the permeability of the blood–brain barrier (BBB). If the logBB values are more than 0.3, they will cross BBB. The logBB value of the screened compounds are less than 0.3, meaning that none of them will be able to cross BBB except for (CID: 10991656). The desired value of the CNS permeability is >−2 and the screen shows good results except for (CID: 10575055), where it was less than the desired value. Seven distinct cytochrome models were used to examine the test drug's metabolism in the body. All of the compounds were tested for their capacity to inhibit CYP1A2, CYP2C19, CYP2D6, CYP2C9, and CYP3A4 as well as their ability to function as a substrate for CYP2D6 and CYP3A4. The total clearance rates of all of the examined compounds were varied, and none of them appeared to be a substrate for organic cation transporter 2 (OCT2). They also failed to anticipate AMES toxicity, showing that these chemicals are neither carcinogenic nor mutagenic, except for: (CID: 129696793), and (CID: 10813589). Three of the selected flavonoids predicted to be negative for hepatotoxicity were (CID: 10813589), (CID: 10991656), and (CID: 10524567), whereas none of the substances tested positive for skin sensitization. Overall, the selected compounds proposed to be safe drug-candidates for human cancer therapy.

Table 7. Structures and chemical properties of the selected flavonoids against MEK1 to predict the drug-likeness: molecular weight, lipophilicity (LogP), number of (#) rotatable bonds, hydrogen acceptors and hydrogen donors, and polar surface area.

Rank	Compound (CID)	Structure	Molecular Weight	LogP	#Rotatable Bonds	#Acceptors	#Donors	Surface Area
1	129696793	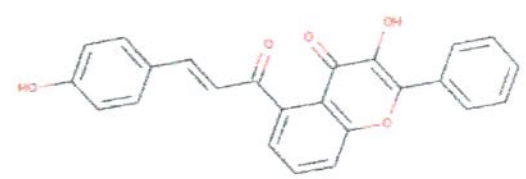	384.387	4.7673	4	5	2	165.349
2	10813589	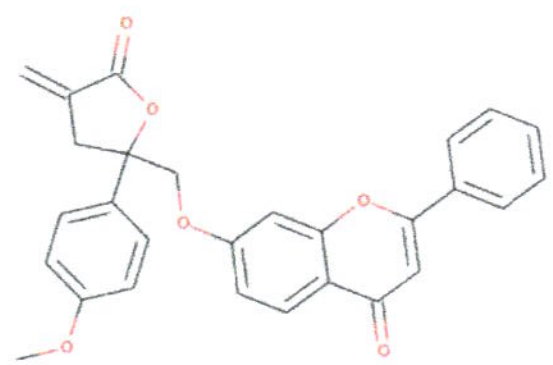	454.478	5.246	6	6	0	195.555
3	10991656	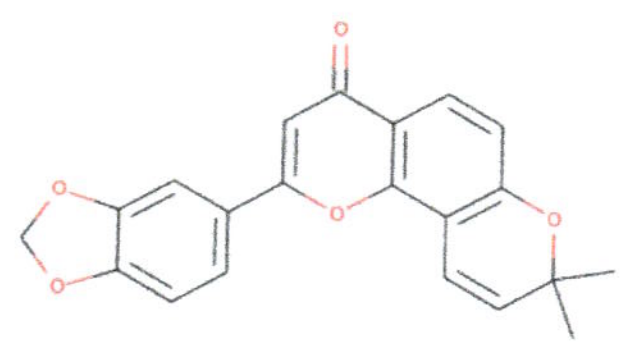	348.354	4.3729	1	5	0	148.966
4	10524567	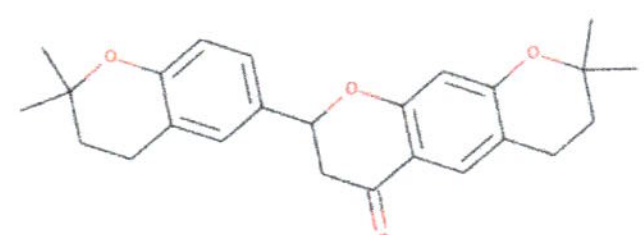	392.495	5.6003	1	4	0	171.717
5	10575055	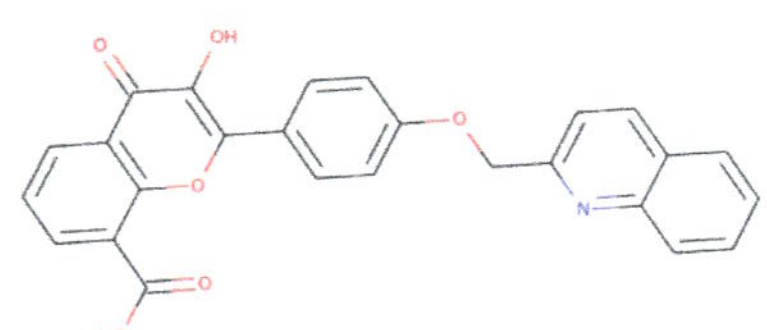	439.423	4.991	5	6	2	186.688

Table 8. ADMET properties of the selected flavonoids against MEK1.

Property	Model Name	Predicted Value					Unit
		129696793	10813589	10991656	10524567	10575055	
Absorption	Water solubility	−4.456	−5.914	−4.984	−5.161	−3.903	Numeric (log mol/L)
	Caco2 permeability	1.089	1.073	1.004	1.116	0.549	Numeric (log Papp in 10^{-6} cm/s)
	Intestinal absorption (human)	90.869	98.737	95.566	96.435	88.343	Numeric (% Absorbed)
	Skin Permeability	−2.735	−2.731	−2.589	−2.729	−2.734	Numeric (log Kp)
	P-glycoprotein substrate	Yes	No	Yes	No	Yes	Categorical (Yes/No)
	P-glycoprotein I inhibitor	Yes	Yes	Yes	Yes	No	Categorical (Yes/No)
	P-glycoprotein II inhibitor	Yes	Yes	Yes	Yes	Yes	Categorical (Yes/No)
Distribution	VDss (human)	−0.696	−0.157	0.121	0.363	−0.432	Numeric (log L/kg)
	Fraction unbound (human)	0.039	0.223	0.195	0.03	0.208	Numeric (Fu)
	BBB permeability	−0.371	−0.749	0.358	−0.005	−0.956	Numeric (log BB)
	CNS permeability	−1.883	−1.903	−1.595	−1.608	−2.889	Numeric (log PS)
Metabolism	CYP2D6 substrate	No	No	No	No	No	Categorical (Yes/No)
	CYP3A4 substrate	Yes	Yes	Yes	Yes	Yes	Categorical (Yes/No)
	CYP1A2 inhibitor	Yes	No	Yes	No	No	Categorical (Yes/No)
	CYP2C19 inhibitor	Yes	Yes	Yes	Yes	No	Categorical (Yes/No)
	CYP2C9 inhibitor	Yes	Yes	Yes	Yes	Yes	Categorical (Yes/No)
	CYP2D6 inhibitor	No	No	No	No	No	Categorical (Yes/No)
	CYP3A4 inhibitor	Yes	Yes	Yes	Yes	No	Categorical (Yes/No)
Excretion	Total Clearance	0.184	0.81	0.345	0.087	0.551	Numeric (log mL/min/kg)
	Renal OCT2 substrate	No	No	No	No	No	Categorical (Yes/No)
Toxicity	AMES toxicity	Yes	Yes	No	No	No	Categorical (Yes/No)
	Max. tolerated dose (human)	0.204	0.64	−0.242	−0.067	0.742	Numeric (log mg/kg/day)
	hERG I inhibitor	No	No	No	No	No	Categorical (Yes/No)
	hERG II inhibitor	Yes	Yes	No	No	Yes	Categorical (Yes/No)
	Oral Rat Acute Toxicity (LD50)	2.767	2.734	2.086	3.018	2.656	Numeric (mol/kg)
	Oral Rat Chronic Toxicity (LOAEL)	0.914	0.805	1.269	1.713	0.755	Numeric (log mg/kg_bw/day)
	Hepatotoxicity	Yes	No	No	No	Yes	Categorical (Yes/No)
	Skin Sensitization	No	No	No	No	No	Categorical (Yes/No)
	T.Pyriformis toxicity	0.29	0.287	0.555	0.491	0.285	Numeric (log ug/L)
	Minnow toxicity	0.09	−2.72	−0.483	−0.22	−1.62	Numeric (log mM)

2.3. MD Simulation

Molecular Dynamic (MD) simulation was performed to refine and assess the stability of protein after adding the missing loop (Supplementary Figure S3) and the binding stability of the protein–ligand complex system. The MD simulation evaluates and delineates the dynamic behavior of the ligand and the binding site residues. The best conformation flavonoids obtained from virtual screening for MEK1 inhibition advanced to MD simulation. The MD results were examined for RMSD of backbone, RMSD of heavy ligand atoms, RMSF values, hydrogen bonds number, and the radius of gyration to assess the stability of the protein–ligand complex. The RMSD value is a measure of how far a protein molecule deviates from its initial conformation over the course of the simulation. The RMSD values for the first rank were found to be within 0.2 nm for the backbone and 0.05–0.1 nm through the simulation, and the protein–ligand complex was considered to be stable (Figure 5A). The second rank RMSD value of backbone began at 0.2 nm then rose to 0.3 nm in 20 ns, then went back to stabilize at 0.2 nm. For the heavy atoms of ligand RMSD (Figure 5B), they showed values in a range of 0.05–0.1 nm then rose up to 0.25 nm after 20 ns to stabilize back at 0.15 nm. RMSD values of interaction of the third rank showed a very stable range, between 0.15 and 0.2 nm, throughout the 30 ns simulation for the backbone and 0.025–0.05 nm for the ligand's heavy atoms. Fourth rank RMSD also showed fluctuations around 0.2–0.3 nm for the backbone and 0.025–0.05 nm for the ligand's heavy atoms. The RMSD values of the fifth rank showed similar fluctuation, between 0.15 and 0.2 for the backbone and between 0.025 and 0.125 nm for the ligand's heavy atoms. The low RMSD fluctuations indicate that the equilibration of a system is achieved through the simulation. The RMSF of the protein coordinates from their beginning positions for each residue was computed to define the flexible areas of the protein during the course of the simulation. The RMSF is a metric that measures how much ligand binding affects the flexibility of MEK1 residues. The RMSF values (Figure 5C) show that the maximum fluctuation was in the amino acid residue region 276–320, which is a proline rich loop and highly flexible region, and the fluctuation was within the range of 1.25 nm. From this perspective also, owing to low fluctuation in RMSF, the protein–ligand complex seems stable (Figure 5C). The radius of gyration is a measure of protein compactness, and the low fluctuations in its value indicate stability of the protein backbone. The radius of gyration fluctuates in a range of around 2.0 nm for all the four simulations and during the entire simulation, which points towards stability of the protein (Figure 5D). The presence of hydrogen bonds is critical for a protein complex's stability. Analysis of the number of hydrogen bonds in (Figure 6) shows them appearing and disappearing during the course of simulation. Looking more closely at (Figure 6), the maximum number of hydrogen bonds and pairs reaches eight for first rank compound, five for the second, two for the third and fourth ranks, and eight for the fifth rank compound. Due to the isolated hydrogen bonds and low average hydrogen-bond number per time frame, the hydrogen-bond network in the complexes appeared to be weak. Other interactions were thought to hold hydrogen bonds in places where they had disappeared. As a result, no notable conformational changes in the complexes were observed across the simulated time period. These findings suggested that the MD simulation trajectory for the complex after equilibrium was reliable enough for future investigation.

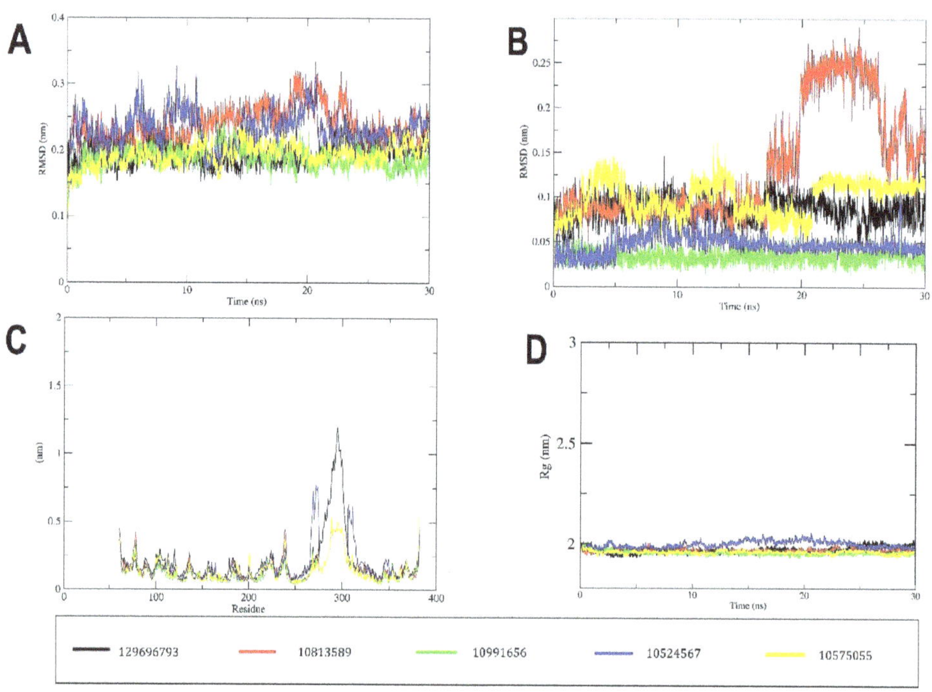

Figure 5. MD simulation results of the docked flavonoids obtained from virtual screening. (**A**) Root–mean-square deviation (RMSD) curve for the protein backbone of the protein–ligand complex. The RMSD plot provides quantification of the overall stability of the protein backbone during 30 ns simulation. (**B**) RMSD curve for ligands' heavy atoms through the simulation. (**C**) Root–mean-square fluctuations' (RMSF) curve of the MEK1 residues for the protein–ligand complexes. (**D**) Radius of gyration (total) of protein–ligand complex.

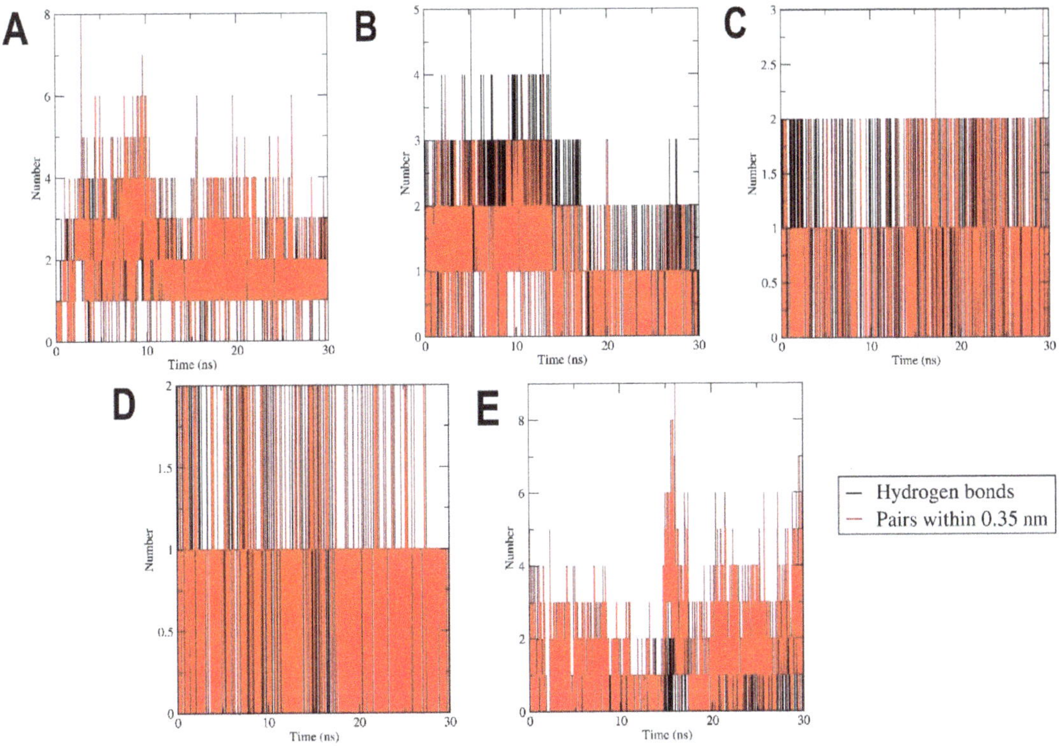

Figure 6. Time evolution of number of hydrogen bonds formed between MEK1 and the compounds. (**A**) 129696793, (**B**) 10813589, (**C**) 10991656, (**D**) 10524567, and (**E**) 10575055.

3. Materials and Methods

3.1. Data Retrieval and Preparation

The three-dimensional structure coordinate of MEK1 complexed with the native inhibitor is retrieved from the protein data bank (PDB, https://www.rcsb.org/, accessed on 19 January 2022). The structure of MEK1 (PDB code: 1S9J) with 2.40 Å resolution [16] was selected and used for the study. MODLLER was used to model the missing prolin-rich loop in the crystallized structure (residues from 276 to 305), which is a high flexible region located after the catalytic site [71]. The unique allosteric site for MEK1 was verified [72], and the coordinates for the grid box covering the catalytic site were prepared using AutoDockTools-1.5.6 [73]. Other preparations included: deleting water, checking for missing atoms, removing heteroatoms, adding polar hydrogens, computing and adding charges, and finally converting the protein into a (pdbqt) file for the molecular docking, also performed using AutoDockTools-1.5.6 software package. A drug-like library prepared from PubChem (www.ncbi.nlm.nih.gov, accessed on 4 April 2021) and 2630 flavonoids were filtered to 1289 by 3D structure availability and the Lipinski rule of five [74]. Ligands were prepared for virtual screening using Open Babel command line [75] and converted from (sdf) file to (pdbqt) after adding charges and hydrogens. The 2-D illustrations for the chemical compounds were prepared using MarvinSketch v18.4, ChemAxon (http://www.chemaxon.com/products/marvin, accessed on 6 June 2021).

3.2. Molecular Docking

Docking was carried out using AutoDock Vina [76] after preparing the configuration file with the details of the grid box coordinates, with energy range of 4 and maximum exhaustiveness of 24. Best mode, least RMSD, and highest docking affinity results were taken and ranked for MEK1. For further analysis of the docking, protein–ligand interaction plots of selected flavonoids with MEK1 was performed using Ligplot+ v1.4.5 [77] and illustrations of the docking were prepared using PyMOL v2.4.0 [78]. Further, calculations of binding energy and dissociation constant were performed by XScore v1.2.11 [79]. The degree of ligand filling the binding site was evaluated by loss in accessible surface area (ASA).

$$\Delta ASA_i = ASA_i^{protien} - ASA_i^{protien-ligand}$$

A residue is said to be taking part in filling the binding site if it loses more than 10 A^2 ASA due to binding [80]. All the ASA calculations of the protein–ligand complexes and the unbound proteins were performed by Naccess v2.1.1 [81].

3.3. Drug-Likeness and Pharmacokinetics Prediction

The "pkCSM-pharmacokinetics" online web-server (http://biosig.unimelb.edu.au/pkcsm/, accessed on 20 July 2021) was used for predictions of drug-likeness and pharmacokinetic properties: Absorption, Distribution, Metabolism, Excretion, and Toxicity (ADMET) [82].

3.4. Molecular Dynamic Simulation

The docked protein–ligand complexes were subjected to energy minimization using Gromacs v2020.5 [83] with the CHARMM36 all atom force field. Ligand and protein were separated to add ligand hydrogen atoms using Avogadro v2018 (https://avogadro.cc/, accessed on 6 January 2022) and then converted for topology using CHARMM force field (https://cgenff.umaryland.edu/, accessed on 6 January 2022), then wrote back with the complex topology file. The models were solvated with a water model in a cubic periodic box with 1 nm distance from the edge of the complex atoms. The solvated system was neutralized by five sodium ions. Energy minimization was carried out through 50,000 steps. An equilibration was conducted by number of particles, volume, and temperature (NVT), and number of particles, pressure, and temperature (NPT) temperature was coupled for ligand, protein, solvent, and ions, separately. Then the system proceeded to the actual MD simulation. The final models obtained at the end of MD were validated and illustrated by VMD (https://www.ks.uiuc.edu/Research/vmd/, accessed on 5 September 2021). For the analysis, Gromacs and Xmgrace (https://plasma-gate.weizmann.ac.il/Grace/, accessed on 7 September 2021) were used.

4. Conclusions

In conclusion, the molecular docking results suggest that some flavonoids could be better inhibitors of MEK1 compared to the native inhibitor based on the binding affinity and ligand interactions. The selected flavonoids could be potential drug candidates after re-engineering to improve the pharmacokinetic properties. Further, MD simulation studies with 100 ns time scale confirm the stability of the first rank flavonoid and MEK1 complex by root–mean–square deviation, root–mean–square fluctuation, and the radius of gyration. Our findings suggest that natural flavonoids are a promising and readily available source of anticancer targeted therapy in the future. However, these interpretations need further confirmatory analysis and validations for the screened molecules to ascertain their efficacy in the illness treatment.

Supplementary Materials: The following supporting information can be downloaded at: https://www.mdpi.com/article/10.3390/ph15020195/s1, Table S1: Autodock Vina docking results of flavonoids against MEK1 binding pocket. Figure S1: Two dimensional sketches of two stereoisomers (R & S) of 4th rank compound (CID: 10524567). Figure S2: S-stereoisomer of (CID: 10524567) interaction with MEK1 compared to the native inhibitor, Table S2: MEK1 interaction residues with S-stereoisomer of (CID: 10524567), Figure S3: Modified MEK1 molecular dynamic (MD) simulation for 2 (ns).

Author Contributions: Conceptualization, W.M.A. and M.R. Data curation, M.R. Formal analysis, W.M.A. Methodology, M.R. Resources, M.R. and T.A.Z. Supervision, S.A.A. and M.R. Visualization, W.M.A. Writing—original draft, W.M.A. Writing—review and editing, M.R., S.A.A. and T.A.Z. All authors have read and agreed to the published version of the manuscript.

Funding: This research received no external funding.

Institutional Review Board Statement: Not applicable.

Informed Consent Statement: Not applicable.

Data Availability Statement: PDB, https://www.rcsb.org/, accessed on 19 January 2022; www.ncbi.nlm.nih.gov, accessed on 4 April 2021.

Acknowledgments: The authors would like to thank King Fahd Medical Research center (KFMRC) for the scientific environment. The MD simulations in this work were performed at King Abdulaziz University's High-Performance Computing Center (Aziz Supercomputer) (http://hpc.kau.edu.sa), Jeddah, Saudi Arabia, and the authors acknowledge the center for technical support.

Conflicts of Interest: The authors declare no conflict of interest.

References

1. Sung, H.; Ferlay, J.; Siegel, R.L.; Laversanne, M.; Soerjomataram, I.; Jemal, A.; Bray, F. Global Cancer Statistics 2020: GLOBOCAN Estimates of Incidence and Mortality Worldwide for 36 Cancers in 185 Countries. *CA Cancer J. Clin.* **2021**, *71*, 209–249. [CrossRef]
2. Haider, T.; Tiwari, R.; Vyas, S.P.; Soni, V. Pharmacology & Therapeutics Molecular Determinants as Therapeutic Targets in Cancer Chemotherapy: An Update. *Pharmacol. Ther.* **2019**, *200*, 85–109. [CrossRef] [PubMed]
3. Braicu, C.; Buse, M.; Busuioc, C.; Drula, R.; Gulei, D.; Raduly, L.; Rusu, A.; Irimie, A.; Atanasov, A.G. A Comprehensive Review on MAPK: A Promising Therapeutic Target in Cancer. *Cancers* **2019**, *11*, 1618. [CrossRef] [PubMed]
4. Oliver, D.; Ji, H.; Liu, P.; Gasparian, A.; Gardiner, E.; Lee, S.; Zenteno, A.; Perinskaya, L.O.; Chen, M.; Buckhaults, P.; et al. Identification of Novel Cancer Therapeutic Targets Using a Designed and Pooled ShRNA Library Screen. *Nat. Sci. Rep.* **2017**, *7*, 43023. [CrossRef]
5. Martino, S.D.I.; Rainone, A.; Troise, A.; Paolo, M.D.I.; Pugliese, S.; Zappavigna, S.; Grimaldi, A.; Valente, D. Overview of FDA-approved anti cancer drugs used for targeted therapy. *World Cancer Res. J.* **2015**, *2*, e553.
6. Trotta, A.P.; Chipuk, J.E. Mitochondrial Dynamics As Regulators Of Cancer Biology. *Physiol. Behav.* **2018**, *176*, 139–148. [CrossRef] [PubMed]
7. Kidger, A.M.; Sipthorp, J.; Cook, S.J. ERK1/2 Inhibitors: New Weapons to Inhibit the RAS-Regulated RAF-MEK1/2-ERK1/2 Pathway. *Pharmacol. Ther.* **2018**, *187*, 45–60. [CrossRef]
8. Lee, S.; Rauch, J.; Kolch, W. Targeting MAPK Signaling in Cancer: Mechanisms of Drug Resistance and Sensitivity. *Int. J. Mol. Sci.* **2020**, *21*, 1102. [CrossRef]
9. Roberts, P.J.; Der, C.J. Targeting the Raf-MEK-ERK Mitogen-Activated Protein Kinase Cascade for the Treatment of Cancer. *Oncogene* **2007**, *26*, 3291–3310. [CrossRef] [PubMed]
10. Courcelles, M.; Lemieux, S.; Voisin, L.; Meloche, S.; Thibault, P. ProteoConnections: A Bioinformatics Platform to Facilitate Proteome and Phosphoproteome Analyses. *Proteomics* **2011**, *11*, 2654–2671. [CrossRef]
11. Roskoski, R. Targeting ERK1/2 Protein-Serine/Threonine Kinases in Human Cancers. *Pharmacol. Res.* **2019**, *142*, 151–168. [CrossRef]
12. Caunt, C.J.; Sale, M.J.; Smith, P.D.; Cook, S.J. MEK1 and MEK2 Inhibitors and Cancer Therapy: The Long and Winding Road. *Nat. Rev. Cancer* **2015**, *15*, 577–592. [CrossRef]
13. Lian, T.; Li, C.; Wang, H. Trametinib in the Treatment of Multiple Malignancies Harboring MEK1 Mutations. *Cancer Treat. Rev.* **2019**, *81*, 101907. [CrossRef]
14. Mccubrey, J.A.; Steelman, L.S.; Chappell, W.H.; Abrams, S.L.; Wong, W.T.; Chang, F.; Lehmann, B.; Terrian, D.M.; Milella, M.; Stivala, F.; et al. Roles of The RAF/MEK/ERK Pathway in Cell Growth, Malignant Transforamtion and Drug Resistance. *Biochim. Biophys. Acta* **2009**, *1773*, 1263–1284. [CrossRef]
15. Zhao, Y.; Adjei, A.A. The Clinical Development of MEK Inhibitors. *Nat. Rev. Clin. Oncol.* **2014**, *11*, 385–400. [CrossRef]
16. Ohren, J.F.; Chen, H.; Pavlovsky, A.; Whitehead, C.; Zhang, E.; Kuffa, P.; Yan, C.; McConnell, P.; Spessard, C.; Banotai, C.; et al. Structures of Human MAP Kinase Kinase 1 (MEK1) and MEK2 Describe Novel Noncompetitive Kinase Inhibition. *Nat. Struct. Mol. Biol.* **2004**, *11*, 1192–1197. [CrossRef] [PubMed]
17. Roskoski, R. MEK1/2 Dual-Specificity Protein Kinases: Structure and Regulation. *Biochem. Biophys. Res. Commun.* **2012**, *417*, 5–10. [CrossRef] [PubMed]
18. Shang, J.; Lu, S.; Jiang, Y.; Zhang, J. Allosteric Modulators of MEK1: Drug Design and Discovery. *Chem. Biol. Drug Des.* **2016**, *88*, 485–497. [CrossRef]

19. Melagraki, G.; Afantitis, A.; Sarimveis, H.; Igglessi-Markopoulou, O.; Koutentis, P.A.; Kollias, G. In Silico Exploration for Identifying Structure-Activity Relationship of MEK Inhibition and Oral Bioavailability for Isothiazole Derivatives. *Chem. Biol. Drug Des.* **2010**, *76*, 397–406. [CrossRef] [PubMed]
20. Cheng, Y.; Tian, H. Current Development Status of MEK Inhibitors. *Molecules* **2017**, *22*, 1551. [CrossRef]
21. Desai, A.; Qazi, G.; Ganju, R.; El-Tamer, M.; Singh, J.; Saxena, A.; Bedi, Y.; Taneja, S.; Bhat, H. Medicinal Plants and Cancer Chemoprevention. *Curr. Drug Metab.* **2008**, *9*, 581–591. [CrossRef] [PubMed]
22. Liu, H.; Jiang, W.; Xie, M. Flavonoids: Recent Advances as Anticancer Drugs. *Recent Pat. AntiCancer Drug Discov.* **2010**, *5*, 152–164. [CrossRef] [PubMed]
23. Newman, D.J.; Cragg, G.M. Natural Products as Sources of New Drugs over the Last 25 Years. *J. Nat. Prod.* **2007**, *70*, 461–477. [CrossRef]
24. Block, G.; Patterson, B.; Subar, A. Fruit, Vegetables, and Cancer Prevention: A Review of the Epidemiological Evidence. *Nutr. Cancer* **1992**, *18*, 1–29. [CrossRef] [PubMed]
25. Garg, S. Flavonoids: Biosynthesis, Metabolism, Mechanism of Antioxidation and Clinical Implications: A Review. *Agric. Rev.* **2020**, *41*, 227–237. [CrossRef]
26. Patil, V.M.; Masand, N. *Anticancer Potential of Flavonoids: Chemistry, Biological Activities, and Future Perspectives*, 1st ed.; Elsevier B.V.: Amsterdam, The Netherlands, 2018; Volume 59. [CrossRef]
27. Ververidis, F.; Trantas, E.; Douglas, C.; Vollmer, G.; Kretzschmar, G.; Panopoulos, N. Biotechnology of Flavonoids and Other Phenylpropanoid-Derived Natural Products. Part I: Chemical Diversity, Impacts on Plant Biology and Human Health. *Biotechnol. J.* **2007**, *2*, 1214–1234. [CrossRef]
28. Abotaleb, M.; Samuel, S.M.; Varghese, E.; Varghese, S.; Kubatka, P.; Liskova, A.; Büsselberg, D. Flavonoids in Cancer and Apoptosis. *Cancers* **2019**, *11*, 28. [CrossRef]
29. Bisol, Â.; de Campos, P.S.; Lamers, M.L. Flavonoids as Anticancer Therapies: A Systematic Review of Clinical Trials. *Phyther. Res.* **2020**, *34*, 568–582. [CrossRef]
30. Dobrzynska, M.; Napierala, M.; Florek, E. Flavonoid Nanoparticles: A Promising Approach for Cancer Therapy. *Biomolecules* **2020**, *10*, 1268. [CrossRef] [PubMed]
31. Stanisic, D.; Costa, A.F.; Cruz, G.; Durán, N.; Tasic, L. Applications of Flavonoids, With an Emphasis on Hesperidin, as Anticancer Prodrugs: Phytotherapy as an Alternative to Chemotherapy. *Stud. Nat. Prod. Chem.* **2018**, *58*, 161–212. [CrossRef]
32. Guerra, B.; Issinger, O.G. Natural Compounds and Derivatives as Ser/Thr Protein Kinase Modulators and Inhibitors. *Pharmaceuticals* **2019**, *12*, 4. [CrossRef] [PubMed]
33. Teillet, F.; Boumendjel, A.; Boutonnat, J.; Ronot, X. Flavonoids as RTK Inhibitors and Potential Anticancer Agents. *Harv. Bus. Rev.* **2008**, *86*, 84–92. [CrossRef] [PubMed]
34. Baby, B.; Antony, P.; Al Halabi, W.; Al Homedi, Z.; Vijayan, R. Structural Insights into the Polypharmacological Activity of Quercetin on Serine/Threonine Kinases. *Drug Des. Dev. Ther.* **2016**, *10*, 3109–3123. [CrossRef]
35. Hsieh, M.H.; Tsai, J.P.; Yang, S.F.; Chiou, H.L.; Lin, C.L.; Hsieh, Y.H.; Chang, H.R. Fisetin Suppresses the Proliferation and Metastasis of Renal Cell Carcinoma through Upregulation of MEK/ERK-Targeting CTSS and ADAM9. *Cells* **2019**, *8*, 948. [CrossRef] [PubMed]
36. Khalaf, H.H.A.; El-Saadani, R.M.A.; El-Amry, H.G.; Abd El-Salam, R.S. Bioactive Compounds of Flaxseed as Natural Antioxidants and Anticancer. *Ann. Agric. Sci. Moshtohor* **2019**, *57*, 395–404. [CrossRef]
37. Tavsan, Z.; Kayali, H.A. Flavonoids Showed Anticancer Effects on the Ovarian Cancer Cells: Involvement of Reactive Oxygen Species, Apoptosis, Cell Cycle and Invasion. *Biomed. Pharmacother.* **2019**, *116*, 109004. [CrossRef] [PubMed]
38. Mateeva, N.; Eyunni, S.V.K.; Redda, K.K.; Ononuju, U.; Hansberry, T.D.; Aikens, C.; Nag, A. Functional Evaluation of Synthetic Flavonoids and Chalcones for Potential Antiviral and Anticancer Properties. *Bioorg. Med. Chem. Lett.* **2017**, *27*, 2350–2356. [CrossRef] [PubMed]
39. Rehan, M.; Mahmoud, M.M.; Tabrez, S.; Hassan, H.M.A.; Ashraf, G.M.D. Exploring Flavonoids for Potential Inhibitors of a Cancer Signaling Protein PI3Kγ Kinase Using Computational Methods. *Anticancer Res.* **2020**, *40*, 4547–4556. [CrossRef] [PubMed]
40. Martínez Medina, J.J.; Naso, L.G.; Pérez, A.L.; Rizzi, A.; Ferrer, E.G.; Williams, P.A.M. Antioxidant and Anticancer Effects and Bioavailability Studies of the Flavonoid Baicalin and Its Oxidovanadium (IV) Complex. *J. Inorg. Biochem.* **2017**, *166*, 150–161. [CrossRef]
41. Jarial, R.; Shard, A.; Thakur, S.; Sakinah, M.; Zularisam, A.W.; Rezania, S.; Kanwar, S.S.; Singh, L. Characterization of Flavonoids from Fern Cheilanthes Tenuifolia and Evaluation of Antioxidant, Antimicrobial and Anticancer Activities. *J. King Saud Univ.-Sci.* **2018**, *30*, 425–432. [CrossRef]
42. de Souza, P.O.; Bianchi, S.E.; Figueiró, F.; Heimfarth, L.; Moresco, K.S.; Gonçalves, R.M.; Hoppe, J.B.; Klein, C.P.; Salbego, C.G.; Gelain, D.P.; et al. Anticancer Activity of Flavonoids Isolated from Achyrocline Satureioides in Gliomas Cell Lines. *Toxicol. Vitr.* **2018**, *51*, 23–33. [CrossRef]
43. Hassan, A.H.E.; Lee, K.T.; Lee, Y.S. Flavone-Based Arylamides as Potential Anticancers: Design, Synthesis and in Vitro Cell-Based/Cell-Free Evaluations. *Eur. J. Med. Chem.* **2020**, *187*, 111965. [CrossRef] [PubMed]
44. Hassan, A.H.E.; Choi, E.; Yoon, Y.M.; Lee, K.W.; Yoo, S.Y.; Cho, M.C.; Yang, J.S.; Kim, H.I.; Hong, J.Y.; Shin, J.S.; et al. Natural Products Hybrids: 3,5,4′-Trimethoxystilbene-5,6,7-Trimethoxyflavone Chimeric Analogs as Potential Cytotoxic Agents against Diverse Human Cancer Cells. *Eur. J. Med. Chem.* **2019**, *161*, 559–580. [CrossRef] [PubMed]

45. Bektic, J.; Guggenberger, R.; Spengler, B.; Christoffel, V.; Pelzer, A.; Berger, A.P.; Ramoner, R.; Bartsch, G.; Klocker, H. The Flavonoid Apigenin Inhibits the Proliferation of Prostatic Stromal Cells via the MAPK-Pathway and Cell-Cycle Arrest in G1/S. *Maturitas* **2006**, *55* (Suppl. 1), 37–46. [CrossRef]
46. Kim, S.H.; Kang, J.G.; Kim, C.S.; Ihm, S.H.; Choi, M.G.; Yoo, H.J.; Lee, S.J. Suppression of AKT Potentiates Synergistic Cytotoxicity of Apigenin with TRAIL in Anaplastic Thyroid Carcinoma Cells. *Anticancer Res.* **2015**, *35*, 6529–6538.
47. Kim, S.H.; Kang, J.G.; Kim, C.S.; Ihm, S.H.; Choi, M.G.; Yoo, H.J.; Lee, S.J. Akt Inhibition Enhances the Cytotoxic Effect of Apigenin in Combination with PLX4032 in Anaplastic Thyroid Carcinoma Cells Harboring BRAF V600E. *J. Endocrinol. Investig.* **2013**, *36*, 1099–1104. [CrossRef]
48. Zhao, G.; Han, X.; Cheng, W.; Ni, J.; Zhang, Y.; Lin, J.; Song, Z. Apigenin Inhibits Proliferation and Invasion, and Induces Apoptosis and Cell Cycle Arrest in Human Melanoma Cells. *Oncol. Rep.* **2017**, *37*, 2277–2285. [CrossRef]
49. Hua, F.; Li, C.H.; Chen, X.G.; Liu, X.P. Daidzein Exerts Anticancer Activity towards SKOV3 Human Ovarian Cancer Cells by Inducing Apoptosis and Cell Cycle Arrest, and Inhibiting the Raf/MEK/ERK Cascade. *Int. J. Mol. Med.* **2018**, *41*, 3485–3492. [CrossRef]
50. Zheng, Z.P.; Yan, Y.; Xia, J.; Zhang, S.; Wang, M.; Chen, J.; Xu, Y. A Phenylacetaldehyde-Flavonoid Adduct, 8-C-(E-Phenylethenyl)-Norartocarpetin, Exhibits Intrinsic Apoptosis and MAPK Pathways-Related Anticancer Potential on HepG2, SMMC-7721 and QGY-7703. *Food Chem.* **2016**, *197*, 1085–1092. [CrossRef]
51. Hseu, Y.C.; Lee, M.S.; Wu, C.R.; Cho, H.J.; Lin, K.Y.; Lai, G.H.; Wang, S.Y.; Kuo, Y.H.; Kumar, K.J.S.; Yang, H.L. The Chalcone Flavokawain b Induces G2/M Cell-Cycle Arrest and Apoptosis in Human Oral Carcinoma HSC-3 Cells through the Intracellular ROS Generation and Downregulation of the Akt/P38 MAPK Signaling Pathway. *J. Agric. Food Chem.* **2012**, *60*, 2385–2397. [CrossRef]
52. Lee, K.W.; Kang, N.J.; Rogozin, E.A.; Kim, H.G.; Cho, Y.Y.; Bode, A.M.; Lee, H.J.; Surh, Y.J.; Bowden, G.T.; Dong, Z. Myricetin Is a Novel Natural Inhibitor of Neoplastic Cell Transformation and MEK1. *Carcinogenesis* **2007**, *28*, 1918–1927. [CrossRef] [PubMed]
53. Hou, D.X.; Kumamoto, T. Flavonoids as Protein Kinase Inhibitors for Cancer Chemoprevention: Direct Binding and Molecular Modeling. *Antioxid. Redox Signal.* **2010**, *13*, 691–719. [CrossRef] [PubMed]
54. Miyata, Y.; Sato, T.; Imada, K.; Dobashi, A.; Yano, M.; Ito, A. A Citrus Polymethoxyflavonoid, Nobiletin, Is a Novel MEK Inhibitor That Exhibits Antitumor Metastasis in Human Fibrosarcoma HT-1080 Cells. *Biochem. Biophys. Res. Commun.* **2008**, *366*, 168–173. [CrossRef]
55. Chen, H.; Yao, K.; Nadas, J.; Bode, A.M.; Malakhova, M.; Oi, N.; Li, H.; Lubet, R.A.; Dong, Z. Prediction of Molecular Targets of Cancer Preventing Flavonoid Compounds Using Computational Methods. *PLoS ONE* **2012**, *7*, e38261. [CrossRef]
56. Alessi, D.R.; Cuenda, A.; Cohen, P.; Dudley, D.T.; Saltiel, A.R. PD 098059 Is a Specific Inhibitor of the Activation of Mitogen-Activated Protein Kinase Kinase in Vitro and in Vivo. *J. Biol. Chem.* **1995**, *270*, 27489–27494. [CrossRef] [PubMed]
57. Ferreira, L.G.; Dos Santos, R.N.; Oliva, G.; Andricopulo, A.D. Molecular Docking and Structure-Based Drug Design Strategies. *Molecules* **2015**, *20*, 13384–13421. [CrossRef]
58. Suhail, M.; Parveen, A.; Husain, A.; Rehan, M. Exploring Inhibitory Mechanisms of Green Tea Catechins as Inhibitors of a Cancer Therapeutic Target, Nuclear Factor-KB (NF-KB). *Biosci. Biotechnol. Res. Asia* **2019**, *16*, 715–723. [CrossRef]
59. Rehan, M. Anticancer Compound XL765 as PI3K/MTOR Dual Inhibitor: A Structural Insight into the Inhibitory Mechanism Using Computational Approaches. *PLoS ONE* **2019**, *14*, e0219180. [CrossRef]
60. Rehan, M.; Mostafa, M. Virtual Screening of 1,4-Naphthoquinone Derivatives for Inhibition of a Key Cancer Signaling Protein, AKT1 Kinase. *Anticancer Res.* **2019**, *39*, 3823–3833. [CrossRef]
61. Rehan, M.; Bajouh, O.S. Virtual Screening of Naphthoquinone Analogs for Potent Inhibitors against the Cancer-Signaling PI3K/AKT/MTOR Pathway. *J. Cell. Biochem.* **2018**, *120*, 1328–1339. [CrossRef]
62. Rehan, M. An Anti-Cancer Drug Candidate OSI-027 and Its Analog as Inhibitors of MTOR: Computational Insights Into the Inhibitory Mechanisms. *J. Cell. Biochem.* **2017**, *118*, 4558–4567. [CrossRef] [PubMed]
63. Rehan, M. A Structural Insight into the Inhibitory Mechanism of an Orally Active PI3K/MTOR Dual Inhibitor, PKI-179 Using Computational Approaches. *J. Mol. Graph. Model.* **2015**, *62*, 226–234. [CrossRef] [PubMed]
64. Rehan, M.; Beg, M.A.; Parveen, S.; Damanhouri, G.A.; Zaher, G.F. Computational Insights into the Inhibitory Mechanism of Human AKT1 by an Orally Active Inhibitor, MK-2206. *PLoS ONE* **2014**, *9*, e109705. [CrossRef] [PubMed]
65. Jamal, M.S.; Parveen, S.; Beg, M.A.; Suhail, M.; Chaudhary, A.G.A.; Damanhouri, G.A.; Abuzenadah, A.M.; Rehan, M. Anticancer Compound Plumbagin and Its Molecular Targets: A Structural Insight into the Inhibitory Mechanisms Using Computational Approaches. *PLoS ONE* **2014**, *9*, e87309. [CrossRef]
66. El-Demerdash, A.; Al-Karmalawy, A.A.; Abdel-Aziz, T.M.; Elhady, S.S.; Darwish, K.M.; Hassan, A.H.E. Investigating the Structure-Activity Relationship of Marine Natural Polyketides as Promising SARS-CoV-2 Main Protease Inhibitors. *RSC Adv.* **2021**, *11*, 31339–31363. [CrossRef]
67. Elkamhawy, A.; Kim, N.Y.; Hassan, A.H.E.; Park, J.E.; Paik, S.; Yang, J.E.; Oh, K.S.; Lee, B.H.; Lee, M.Y.; Shin, K.J.; et al. Thiazolidine-2,4-Dione-Based Irreversible Allosteric IKK-β Kinase Inhibitors: Optimization into in Vivo Active Anti-Inflammatory Agents. *Eur. J. Med. Chem.* **2020**, *188*, 111955. [CrossRef]
68. Elkamhawy, A.; Kim, N.Y.; Hassan, A.H.E.; Park, J.E.; Yang, J.E.; Elsherbeny, M.H.; Paik, S.; Oh, K.S.; Lee, B.H.; Lee, M.Y.; et al. Optimization Study towards More Potent Thiazolidine-2,4-Dione IKK-β Modulator: Synthesis, Biological Evaluation and in Silico Docking Simulation. *Bioorg. Chem.* **2019**, *92*, 103261. [CrossRef]

69. Hashemzadeh, S.; Ramezani, F.; Rafii-Tabar, H. Study of Molecular Mechanism of the Interaction Between MEK1/2 and Trametinib with Docking and Molecular Dynamic Simulation. *Interdiscip. Sci. Comput. Life Sci.* **2019**, *11*, 115–124. [CrossRef]
70. Ya'u Ibrahim, Z.; Uzairu, A.; Shallangwa, G.; Abechi, S. Molecular Docking Studies, Drug-Likeness and in-Silico ADMET Prediction of Some Novel β-Amino Alcohol Grafted 1,4,5-Trisubstituted 1,2,3-Triazoles Derivatives as Elevators of P53 Protein Levels. *Sci. Afr.* **2020**, *10*, e00570. [CrossRef]
71. Eswar, N.; Webb, B.; Marti-Renom, M.A.; Madhusudhan, M.S.; Eramian, D.; Shen, M.; Pieper, U.; Sali, A. Comparative Protein Structure Modeling Using Modeller. *Curr. Protoc. Bioinform.* **2006**, *15*, 5–6. [CrossRef]
72. Zhao, Z.; Xie, L.; Bourne, P.E. Insights into the Binding Mode of MEK Type-III Inhibitors. A Step towards Discovering and Designing Allosteric Kinase Inhibitors across the Human Kinome. *PLoS ONE* **2017**, *12*, e0179936. [CrossRef] [PubMed]
73. Morris, G.M.; Ruth, H.; Lindstrom, W.; Sanner, M.F.; Belew, R.K.; Goodsell, D.S.; Olson, A.J. Software News and Updates AutoDock4 and AutoDockTools4: Automated Docking with Selective Receptor Flexibility. *J. Comput. Chem.* **2009**, *30*, 2785–2791. [CrossRef] [PubMed]
74. Lipinski, C.A.; Lombardo, F.; Dominy, B.W.; Feeney, P.J. Experimental and Computational Approaches to Estimate Solubility and Permeability in Drug Discovery and Development Settings. *Adv. Drug Deliv. Rev.* **1996**, *64*, 4–17. [CrossRef]
75. O'Boyle, N.M.; Banck, M.; James, C.A.; Morley, C.; Vandermeersch, T.; Hutchison, G.R. Open Babel: An open chemical toolbox. *J. Cheminform.* **2011**, *3*, 33. [CrossRef] [PubMed]
76. Trott, O.; Olson, A.J. AutoDock Vina: Improving the Speed and Accuracy of Docking with a New Scoring Function, Efficient Optimization, and Multithreading. *J. Comput. Chem.* **2009**, *31*, 455–461. [CrossRef]
77. Laskowski, R.A.; Swindells, M.B. LigPlot+: Multiple Ligand-Protein Interaction Diagrams for Drug Discovery. *J. Chem. Inf. Model.* **2011**, *51*, 2778–2786. [CrossRef]
78. Lua, R.C.; Lichtarge, O. PyETV: A PyMOL Evolutionary Trace Viewer to Analyze Functional Site Predictions in Protein Complexes. *Bioinformatics* **2010**, *26*, 2981–2982. [CrossRef]
79. Wang, R.; Lai, L.; Wang, S. Further Development and Validation of Empirical Scoring Functions for Structure-Based Binding Affinity Prediction. *J. Comput. Aided Mol. Des.* **2002**, *16*, 11–26. [CrossRef]
80. Roy, A.S.; Tripathy, D.R.; Chatterjee, A.; Dasgupta, S. A Spectroscopic Study of the Interaction of the Antioxidant Naringin with Bovine Serum Albumin. *J. Biophys. Chem.* **2010**, *1*, 141–152. [CrossRef]
81. Hubbard, S.J.; Thornton, J.M. *'NACCESS', Computer Program*; Department of Biochemistry and Molecular Biology, University College London: London, UK, 1993.
82. Pires, D.E.V.; Blundell, T.L.; Ascher, D.B. PkCSM: Predicting Small-Molecule Pharmacokinetic and Toxicity Properties Using Graph-Based Signatures. *J. Med. Chem.* **2015**, *58*, 4066–4072. [CrossRef]
83. Abraham, M.J.; Murtola, T.; Schulz, R.; Páll, S.; Smith, J.C.; Hess, B.; Lindah, E. Gromacs: High Performance Molecular Simulations through Multi-Level Parallelism from Laptops to Supercomputers. *SoftwareX* **2015**, *1*, 19–25. [CrossRef]

pharmaceuticals

Article

Structure-Based Identification and Biological Characterization of New NAPRT Inhibitors

Jorge Franco [1,2], Francesco Piacente [3], Melanie Walter [4], Simone Fratta [5], Moustafa Ghanem [2], Andrea Benzi [3], Irene Caffa [2], Alexander V. Kurkin [6], Andrea Altieri [6,7], Patrick Herr [4], Macarena Martínez-Bailén [5], Inmaculada Robina [5], Santina Bruzzone [3], Alessio Nencioni [2,8] and Alberto Del Rio [1,9,*]

1 Innovamol Consulting Srl, 41126 Modena, Italy; jorfrasus1995@gmail.com
2 Department of Internal Medicine and Medical Specialties, University of Genoa, 16132 Genoa, Italy; moustafa.ghanem@edu.unige.it (M.G.); irene.caffa@libero.it (I.C.); alessio.nencioni@unige.it (A.N.)
3 Department of Experimental Medicine, University of Genoa, 16132 Genoa, Italy; francesco.piacente@unige.it (F.P.); andreeabenzi@gmail.com (A.B.); santina.bruzzone@unige.it (S.B.)
4 Department of Oncology and Metabolism, University of Sheffield, Sheffield S10 2TN, UK; mwalter3@sheffield.ac.uk (M.W.); p.herr@sheffield.ac.uk (P.H.)
5 Department of Organic Chemistry, University of Seville, 41012 Seville, Spain; sfratta@us.es (S.F.); mmartinez45@us.es (M.M.-B.); robina@us.es (I.R.)
6 Department of Chemistry, Lomonosov Moscow State University, 119991 Moscow, Russia; indolester@gmail.com (A.V.K.); aaltieri@edasascientific.com (A.A.)
7 EDASA Scientific Srls, 66050 San Salvo, Italy
8 IRCCS Ospedale Policlinico San Martino, 16132 Genoa, Italy
9 Institute of Organic Synthesis and Photoreactivity (ISOF), National Research Council (CNR), 40129 Bologna, Italy
* Correspondence: alberto.delrio@gmail.com

Citation: Franco, J.; Piacente, F.; Walter, M.; Fratta, S.; Ghanem, M.; Benzi, A.; Caffa, I.; Kurkin, A.V.; Altieri, A.; Herr, P.; et al. Structure-Based Identification and Biological Characterization of New NAPRT Inhibitors. *Pharmaceuticals* **2022**, *15*, 855. https://doi.org/10.3390/ph15070855

Academic Editors: Marialuigia Fantacuzzi and Mariangela Agamennone

Received: 21 June 2022
Accepted: 9 July 2022
Published: 12 July 2022

Abstract: NAPRT, the rate-limiting enzyme of the Preiss–Handler NAD biosynthetic pathway, has emerged as a key biomarker for the clinical success of NAMPT inhibitors in cancer treatment. Previous studies found that high protein levels of NAPRT conferred resistance to NAMPT inhibition in several tumor types whereas the simultaneous blockade of NAMPT and NAPRT results in marked anti-tumor effects. While research has mainly focused on NAMPT inhibitors, the few available NAPRT inhibitors (NAPRTi) have a low affinity for the enzyme and have been scarcely characterized. In this work, a collection of diverse compounds was screened in silico against the NAPRT structure, and the selected hits were tested through cell-based assays in the NAPRT-proficient OVCAR-5 ovarian cell line and on the recombinant hNAPRT. We found different chemotypes that efficiently inhibit the enzyme in the micromolar range concentration and for which direct engagement with the target was verified by differential scanning fluorimetry. Of note, the therapeutic potential of these compounds was evidenced by a synergistic interaction between the NAMPT inhibitor FK866 and the new NAPRTi in terms of decreasing OVCAR-5 intracellular NAD levels and cell viability. For example, compound IM29 can potentiate the effect of FK866 of more than two-fold in reducing intracellular NAD levels. These results pave the way for the development of a new generation of human NAPRTi with anticancer activity.

Keywords: NAPRT inhibitors; NAD biosynthesis; Preiss–Handler pathway; OVCAR-5; bioactive molecules; NAMPT; molecular design; virtual screening

1. Introduction

Nicotinamide adenine dinucleotide (NAD) is a vital pyridine nucleotide. The first role that was discovered for NAD^+ and its phosphorylated form ($NADP^+$) was as an essential coenzyme in redox reactions that are involved in cell energy and anabolic metabolism. By exchanging hydride, $NAD(P)^+$ is constantly shuttling between its oxidized and reduced forms in hundreds of enzymatic reactions that take part in key pathways in mammalian cells, such as glycolysis, tricarboxylic acid cycle (TCA cycle), oxidative phosphorylation, and serine biosynthesis [1–3]. In addition, marked cell regulatory properties have been ascribed to NAD by acting as a substrate for several families of enzymes, which always release nicotinamide (Nam) as a result of NAD degradation [1,4–6]. Indeed, NAD is consumed in post-translational modifications of target proteins by mono- and poly-(ADP–ribose) polymerases (PARPs) and sirtuins (SIRT1-7), the last endowed with protein deac(et)ylase activity [7,8]. NAD is also the precursor of the Ca^{2+}-mobilizing second messenger cyclic ADP-ribose (cADPR), produced by the ectoenzymes CD38 and CD157 [9]. The enzyme sterile alpha and TIR motif-containing 1 (SARM1) exerts NAD-cleavage activity in neurons and represents a new family of NAD-consuming enzymes [10]. The activities of these enzymes are modulated by NAD availability and regulate a series of fundamental cellular processes including DNA repair, apoptosis, cell metabolism, cell cycle progression, and immune responses [11].

Unlike a redox cofactor, NAD is consumed when acting as a substrate. Therefore, continuous NAD biosynthesis is required in normal human tissues to preserve NAD homeostasis and thus health [12]. Due to aberrant metabolism, cell growth, and proliferation, tumor cells require higher NAD production with respect to healthy tissues to support the increased activity of NAD-degrading enzymes [13]. Therefore, interfering with the NAD biosynthetic machinery was conceived as a promising therapeutic strategy against cancer [11,14]. The depletion of NAD strongly affects multiple cellular metabolic pathways, leads to a rapid decline in adenosine triphosphate (ATP) levels, and ultimately causes cancer cell death [15]. For a more detailed overview of the in vitro and in vivo effects of chemical agents targeting NAD biosynthesis in cancer cells, we refer the readers to recent review articles [16].

There are three main pathways contributing to NAD biosynthesis in mammals: the de novo pathway with tryptophan as a NAD precursor, the Preiss–Handler pathway, which utilizes nicotinic acid (NA) as a starting block, and the nicotinamide (Nam) salvage pathway (Figure 1) [17]. In addition, the ribosylated precursors nicotinamide riboside (NR) and nicotinic acid riboside (NAR) represent additional forms of vitamin B3 that can be converted into NAD.

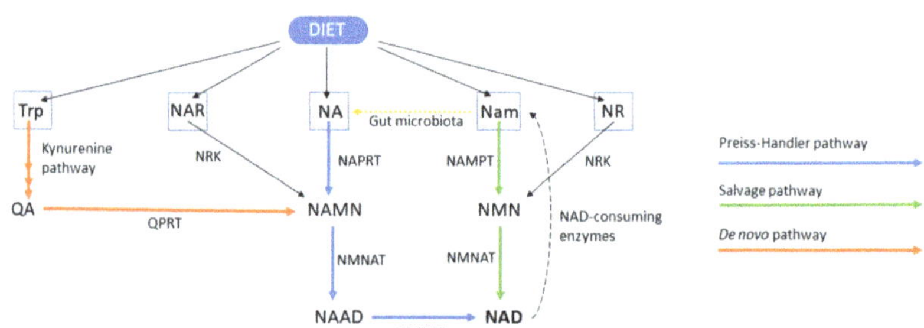

Figure 1. Graphical depiction of NAD biosynthesis in mammals. Trp, tryptophan; NA, nicotinic acid; NAR, nicotinic acid riboside; NR, nicotinamide riboside; Nam, nicotinamide; QA, quinolinic acid; NAMN, nicotinic acid mononucleotide; NMN, nicotinamide mononucleotide; NAAD, nicotinic acid adenine dinucleotide; NAD, nicotinamide adenine dinucleotide; QPRT, quinolinate phosphoribosyltransferase; NAPRT, nicotinate phosphoribosyltransferase; NAMPT, nicotinamide phosphoribosyltransferase; NRK, nicotinamide riboside kinase; NMNAT, nicotinamide mononucleotide adenylyltransferase; NADSYN, nicotinamide adenine dinucleotide synthetase.

The Nam salvage pathway plays a key role in maintaining NAD homeostasis in mammalian cells [18]. Duarte-Pereira and colleagues observed that the gene of the rate-limiting enzyme of this pathway, nicotinamide phosphoribosyltransferase (NAMPT), was ubiquitously expressed at the mRNA level in all human normal tissues and tumors that were studied [19]. Moreover, NAD-consuming enzymes release nicotinamide as a by-product, which makes Nam the most accessible NAD precursor. Accordingly, research focusing on NAD-lowering as an anticancer strategy has largely focused on NAMPT and led to the identification of numerous inhibitors, including FK866 and CHS-828 [20–22]. Despite the potent antitumor activity of these agents in preclinical models, the efficacy of these compounds in clinical trials has been disappointing. In addition, thrombocytopenia and gastrointestinal symptoms arose as dose-limiting toxicities [23,24]. Studies show that the limited clinical activity of these agents reflects the overexpression of enzymes from the Preiss–Handler pathway, at least in a subset of human malignancies [25,26].

Human nicotinate phosphoribosyltransferase (hNAPRT, Uniprot: Q6XQN6) is the rate-limiting enzyme of the Preiss–Handler pathway. The enzyme catalyzes the conversion of NA and 5-phosphoribosyl-1-pyrophosphate (PRPP) to nicotinic acid mononucleotide (NAMN) and pyrophosphate (PPi) in an ATP-dependent manner [27]. The Preiss–Handler pathway continues with the adenylation of NAMN catalyzed by NMNAT1-3 and ends with the amidation of NAAD into NAD, which is catalyzed by NAD synthetase (NADSYN). Human NAPRT belongs to the Type II phosphoribosyltransferase family of functional dimeric proteins that are involved in NAD biosynthesis, together with quinolinate phosphoribosyltransferase (QPRT) and NAMPT. The crystal structure of hNAPRT was solved by Marletta et al. and evidenced that the NAPRT monomer consists of an irregular α/β barrel domain and of a second open-faced sandwich domain [28]. Despite the low sequence similarity between hNAPRT and its bacterial homologs, the main amino acids that are involved in the recognition and stabilization of NAPRT substrates are strictly conserved, as demonstrated through site-directed mutagenesis experiments [29].

NAPRT displays marked tissue and tumor specificity in terms of expression and its regulation mechanisms and is mostly present in several catabolic healthy mammalian tissues including the heart, kidney, liver, and small intestine [19,25,30–33]. In tissues that express the NAPRT protein, NA is the preferred precursor of NAD. Accordingly, the NAD pool of HEK293 cells was dramatically increased when these were cultured in NA-rich conditions, whereas similar doses of Nam resulted in a much lower effect [33]. This observation is likely related to the fact that, unlike NAMPT, NAPRT activity is not inhibited by NAD [33]. Interestingly, in addition to the role of NAMPT and NAPRT as intracellular NAD-producing enzymes (mostly located in the nucleus and the cytoplasm), NAMPT and NAPRT also exist as extracellular proteins, which exert pro-inflammatory and pro-tumorigenic effects [34,35].

With respect to the amplification of the NAPRT gene and to its expression, these are very variable in human tumors [19,25,31]. Tumors originating from normal tissues that highly express NAPRT were found to amplify the NAPRT gene at a high frequency (and to express it at high levels as a result), whereas tumors arising from tissues that do not express NAPRT will strongly rely on NAMPT activity for NAD biosynthesis and for cell survival [31]. In ovarian, prostate, breast, and pancreatic cancers, NAPRT was to be found upregulated [25]. On the contrary, gastric, renal, and colorectal carcinoma, as well as several leukemia cell lines, were reported to have low or no NAPRT expression [19]. Among the mechanisms that regulate NAPRT expression, hypermethylation of the NAPRT gene promoter leads to gene silencing and has been found in several NAPRT-negative tumors. NAPRT promoter hypermethylation in cancer is frequently associated with mutations in the protein phosphatase Mg^{2+}/Mn^{2+}-dependent 1D (PPM1D) or isocitrate dehydrogenase 1 (IDH1) genes, as well as with the epithelial-mesenchymal transition (EMT)-subtype of gastric cancer [36–38]. Due to their dependency on the Nam salvage pathway for survival, NAPRT-deficient cancers are extremely sensitive to treatment with NAMPT inhibitors [39,40]. On the other hand, NAPRT upregulation confers resistance to NAMPT inhibitors to many malignancies. Piacente et al. demonstrated that NAPRT-proficient ovarian and pancreatic cancers are resistant to FK866, whereas NAPRT downregulation through gene silencing or chemical inhibition with 2-hydroxynicotinic acid (2-HNA) sensitized tumor cells to NAMPT inhibitors both in vitro and in vivo [25]. In line with these results, another study highlighted that sensitivity to NAMPT inhibition in several ovarian cancer cell lines was inversely proportional to NAPRT expression [41]. Despite its usefulness in basic research as NAPRTi, the low molecular weight and high functionality of 2-hydroxynicotinic acid suggest that this compound could have non-specific effects. Together with its low potency, such features make it unlikely that 2-HNA could be utilized as a NAPRTi in the clinic.

To the best of our knowledge, additional NAPRT inhibitors were reported in the 1970s, including nicotinic acid analogs and non-steroidal anti-inflammatory agents such as flufenamic acid, salicylic acid, mefenamic acid, and phenylbutazone [42–44]. Unfortunately, the available data, which were mostly obtained by monitoring NAPRT activity in human platelet lysates, suggest a weak inhibitory activity for these agents. Thus, the expected low specificity of 2-HNA combined with the limited activity and characterization of the other reported NAPRTi mandate the search for new, potent NAPRTi. In this work, we identified several new small-molecule NAPRTi which, upon further compound optimization, could lead to a new generation of molecules that sensitize NAPRT-proficient malignancies to NAMPT inhibitors and, possibly, to other anticancer agents.

2. Results

2.1. Sensitization of OVCAR-5 Cells to FK866

As described in the experimental section, a structure-based virtual screening was run on NAPRT to identify putative inhibitors. The docking results were thoroughly inspected and resulted in a selection of 62 compounds (Table 1 and Table S1 of Supplementary Information) to be tested employing in vitro assays on the ovarian cancer cell line OVCAR-5 that expresses high levels of NAPRT.

Table 1. The number of in silico screening hit compounds that were obtained from each provider after visual inspection of the top-score docking binding poses.

Provider	Number of Compounds
University of Seville	22
MCULE	16
EDASA	13
Prestwick	11

As noted in the methodology section, a 50 mM stock solution in DMSO was prepared for each compound. Compound IM 28 could not be solubilized in DMSO at the specified concentration and was excluded from the study.

The first assay that we performed to test the putative NAPRT inhibitors was based on the principle that NAPRT-expressing OVCAR-5 cells are resistant to FK866 whereas the simultaneous chemical inhibition of NAPRT with 2-HNA renders the cells sensitive to the NAMPT inhibitor due to the cooperation between FK866 and 2-HNA in depleting the intracellular NAD pool [25]. We observed that, in line with the work that was conducted by Piacente et al., OVCAR-5 cells withstood 72 h treatments with 100 nM FK866 or 1 mM 2-HNA, whereas co-treatment with both compounds at the specified concentrations resulted in a marked synergistic effect exerting pronounced cell death (Figure 2). Thus, 2-HNA was used throughout the study as a positive control in our cell viability assay and any test compound resembling its activity was considered a potential NAPRT inhibitor. It is worth mentioning that, to reflect the physiological contribution of the Preiss–Handler pathway to NAD biosynthesis, all cell-based assays that are presented in this study were performed under a nicotinic acid concentration of 0.3 μM in the cell culture medium (Figure S1).

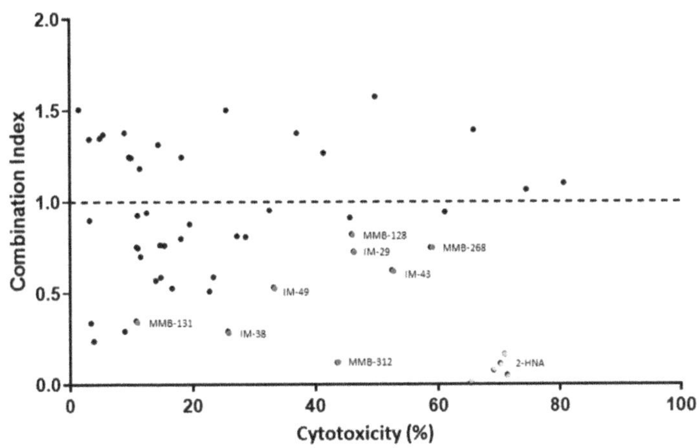

Figure 2. CI of compounds that were tested in combination with FK866 on OVCAR-5 cells. The cells were cultured in RPMI 1640 medium containing test compounds at a 100 μM concentration with and without 100 nM FK866 and the cell viability was determined following 72 h treatments. The data are shown as CI vs. cytotoxicity exerted on cells. Blue dots represent the reference NAPRT inhibitor, 2-HNA. Red dots belong to the best performing test compounds.

We decided to use combination index (CI) as a measure to determine the degree of interaction between FK866 and the test compounds i.e., to identify potential new NAPRT inhibitors. A CI < 1 indicates a synergistic interaction between the test compound and FK866 in decreasing cell viability. A CI = 1 and a CI > 1 are indicative of an additive and of an antagonistic effect, respectively.

Figure 2 shows that the simultaneous administration of FK866 and 2-HNA produced the lowest CI out of all the combinations that were tested (CI = 0.08) and marked cytotoxicity. Among all the tested compounds, eight (represented by red dots in Figure 2) were considered to be especially interesting for their low CI or their high cytotoxicity when they were combined with FK866. Compounds IM 29, IM 43, MMB-128, and MMB-268 exhibited a CI ranging between 0.62 and 0.82 and high cytotoxicity in combination with FK866 and these results might suggest other interactions within the cells while some degree of synergism (CI < 1) with FK866 was evident, suggesting the inhibition of the Preiss–Handler NAD biosynthetic pathway.

Compounds IM 38, IM 49, MMB-131, and MMB-312 produced less pronounced cytotoxicity when they were combined with FK866. However, this subset of compounds showed especially low combination indexes ($0.12 \leq CI \leq 0.53$) and, therefore, a larger synergism with FK866.

These eight compounds were all considered promising candidates for NAPRT inhibition and were selected for further in vitro assays.

2.2. *Effect of Test Compounds on Intracellular NAD Levels*

The assay that was performed on OVCAR-5 cell viability led to a set of potential NAPRT inhibitors, namely IM 29, IM 38, IM 43, IM 49, MMB-128, MMB-131, MMB-268, and MMB-312. In order to confirm that the cited compounds sensitized OVCAR-5 cells to the NAMPT inhibitor FK866 by interfering with NAD biosynthesis, we quantified the intracellular NAD^+ levels in OVCAR-5 cells after 20-h treatments with these compounds with or without FK866.

As Figure 3 shows, treatment with 30 nM FK866 alone decreased the NAD^+ levels by 24%. The effect of FK866 was strongly enhanced when the drug was combined with 1 mM 2-HNA, yielding an 86 % drop in the intracellular NAD^+ concentration, thus evidencing marked synergism between the two compounds in depleting NAD+. This observation is in line with the fact that the combination of FK866 and 2-HNA achieved the lowest CI and the highest cytotoxicity in the cell viability studies.

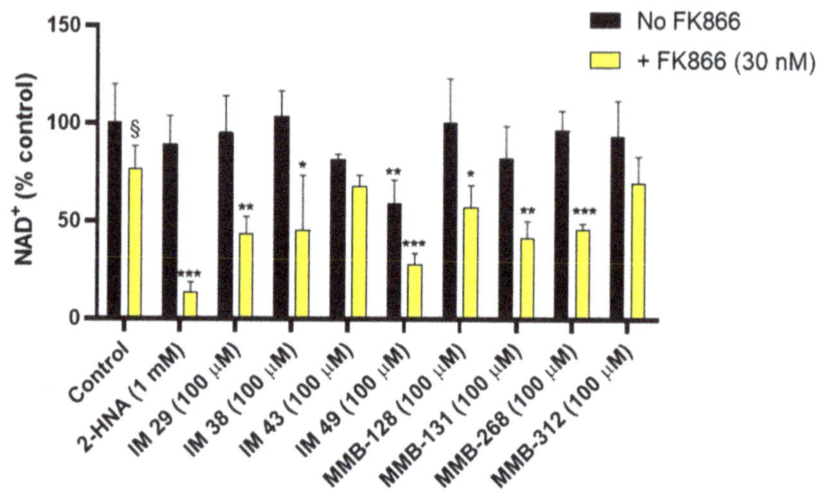

Figure 3. A subset of test compounds cooperates synergistically with FK866 in decreasing intracellular NAD+ levels in OVCAR-5 cells. The compounds were administered to cells alone or in combination with FK866 for 20 h. Quantification of NAD+ was performed via an enzyme cycling assay and normalized to cell lysate protein content. The results are the mean ± SD of two technical replicates with two biological replicates each. *, $p < 0.05$; **, $p < 0.01$; ***, $p < 0.001$ vs. the respective control, i.e., FK866-untreated control for the test compounds that were administered alone to cells, and FK866-treated control for the combinations of test compound and FK866; §, $p < 0.05$ versus FK866-untreated, control cells.

In keeping with the cell viability experiments, the test compounds IM 29, IM 38, MMB-128, MMB-131, and MMB-268 were found to mimic the effect of 2-HNA on intracellular NAD+ levels and thus represent a set of potential NAPRT inhibitors. Indeed, co-treatments with the cited compounds at a 100 µM concentration and 30 nM FK866 produced a marked drop in the NAD+ levels (42.9–56.4%). A similar decrease in NAD+ was obtained in response to the NAMPT inhibitor combined with compound IM 49. Nevertheless, compound IM 49 by itself strongly reduced the OVCAR-5 intracellular NAD+ levels. Therefore, a lower degree of synergism was observed between this test compound and FK866. Possibly, IM 49 can interfere with other enzyme(s) that are involved in NAD synthesis in addition to hNAPRT.

Besides the test compounds that effectively cooperate with FK866 in decreasing NAD+ levels, the co-administration of compounds IM 43 and MMB-312 with FK866 only yielded additive effects in terms of NAD+ reduction in OVCAR-5 cells. As depicted in Figure 3, cotreatments with the cited compounds and FK866 did not cause a major reduction of the NAD+ levels. These results appear not to be in line with our cell viability assay, where IM 43 exerted high cytotoxicity in combination with FK866 in OVCAR-5 cells, while MMB-312 achieved the lowest CI value (CI = 0.12) out of all the tested compounds. These results suggest that other mechanisms may come into play to justify the synergy between IM 43 or MMB-312 and FK866, which need to be addressed in further studies.

2.3. Inhibition of the Recombinant hNAPRT

The six test compounds that were found to exert a synergistic effect with FK866 on OVCAR-5 intracellular NAD+ levels were tested in a biochemical assay with recombinant hNAPRT as inhibitors of the enzyme catalytic activity. These inhibition studies resulted in the identification of five new hNAPRT inhibitors with IC_{50} values in the micromolar range. Compound IM 29 exhibited the highest activity against the formation of nicotinic acid mononucleotide with an IC_{50} of 160 μM (Figure 4). Comparable results were obtained for compounds IM 38, IM 49, MMB-128, and MMB-131, with estimated IC_{50} in the 200–300 μM range. This enzymatic assay confirmed that the observed cooperation between FK866 and the tested compounds in decreasing OVCAR-5 cell viability and intracellular NAD^+ levels was due to the ability of the compounds to inhibit hNAPRT. Noteworthy, compound MMB-268 showed negligible inhibition of hNAPRT (data not shown).

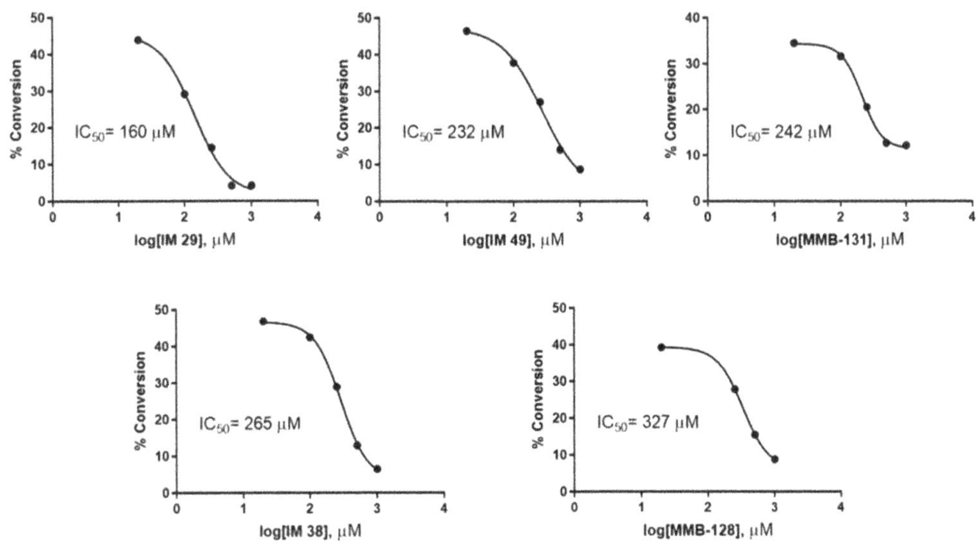

Figure 4. Compounds IM 29, IM 49, MMB-131, IM 38, and MMB-128 inhibit recombinant hNAPRT in the μM range. The test compounds were added at different concentrations to reaction mixtures containing hNAPRT and substrates and the half maximal inhibitory concentration (IC_{50}) was obtained for each compound by measuring the amounts of NA and NAMN that was present after the reactions.

2.4. hNAPRT Melting Temperature Experiment

With the purpose to provide support to the results that were obtained in the enzymatic activity assay, the ability of the new hNAPRT inhibitors to stabilize the protein through binding was evaluated via the measurement of the hNAPRT melting temperature in the presence or absence of the inhibitors at a 100 μM concentration. The resulting Tm shifts are depicted in Figure 5. Interestingly, a significant positive shift by approximately 0.5 °C in the Tm of hNAPRT was induced by compounds MMB-128 and MMB-131, suggesting a certain affinity of the inhibitors towards the protein that results in thermal stabilization. On the contrary, the melting temperature of hNAPRT remained unchanged when the enzyme was exposed to compounds IM 29, IM 38, or IM 49. A plausible explanation for this outcome is that larger compounds are more likely to establish a higher number of interactions and thus result in greater stabilization of a given protein. Indeed, significant Tm shifts were reached with the two largest inhibitors (MMB-128 and MMB-131).

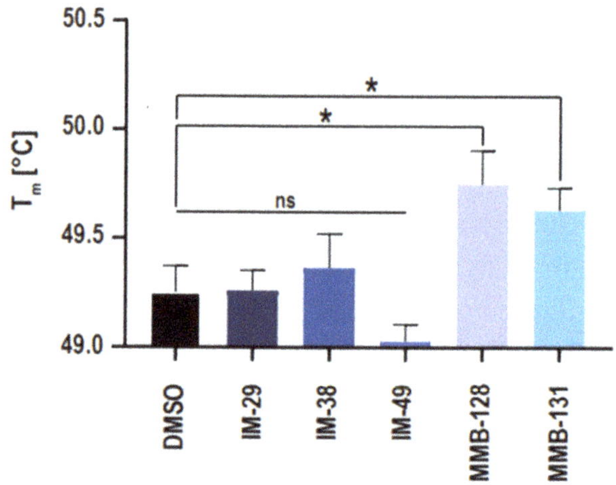

Figure 5. NAPRT inhibitors MMB-128 and MMB-131 directly engage recombinant hNAPRT. Differential scanning fluorimetry (DSF) was used to determine the thermal stabilization of hNAPRT protein upon inhibitor binding. Recombinant hNAPRT protein was exposed to 100 μM hNAPRT inhibitors over a defined temperature gradient and the melting temperature Tm was calculated for each compound. The data are shown as means ± standard error of the means (SEM) of two technical replicates with three biological replicates each. * $p < 0.05$; unpaired *t*-test.

2.5. Analysis of Inhibitor Binding Pose

Following the in vitro characterization of the new hNAPRT inhibitors, a molecular docking on the hNAPRT structure was performed to shed light on the binding mode features that are likely related to the enzyme inhibition. Docking of compound IM 29 shows that the small and rigid molecule binds deep into the active site pocket of hNAPRT and it is stabilized by several interactions with protein residues (Figure 6A,B). Remarkably, IM 29 establishes two pi-cation contacts with ARG318A, a residue that is reported to exert a key role in catalysis that is potentially hampered by IM 29 binding [28,29]. The ionization at a physiological pH of the amino group of IM 29 appears relevant for engagement and inhibition of hNAPRT. Indeed, the carboxylate side chain of GLU167A in proximity to the ammonium moiety of IM 29 indicates suitable ligand-receptor electrostatic complementarity and results in a salt bridge and short-distance hydrogen bond between the cited amino acid and the hNAPRT inhibitor.

As for compound IM 29, docking into the hNAPRT active site provides a favorable binding mode for IM 38. The ligand fits in the active site maintaining the inherent planarity of its amide conjugated system, resulting in negligible ligand strain. Electrostatic potential mapping of the protein (Figure 6C) shows that the most electronegative atoms of the inhibitor are surrounded by positively charged active site residues. As for many drug candidates in medicinal chemistry, the affinity of hNAPRT towards IM 38 is largely explainable by the occurrence of numerous hydrogen bonds upon ligand binding. The carboxylate moiety of the ligand is key for its bioactivity, participating in two H-bonds with the backbone of hNAPRT residues HIS213A and SER214A (Figure 6D). Furthermore, through its amide and ether groups, IM 38 establishes several hydrogen bonds with the amino acids ARG318A and LEU170A. The latter residue also interacts with the chlorine atom of IM 38, albeit the poor directionality of the halogen bond suggests this interaction to be of limited relevance.

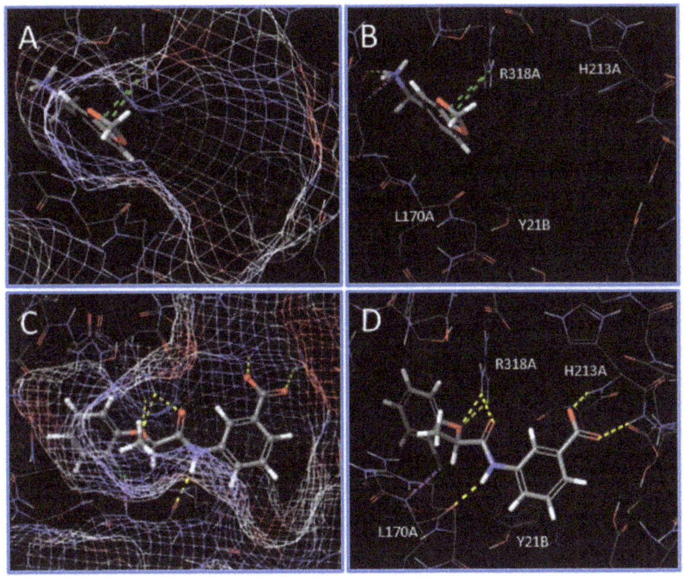

Figure 6. Binding mode that is predicted by molecular docking of compounds IM 29 (**A**,**B**) and IM 38 (**C**,**D**) in the hNAPRT active site. The analysis of the docking binding poses was performed on the academic version of Schrodinger Maestro v.2017-4. Protein is represented in thin sticks whereas ligands are depicted in thick tubes. Hydrogen bonds appear as yellow dotted lines. Salt bridges are represented by pink dotted lines. Pi-cation interactions are depicted as green dotted lines and halogen bonds are shown as purple dotted lines.

3. Discussion

Since NAD depletion emerged as a promising anticancer strategy, most of the research has focused on disabling the nicotinamide salvage pathway, which overall represents the main route to NAD biosynthesis in mammals. Indeed, several families of potent NAMPT inhibitors have emerged over the last decades with encouraging preclinical antitumor efficacy. Unfortunately, the expectations on some of these NAMPT inhibitors were not met later on in clinical trials, which showed limited clinical activity for these compounds. Piacente et al. hypothesized that alternative NAD production routes could represent a mechanism of tumor resistance to NAMPT inhibitors and demonstrated that the Preiss–Handler pathway gene, NAPRT, is frequently amplified and overexpressed in a subset of human tumors such as ovarian, breast, pancreatic, and prostate cancer. In the cited study, targeting NAPRT through silencing or chemical inhibition effectively sensitized NAPRT-expressing cancer cells to FK866 both in vitro and in vivo. In addition, the authors highlighted the need to discover new NAPRT inhibitors with increased potency with respect to 2-hydroxynicotinic acid and other active compounds that date back to the last century.

Here we report the first study that is aimed at identifying new hNAPRT inhibitors through a structure-based drug design approach with the evaluation of bioactivity with state-of-the-art in vitro assays.

As we were particularly interested in identifying active compounds in NAPRT-expressing cells, our first assay was designed as an in vitro screening to study which test compounds successfully sensitized OVCAR-5 cells to the NAMPT inhibitor FK866, essentially recreating the results that were obtained with the reference NAPRT inhibitor, 2-HNA. A total of eight compounds demonstrated synergism along with FK866 in decreasing cell viability, suggesting an impairment of the Preiss–Handler pathway. Most of these hits also showed the ability to cooperate with FK866 synergistically to decrease intracellular NAD^+ levels in OVCAR-5 cells. Specifically, such an effect on intracellular NAD^+ was observed in response to six compounds (IM 29, IM 38, IM 49, MMB-128, MMB-131, and MMB-268) when these were combined with FK866. Conversely, two compounds (IM 43 and MMB-312) showed promising results in the cell viability assay but failed to cooperate with FK866 to lower the NAD^+ levels, suggesting that their antitumor effect may reflect an off-target activity.

Following cell assays, we proceeded to elucidate the regulatory properties of the best compounds on recombinant hNAPRT. A total of five compounds (IM 29, IM 38, IM 49, MMB-128, and MMB-131) showed inhibitory properties against hNAPRT and IC_{50} in the micromolar range. The suitability of our assays is supported by the fact that only one compound (MMB-268) out of the pool of candidates that exerted cytotoxicity and intracellular NAD^+ decrease in combination with FK866 failed to inhibit hNAPRT enzymatic activity. Therefore, MMB-268 that showed promising activities on the OVCAR-5 cell line may be related to a biological target other than hNAPRT. For instance, one possibility is that MMB-268 could impair NAD synthetase (NADSYN) activity, catalyzing the conversion of NAAD into NAD independently of the Nam salvage biosynthetic pathway. NADSYN inhibition could lead to results that are comparable to hNAPRT inhibition in our in vitro cell assays. Finally, we decided to study the capability of the hNAPRT inhibitors that were herein identified to confer thermal stability to hNAPRT by DSF. Direct engagement of compounds MMB-128 and MMB-131 to hNAPRT was confirmed by positive shifts in the protein melting temperature when exposed to the inhibitors.

In summary, our findings broaden the chemical space of known hNAPRT inhibitors and pave the way for the identification of new NAD-lowering anticancer drugs. The new hNAPRT inhibitors that we report are characterized by a low molecular weight and, thus, are susceptible to undergo optimization studies shortly to increase their potency through chemical structure modifications.

4. Materials and Methods

4.1. High Throughput Virtual Screening

A high-throughput molecular docking screening was performed to discover new inhibitors of hNAPRT. The crystal structure of hNAPRT (PDB code: 4YUB) was retrieved from The Protein Data Bank [45]. This structure was prepared with the academic version of Schrodinger Maestro v.2017-4 with standard preparation procedures, that include the removal of water molecules, correct assignment of bond orders, the addition of hydrogens, and the optimization of protonation states and restrained energy minimization. A 15 Å grid centered on the enzymatically relevant active site residues LEU170A, ARG318A, and TYR21B was generated and docking virtual screening was run with the modeling tools that are offered by the Mcule platform, performing ligand preparation with Gypsum-DL using the default settings [28,29,46,47].

The docking results were thoroughly visually inspected considering widely accepted aspects in the scientific community such as quality of ligand-receptor interactions, docking score, fitting in the active site, ligand strain, and drug-likeness [48]. The most promising compounds were purchased or provided by collaborators and tested in vitro.

4.2. Virtual Screening Libraries

The chemical compounds that were subjected to docking screening were obtained from several sources. The list of libraries and providers is numbered below.

(1) Approximately 1×10^5 molecules from the "Potentially purchasable compounds" Mcule database, selected based on simple physicochemical properties.
(2) The Prestwick Chemical Library® [49].
(3) Approximately 2000 in-stock compounds from EDASA Scientific [50].
(4) A small selection of approximately 100 molecules that were synthesized by the University of Seville, some of which have already been described in the literature (see Supplementary Information for synthetic procedures and compounds characterization) [51–54].

Compounds were obtained in mg quantity and subsequently dissolved in DMSO to prepare a 50 mM stock solution. The purity of all compounds was > 95%, as declared by vendors or collaborators.

4.3. Cell Lines and Reagents

OVCAR-5 cells were obtained from the NCI-60 panel in 2015 as a kind gift from Prof. Zoppoli. The cells were passaged for less than six months before resuscitation for the experiments. Testing for mycoplasma was routinely done with the MycoAlert Mycoplasma Detection Kit (Lonza Group, Basel, Switzerland). The cells were maintained and treated with RPMI 1640 medium that was supplemented with 10% heat-inactivated FBS, penicillin (50 units/mL), and streptomycin (50 µg/mL) (Life Technologies, Monza, Italy). FK866 was kindly provided by the NIMH Chemical Synthesis and Drug Supply Program. 2-hydroxynicotinic acid was purchased (Sigma Aldrich S.r.l., Milan, Italy).

4.4. Cell Viability Assay

A total of 3×10^3 OVCAR-5 cells/well were plated in 96-well plates in the regular culture medium. 24 h later the cell medium was removed and the cells were subsequently incubated either in the regular medium that was supplemented with 0.3 µM nicotinic acid (control wells) or in the treatment medium which contained 0.3 µM nicotinic acid and combinations of 100 nM FK866, 1 mM 2-HNA, and 100 µM of test compounds. Each condition was performed in triplicate and the cells remained under treatment for a total of 72 h at 37 °C. Thereafter, the culture plates were fixed with 10% trichloroacetic acid at 4 °C for 20 min, washed with cold water, and dried overnight. The plates were stained with 0.04% sulforhodamine B (SRB) in 1% acetic acid, washed four times with 1% acetic acid to remove the unbound dye, and dried overnight. Lastly, Trizma®-base 10 mM was added to the plates and cell viability was quantified by absorbance measurements on a Tecan Infinite® 200 PRO instrument.

To quantify the extent of the interaction between FK866 and the putative NAPRT inhibitors or 2-HNA, we applied the combination index (CI) equation that is depicted below:

$$CI = \frac{mortality\ \%\ A + mortality\ \%\ B}{mortality\ \%\ (A + B)} \tag{1}$$

where: '*A*' refers to the treatment of cells with FK866 100 nM; '*B*' refers to the treatment of cells with the test compound 100 µM or 2-HNA 1 mM; '*A* + *B*' refers to the coadministration of FK866 100 nM and the test compound at 100 µM or 2-HNA 1 mM. A synergistic interaction is evidenced by CI < 1, additive effect produces CI close to 1, whereas CI > 1 corresponds to antagonistic effect.

4.5. Intracellular NAD Quantification

A total of 1×10^5 OVCAR-5 cells/well were plated in 24-well plates in the regular culture medium. 24 h later the cell medium was removed and the cells were cultured either in the regular medium that was supplemented with 0.3 µM nicotinic acid (control wells) or in the treatment medium (regular medium with 0.3 µM nicotinic acid and combinations of 30 nM FK866, 1 mM 2-HNA and 100 µM of test compounds). Each condition was performed in duplicate and the cells remained under treatment for a total of 20 h at 37 °C. Thereafter, the cell medium was removed and the cells were harvested and lysed with 0.6 M perchloric acid (PCA). Samples in PCA were neutralized by diluting the extracts in 100 mM sodium phosphate buffer (pH 8) and the total intracellular NAD^+ content was determined with a sensitive enzyme cycling assay that exploits the use of alcohol dehydrogenase [55]. The obtained NAD^+ values were normalized to cell lysate protein content that was quantified by the Bradford method.

4.6. Recombinant hNAPRT Production

The production of recombinant hNAPRT employed *E. coli* bacteria as an expression system. The human coding sequence for NAPRT was codon optimized for the expression in *E. coli* and cloned in a pET-23a vector containing a C-terminal His-tag by GenScript. A total of 5 ml of bacterial culture was grown overnight at 37 °C in the Luria–Bertani medium that was supplemented with 100 μg/mL ampicillin. The day after the culture was diluted at 1:100 in fresh medium and incubated at 25 °C. When a 0.3–0.4 OD_{600} was reached, protein expression was started at 20 °C by adding 1 mM IPTG, followed by overnight incubation.

The induced cells were harvested by mild centrifugation (5000 rpm, 10 min) in a Beckman Coulter J6-HC centrifuge with JA-10 rotor and resuspended in 1:50 original volume with equilibrium buffer (100 mM K_2HPO_4, 300 mM KCl, and 5 mM imidazole, pH 7.4). After sonication (18 × 10 s), the crude extract was clarified by centrifugation (6000 rpm, 15 min) with JA-20 rotor, and purified by His-tag affinity chromatography as follows. The supernatant was batch-mixed (1 h) with a HisPur Cobalt resin (Thermo Fisher Scientific, Pittsburg, PA, USA), previously equilibrated in the above equilibrium buffer, and then packed into a chromatographic column. The flow-through and the subsequent 10 mM imidazole wash buffer were discarded. The recombinant protein was eluted by equilibration buffer containing 150 mM imidazole.

After centrifuge-assisted protein concentration in a Protein Concentrator 10K (Pierce-Thermo Fisher Scientific, Pittsburg, PA, USA), the hNAPRT amount was quantified by absorbance measurement at 280 nm. The solution containing the protein was dialyzed overnight against 50 mM Tris/HCl, pH 7.4, 10 mM KCl, and 1 mM DTT, to remove imidazole and change the buffer with the reaction buffer. To improve enzyme stability, after dialysis, 0.5 mM PRPP and 20% glycerol were added and the protein was aliquoted and kept at −20 °C. All the steps were performed at 4 °C.

4.7. hNAPRT Inhibition Assay

hNAPRT was expressed in *E. coli* with an *N*-terminal His-tag and purified as described above. The enzymatic reactions were carried out at 37 °C in standard reaction mixtures containing 50 mM Tris-HCl, pH 7.4, 10 mM KCl, 2 mM $MgCl_2$, 100 μM nicotinic acid, 200 μM PRPP, 0.1 mg/mL purified recombinant hNAPRT, and the test compound at five different concentrations ranging from 20 μM to 1000 μM.

Blank mixtures without compounds but with equal amounts of DMSO were set in parallel and their rates fixed as 100% activity. NAPRT was pre-incubated with the compounds for 5 min and reactions were started by the addition of the enzyme substrates and stopped after 45 min of incubation by heating to 85 °C for 3 min. Following centrifugation of the reaction mix, the supernatant was analyzed by HPLC by injection into a reverse-phase column (XTerra MS C18 Column, 125Å, 5 μm, 4.6 × 150 mm, Waters). The eluted species were monitored at 260 nm and the peaks of nicotinic acid and nicotinic acid mononucleotide were quantified with reference to standard curves. The percentage of nicotinic acid conversion was calculated for each reaction and the IC_{50} values for the active compounds were determined with GraphPad Prism 8 software (GraphPad Software, S. Diego, CA, USA).

4.8. NAPRT Melting Temperature Determination

The melting temperature ™ of hNAPRT was determined through the differential scanning fluorimetry (DSF) technique. Triplicates of 100 μM of each test compound and DMSO were added to a 96-well clear bottom BioRad PCR plate. Subsequently, DSF protein buffer, containing 5 μM of recombinant hNAPRT protein and 5X SPYRO® Orange (S5692, Sigma-Aldrich) in 20 mM HEPES pH 7.5, 100 mM NaCl, 10 mM Mg-acetate, and 1 mM DTT was added. The plate was sealed and exposed to a temperature gradient from 20 to 95 °C in a BioRad CFX96 Real-Time System. The fluorescence for each temperature increment was measured at 465–580 nm with an excitation wavelength of 465 nm. The DSF templates that were provided by Niesen et al. were used for data analysis followed by GraphPad Prism 8 for statistical analysis of the generated data [56].

4.9. Statistical Analyses

All the experiments were repeated at least 3 times. Statistics were performed with GraphPad Prism v.8 software (GraphPad Software, S. Diego, CA, USA). All the parameters were tested by paired *t*-test or one-way ANOVA followed by the Tukey test. p-values < 0.05 were considered significant.

Supplementary Materials: The following supporting information can be downloaded at: https://www.mdpi.com/article/10.3390/ph15070855/s1, Table S1: Details of chemical structures of tested compounds as putative hNAPRT inhibitors. Figure S1: Nicotinic acid in culture medium determines the susceptibility of the OVCAR-5 cell line to treatments with FK866 and 2-hydroxynicotinic acid.

Author Contributions: Conceptualization, J.F. and A.D.R.; methodology, A.A., A.V.K., P.H., I.R., S.B., A.N. and A.D.R.; validation, J.F., F.P., M.W., S.F. and I.C.; investigation, J.F., M.W. and S.F.; resources, A.A., A.V.K., M.M.-B., P.H., I.R., S.B., A.N. and A.D.R.; data curation, J.F., F.P., A.B., M.G., M.W., I.C., S.B., A.N. and A.D.R.; writing—original draft preparation, J.F. and A.D.R.; writing—review and editing, J.F. and A.D.R.; supervision, A.D.R.; project administration, A.D.R.; funding acquisition, A.D.R. All authors have read and agreed to the published version of the manuscript.

Funding: This work received funding from the European Union's Horizon 2020 research and innovation programme under the Marie Skłodowska-Curie grant agreement no. 813284. In addition, it was supported in part by the Associazione Italiana per la Ricerca sul Cancro (AIRC; IG#22098; to A. Nencioni) and by the Italian Ministry of Health (PE-2016-02362694 and PE-2016-02363073) and by the Spanish Ministry of Science and Innovation (PID2020-116460RB-I00).

Institutional Review Board Statement: Not applicable.

Informed Consent Statement: Not applicable.

Data Availability Statement: Not applicable.

Conflicts of Interest: The authors declare no conflict of interest.

References

1. Yang, Y.; Sauve, A.A. NAD+ Metabolism: Bioenergetics, Signaling and Manipulation for Therapy. *Biochim. Biophys. Acta-Proteins Proteom.* **2016**, *1864*, 1787–1800. [CrossRef] [PubMed]
2. Cantó, C.; Menzies, K.J.; Auwerx, J. NAD+ Metabolism and the Control of Energy Homeostasis: A Balancing Act between Mitochondria and the Nucleus. *Cell Metab.* **2015**, *22*, 31–53. [CrossRef] [PubMed]
3. Navas, L.E.; Carnero, A. NAD+ Metabolism, Stemness, the Immune Response, and Cancer. *Signal Transduct. Target. Ther.* **2021**, *6*, 2. [CrossRef] [PubMed]
4. Lin, H. Nicotinamide Adenine Dinucleotide: Beyond a Redox Coenzyme. *Org. Biomol. Chem.* **2007**, *5*, 2541–2554. [CrossRef]
5. Covarrubias, A.J.; Perrone, R.; Grozio, A.; Verdin, E. NAD+ Metabolism and Its Roles in Cellular Processes during Ageing. *Nat. Rev. Mol. Cell Biol.* **2021**, *22*, 119–141. [CrossRef]
6. Houtkooper, R.H.; Cantó, C.; Wanders, R.J.; Auwerx, J. The Secret Life of NAD+: An Old Metabolite Controlling New Metabolic Signaling Pathways. *Endocr. Rev.* **2010**, *31*, 194–223. [CrossRef]
7. Berger, N.A.; Besson, V.C.; Boulares, A.H.; Bürkle, A.; Chiarugi, A.; Clark, R.S.; Curtin, N.J.; Cuzzocrea, S.; Dawson, T.M.; Dawson, V.L.; et al. Opportunities for the Repurposing of PARP Inhibitors for the Therapy of Non-Oncological Diseases. *Br. J. Pharmacol.* **2018**, *175*, 192–222. [CrossRef]
8. Houtkooper, R.H.; Pirinen, E.; Auwerx, J. Sirtuins as Regulators of Metabolism and Healthspan. *Nat. Rev. Mol. Cell Biol.* **2012**, *13*, 225–238. [CrossRef]
9. Quarona, V.; Zaccarello, G.; Chillemi, A.; Brunetti, E.; Singh, V.K.; Ferrero, E.; Funaro, A.; Horenstein, A.L.; Malavasi, F. CD38 and CD157: A Long Journey from Activation Markers to Multifunctional Molecules. *Cytom. Part B-Clin. Cytom.* **2013**, *84*, 207–217. [CrossRef]
10. Essuman, K.; Summers, D.W.; Sasaki, Y.; Mao, X.; DiAntonio, A.; Milbrandt, J. The SARM1 Toll/Interleukin-1 Receptor Domain Possesses Intrinsic NAD+ Cleavage Activity That Promotes Pathological Axonal Degeneration. *Neuron* **2017**, *93*, 1334–1343.e5. [CrossRef]
11. Chiarugi, A.; Dölle, C.; Felici, R.; Ziegler, M. The NAD Metabolome—A Key Determinant of Cancer Cell Biology. *Nat. Rev. Cancer* **2012**, *12*, 741–752. [CrossRef] [PubMed]

12. Zapata-Pérez, R.; Wanders, R.J.A.; Karnebeek, C.D.M.; Houtkooper, R.H. NAD + Homeostasis in Human Health and Disease. *EMBO Mol. Med.* **2021**, *13*, e13943. [CrossRef] [PubMed]
13. Pramono, A.A.; Rather, G.M.; Herman, H.; Lestari, K.; Bertino, J.R. NAD-and NADPH-Contributing Enzymes as Therapeutic Targets in Cancer: An Overview. *Biomolecules* **2020**, *10*, 358. [CrossRef] [PubMed]
14. Yaku, K.; Okabe, K.; Hikosaka, K.; Nakagawa, T. NAD Metabolism in Cancer Therapeutics. *Front. Oncol.* **2018**, *8*, 622. [CrossRef] [PubMed]
15. Del Nagro, C.; Xiao, Y.; Rangell, L.; Reichelt, M.; O'Brien, T. Depletion of the Central Metabolite NAD Leads to Oncosismediated Cell Death. *J. Biol. Chem.* **2014**, *289*, 35182–35192. [CrossRef]
16. Ghanem, M.S.; Monacelli, F.; Nencioni, A. Advances in Nad-Lowering Agents for Cancer Treatment. *Nutrients* **2021**, *13*, 1665. [CrossRef]
17. Nikiforov, A.; Dölle, C.; Niere, M.; Ziegler, M. Pathways and Subcellular Compartmentation of NAD Biosynthesis in Human Cells: From Entry of Extracellular Precursors to Mitochondrial NAD Generation. *J. Biol. Chem.* **2011**, *286*, 21767–21778. [CrossRef]
18. Rongvaux, A.; Andris, F.; van Gool, F.; Leo, O. Reconstructing Eukaryotic NAD Metabolism. *BioEssays* **2003**, *25*, 683–690. [CrossRef]
19. Duarte-Pereira, S.; Pereira-Castro, I.; Silva, S.S.; Correia, M.G.; Neto, C.; da Costa, L.T.; Amorim, A.; Silva, R.M. Extensive Regulation of Nicotinate Phosphoribosyltransferase (NAPRT) Expression in Human Tissues and Tumors. *Oncotarget* **2016**, *7*, 1973–1983. [CrossRef]
20. Sampath, D.; Zabka, T.S.; Misner, D.L.; O'Brien, T.; Dragovich, P.S. Inhibition of Nicotinamide Phosphoribosyltransferase (NAMPT) as a Therapeutic Strategy in Cancer. *Pharmacol. Ther.* **2015**, *151*, 16–31. [CrossRef]
21. Olesen, U.H.; Christensen, M.K.; Björkling, F.; Jäättelä, M.; Jensen, P.B.; Sehested, M.; Nielsen, S.J. Anticancer Agent CHS-828 Inhibits Cellular Synthesis of NAD. *Biochem. Biophys. Res. Commun.* **2008**, *367*, 799–804. [CrossRef] [PubMed]
22. Hasmann, M.; Schemainda, I. FK866, a Highly Specific Noncompetitive Inhibitor of Nicotinamide Phosphoribosyltransferase, Represents a Novel Mechanism for Induction of Tumor Cell Apoptosis. *Cancer Res.* **2003**, *63*, 7436–7442. [PubMed]
23. Goldinger, S.M.; Bischof, S.G.; Fink-Puches, R.; Klemke, C.D.; Dréno, B.; Bagot, M.; Dummer, R. Efficacy and Safety of Apo866 in Patients with Refractory or Relapsed Cutaneous T-Cell Lymphoma: A Phase 2 Clinical Trial. *JAMA Dermatol.* **2016**, *152*, 837–839. [CrossRef] [PubMed]
24. Von Heideman, A.; Berglund, Å.; Larsson, R.; Nygren, P.; Larsson, R. Safety and Efficacy of NAD Depleting Cancer Drugs: Results of a Phase i Clinical Trial of CHS 828 and Overview of Published Data. *Cancer Chemother. Pharmacol.* **2010**, *65*, 1165–1172. [CrossRef] [PubMed]
25. Piacente, F.; Caffa, I.; Ravera, S.; Sociali, G.; Passalacqua, M.; Vellone, V.G.; Becherini, P.; Reverberi, D.; Monacelli, F.; Ballestrero, A.; et al. Nicotinic Acid Phosphoribosyltransferase Regulates Cancer Cell Metabolism, Susceptibility to NAMPT Inhibitors, and DNA Repair. *Cancer Res.* **2017**, *77*, 3857–3869. [CrossRef]
26. Piacente, F.; Caffa, I.; Nencioni, A. Nicotinic Acid: A Case for a Vitamin That Moonlights for Cancer? *Cell Cycle* **2017**, *16*, 1635–1636. [CrossRef]
27. Hara, N.; Yamada, K.; Shibata, T.; Osago, H.; Tsuchiya, M. Nicotinamide Phosphoribosyltransferase/Visfatin Does Not Catalyze Nicotinamide Mononucleotide Formation in Blood Plasma. *PLoS ONE* **2011**, *6*, e22781. [CrossRef]
28. Marletta, A.S.; Massarotti, A.; Orsomando, G.; Magni, G.; Rizzi, M.; Garavaglia, S. Crystal Structure of Human Nicotinic Acid Phosphoribosyltransferase. *FEBS Open Biol.* **2015**, *5*, 419–428. [CrossRef]
29. Galassi, L.; di Stefano, M.; Brunetti, L.; Orsomando, G.; Amici, A.; Ruggieri, S.; Magni, G. Characterization of Human Nicotinate Phosphoribosyltransferase: Kinetic Studies, Structure Prediction and Functional Analysis by Site-Directed Mutagenesis. *Biochimie* **2012**, *94*, 300–309. [CrossRef]
30. Duarte-Pereira, S.; Fajarda, O.; Matos, S.; Oliveira, J.L.; Silva, R.M. Naprt Expression Regulation Mechanisms: Novel Functions Predicted by a Bioinformatics Approach. *Genes* **2021**, *12*, 2022. [CrossRef]
31. Chowdhry, S.; Zanca, C.; Rajkumar, U.; Koga, T.; Diao, Y.; Raviram, R.; Liu, F.; Turner, K.; Yang, H.; Brunk, E.; et al. NAD Metabolic Dependency in Cancer Is Shaped by Gene Amplification and Enhancer Remodelling. *Nature* **2019**, *569*, 570–575. [CrossRef] [PubMed]
32. Olesen, U.H.; Hastrup, N.; Sehested, M. Expression Patterns of Nicotinamide Phosphoribosyltransferase and Nicotinic Acid Phosphoribosyltransferase in Human Malignant Lymphomas. *APMIS* **2011**, *119*, 296–303. [CrossRef] [PubMed]
33. Hara, N.; Yamada, K.; Shibata, T.; Osago, H.; Hashimoto, T.; Tsuchiya, M. Elevation of Cellular NAD Levels by Nicotinic Acid and Involvement of Nicotinic Acid Phosphoribosyltransferase in Human Cells. *J. Biol. Chem.* **2007**, *282*, 24574–24582. [CrossRef]
34. Audrito, V.; Messana, V.G.; Deaglio, S. NAMPT and NAPRT: Two Metabolic Enzymes With Key Roles in Inflammation. *Front. Oncol.* **2020**, *10*, 358. [CrossRef] [PubMed]
35. Managò, A.; Audrito, V.; Mazzola, F.; Sorci, L.; Gaudino, F.; Gizzi, K.; Vitale, N.; Incarnato, D.; Minazzato, G.; Ianniello, A.; et al. Extracellular Nicotinate Phosphoribosyltransferase Binds Toll like Receptor 4 and Mediates Inflammation. *Nat. Commun.* **2019**, *10*, 4116. [CrossRef]
36. Tateishi, K.; Wakimoto, H.; Iafrate, A.J.; Tanaka, S.; Loebel, F.; Lelic, N.; Wiederschain, D.; Bedel, O.; Deng, G.; Zhang, B.; et al. Extreme Vulnerability of IDH1 Mutant Cancers to NAD+ Depletion. *Cancer Cell* **2015**, *28*, 773–784. [CrossRef] [PubMed]

37. Lee, J.; Kim, H.; Lee, J.E.; Shin, S.J.; Oh, S.; Kwon, G.; Kim, H.; Choi, Y.Y.; White, M.A.; Paik, S.; et al. Selective Cytotoxicity of the NAMPT Inhibitor FK866 Toward Gastric Cancer Cells With Markers of the Epithelial-Mesenchymal Transition, Due to Loss of NAPRT. *Gastroenterology* **2018**, *155*, 799–814.e13. [CrossRef]
38. Fons, N.R.; Sundaram, R.K.; Breuer, G.A.; Peng, S.; McLean, R.L.; Kalathil, A.N.; Schmidt, M.S.; Carvalho, D.M.; Mackay, A.; Jones, C.; et al. PPM1D Mutations Silence NAPRT Gene Expression and Confer NAMPT Inhibitor Sensitivity in Glioma. *Nat. Commun.* **2019**, *10*, 3790. [CrossRef]
39. Franceschini, N.; Oosting, J.; Tamsma, M.; Niessen, B.; Briaire-De Bruijn, I.; van den Akker, B.; Kruisselbrink, A.B.; Palubeckait, I.; Bovée, J.V.M.G.; Cleton-Jansen, A.-M.; et al. Targeting the NAD Salvage Synthesis Pathway as a Novel Therapeutic Strategy for Osteosarcomas with Low NAPRT Expression. *Int. J. Mol. Sci.* **2021**, *22*, 6273. [CrossRef]
40. Watson, M.; Roulston, A.; Bélec, L.; Billot, X.; Marcellus, R.; Bédard, D.; Bernier, C.; Branchaud, S.; Chan, H.; Dairi, K.; et al. The Small Molecule GMX1778 Is a Potent Inhibitor of NAD + Biosynthesis: Strategy for Enhanced Therapy in Nicotinic Acid Phosphoribosyltransferase 1-Deficient Tumors. *Mol. Cell. Biol.* **2009**, *29*, 5872–5888. [CrossRef]
41. Kudo, K.; Nomura, M.; Sakamoto, Y.; Ito, S.; Morita, M.; Kawai, M.; Yamashita, Y.; Ito, K.; Yamada, H.; Shima, H.; et al. Divergent Metabolic Responses Dictate Vulnerability to NAMPT Inhibition in Ovarian Cancer. *FEBS Lett.* **2020**, *594*, 1379–1388. [CrossRef] [PubMed]
42. Gaut, Z.N.; Solomon, H.M. Inhibition of 7- 14 C-Nicotinic Acid Metabolism in the Human Blood Platelet by Anti-Inflammatory Drugs. *Res. Commun. Chem. Pathol. Pharm.* **1970**, *1*, 547–552.
43. Gaut, Z.N.; Solomon, H.M. Inhibition of Nicotinate Phosphoribosyl Transferase by Nonsteroidal Anti-Inflammatory Drugs: A Possible Mechanism of Action. *J. Pharm. Sci.* **1971**, *60*, 1887–1888. [CrossRef]
44. Gaut, Z.N.; Solomon, H.M. Inhibition of Nicotinate Phosphoribosyltransferase in Human Platelet Lysate by Nicotinic Acid Analogs. *Biochem. Pharmacol.* **1971**, *20*, 2903–2906. [CrossRef]
45. The Protein Data Bank. Available online: www.rcsb.org (accessed on 8 June 2022).
46. Mcule. Available online: www.mcule.com (accessed on 8 June 2022).
47. Ropp, P.J.; Spiegel, J.O.; Walker, J.L.; Green, H.; Morales, G.A.; Milliken, K.A.; Ringe, J.J.; Durrant, J.D. GypSUm-DL: An Open-Source Program for Preparing Small-Molecule Libraries for Structure-Based Virtual Screening. *J. Cheminform.* **2019**, *11*, 34. [CrossRef]
48. Fischer, A.; Smieško, M.; Sellner, M.; Lill, M.A. Decision Making in Structure-Based Drug Discovery: Visual Inspection of Docking Results. *J. Med. Chem.* **2021**, *64*, 2489–2500. [CrossRef]
49. Prestwick Chemical. Available online: www.prestwickchemical.com (accessed on 8 June 2022).
50. EDASA Scientific. Available online: www.edasascientific.com (accessed on 8 June 2022).
51. Martínez-Bailén, M.; Galbis, E.; Carmona, A.T.; de-Paz, M.V.; Robina, I. Preparation of Water-Soluble Glycopolymers Derived from Five-Membered Iminosugars. *Eur. Polym. J.* **2019**, *119*, 213–221. [CrossRef]
52. Martínez-Bailén, M.; Carmona, A.T.; Moreno-Clavijo, E.; Robina, I.; Ide, D.; Kato, A.; Moreno-Vargas, A.J. Tuning of β-Glucosidase and α-Galactosidase Inhibition by Generation and in Situ Screening of a Library of Pyrrolidine-Triazole Hybrid Molecules. *Eur. J. Med. Chem.* **2017**, *138*, 532–542. [CrossRef]
53. Martínez-Bailén, M.; Carmona, A.T.; Patterson-Orazem, A.C.; Lieberman, R.L.; Ide, D.; Kubo, M.; Kato, A.; Robina, I.; Moreno-Vargas, A.J. Exploring Substituent Diversity on Pyrrolidine-Aryltriazole Iminosugars: Structural Basis of β-Glucocerebrosidase Inhibition. *Bioorg. Chem.* **2019**, *86*, 652–664. [CrossRef]
54. Popowycz, F.; Gerber-Lemaire, S.; Demange, R.; Rodriguez-García, E.; Carmona Asenjo, A.T.; Robina, I.; Vogel, P. Derivatives of (2R,3R,4S)-2-Aminomethylpyrrolidine-3,4-diol are Selective α-Mannosidase Inhibitors. *Bioorg. Med. Chem. Lett.* **2001**, *11*, 2489–2493. [CrossRef]
55. Bruzzone, S.; de Flora, A.; Usai, C.; Graeff, R.; Lee, H.C. Cyclic ADP-Ribose Is a Second Messenger in the Lipopolysaccharide-Stimulated Proliferation of Human Peripheral Blood Mononuclear Cells. *Biochem. J.* **2003**, *375*, 395–403. [CrossRef] [PubMed]
56. Niesen, F.H.; Berglund, H.; Vedadi, M. The Use of Differential Scanning Fluorimetry to Detect Ligand Interactions That Promote Protein Stability. *Nat. Protoc.* **2007**, *2*, 2212–2221. [CrossRef] [PubMed]

pharmaceuticals

MDPI

Article

Identification of NAPRT Inhibitors with Anti-Cancer Properties by In Silico Drug Discovery

Moustafa S. Ghanem [1], Irene Caffa [1], Alberto Del Rio [2,3], Jorge Franco [1,2], Marco Daniele Parenti [3], Fiammetta Monacelli [1,4], Michele Cea [1], Amr Khalifa [1], Aimable Nahimana [5], Michel A. Duchosal [5], Silvia Ravera [6], Nadia Bertola [6], Santina Bruzzone [7], Alessio Nencioni [1,4,*] and Francesco Piacente [7]

1 Department of Internal Medicine and Medical Specialties, University of Genoa, 16132 Genoa, Italy; moustafa.ghanem@edu.unige.it (M.S.G.); irene.caffa@unige.it (I.C.); jorge.franco@innovamol.com (J.F.); fiammetta.monacelli@unige.it (F.M.); michele.cea@unige.it (M.C.); amr.khalifa@edu.unige.it (A.K.)

2 Innovamol Consulting Srl, 41126 Modena, Italy; alberto.delrio@innovamol.com

3 National Research Council (CNR), Institute of Organic Synthesis and Photoreactivity (ISOF), 40129 Bologna, Italy; marcodaniele.parenti@isof.cnr.it

4 IRCCS Ospedale Policlinico San Martino, 16132 Genoa, Italy

5 Service and Central Laboratory of Hematology, University Hospital of Lausanne, 1015 Lausanne, Switzerland; aimable.nahimana@chuv.ch (A.N.); michel.duchosal@chuv.ch (M.A.D.)

6 Department of Pharmacy, Biochemistry Lab, 16132 Genoa, Italy; silvia.ravera@unige.it (S.R.); nadia.bertola@gmail.com (N.B.)

7 Department of Experimental Medicine, University of Genoa, 16132 Genoa, Italy; santina.bruzzone@unige.it (S.B.); francesco.piacente@unige.it (F.P.)

* Correspondence: alessio.nencioni@unige.it

Abstract: Depriving cancer cells of sufficient NAD levels, mainly through interfering with their NAD-producing capacity, has been conceived as a promising anti-cancer strategy. Numerous inhibitors of the NAD-producing enzyme, nicotinamide phosphoribosyltransferase (NAMPT), have been developed over the past two decades. However, their limited anti-cancer activity in clinical trials raised the possibility that cancer cells may also exploit alternative NAD-producing enzymes. Recent studies show the relevance of nicotinic acid phosphoribosyltransferase (NAPRT), the rate-limiting enzyme of the Preiss–Handler NAD-production pathway for a large group of human cancers. We demonstrated that the NAPRT inhibitor 2-hydroxynicotinic acid (2-HNA) cooperates with the NAMPT inhibitor FK866 in killing NAPRT-proficient cancer cells that were otherwise insensitive to FK866 alone. Despite this emerging relevance of NAPRT as a potential target in cancer therapy, very few NAPRT inhibitors exist. Starting from a high-throughput virtual screening approach, we were able to identify and annotate two additional chemical scaffolds that function as NAPRT inhibitors. These compounds show comparable anti-cancer activity to 2-HNA and improved predicted aqueous solubility, in addition to demonstrating favorable drug-like profiles.

Keywords: NAPRT inhibitors; cancer metabolism; NAD; anti-cancer agents; NAMPT; NAD synthesis; in silico drug design

Citation: Ghanem, M.S.; Caffa, I.; Del Rio, A.; Franco, J.; Parenti, M.D.; Monacelli, F.; Cea, M.; Khalifa, A.; Nahimana, A.; Duchosal, M.A.; et al. Identification of NAPRT Inhibitors with Anti-Cancer Properties by In Silico Drug Discovery. *Pharmaceuticals* **2022**, *15*, 848. https://doi.org/10.3390/ph15070848

Academic Editors: Marialuigia Fantacuzzi and Mariangela Agamennone

Received: 30 May 2022
Accepted: 6 July 2022
Published: 10 July 2022

1. Introduction

Nicotinamide adenine dinucleotide (NAD) is broadly involved in fundamental biological processes inside the cell. It is particularly unique in its ability to function not only as a cofactor in redox reactions that are intimately involved in energy metabolism but also as a substrate for NAD-consuming enzymes, including poly(ADP-ribose) polymerases (PARPs), sirtuins, CD38, and CD157 [1–5]. Since NAD gets degraded by the catalytic activity of these enzymes, continuous NAD production is required. Most mammalian tissues generate NAD starting from nicotinamide (NAM) through the salvage pathway (also known as the "amidated" pathway). A parallel NAD-generating route named the Preiss–Handler (PH) pathway (or "deamidated" pathway) utilizes nicotinic acid (NA) as its building block and also operates in many tissues [6–8]. In addition, NAD can be synthesized from the amino acid tryptophan through the de novo pathway, which is mainly active in hepatic and renal tissues [9,10]. It is also worth mentioning that nicotinamide riboside (NR) and its reduced form (NRH) were recently recognized as additional NAD precursors that boost NAD production through alternative salvage pathways [11–13].

Cancer cells extensively rely on these NAD-biosynthetic routes (summarized in Figure 1) in order to keep adequate NAD levels, which, in turn, are necessary to fuel their reprogrammed metabolism [14] and to compensate for the extensive NAD breakdown caused by vital, NAD-consuming, enzymatic activities, such as PARP-mediated DNA repair [1–4]. Accordingly, interfering with the NAD biosynthetic machinery has been put forward as an appealing therapeutic approach against cancer. Most studies in this field focused on interrupting the NAM salvage pathway by targeting its rate-limiting enzyme, nicotinamide phosphoribosyltransferase (NAMPT). This reflects the fact that potent and highly active (at least in preclinical models) NAMPT inhibitors, such as FK866 and CHS828, were among the first NAD-lowering agents to be reported [15–17] and the observation that NAMPT is commonly overexpressed in a variety of human cancers [17,18]. Regrettably, despite their efficacy in preclinical models, NAMPT inhibitors showed poor efficacy in clinical trials [19–22], indicating that tumor cells exploit surrogate NAD-producing routes, in particular the PH pathway, to circumvent NAMPT blockade [23,24].

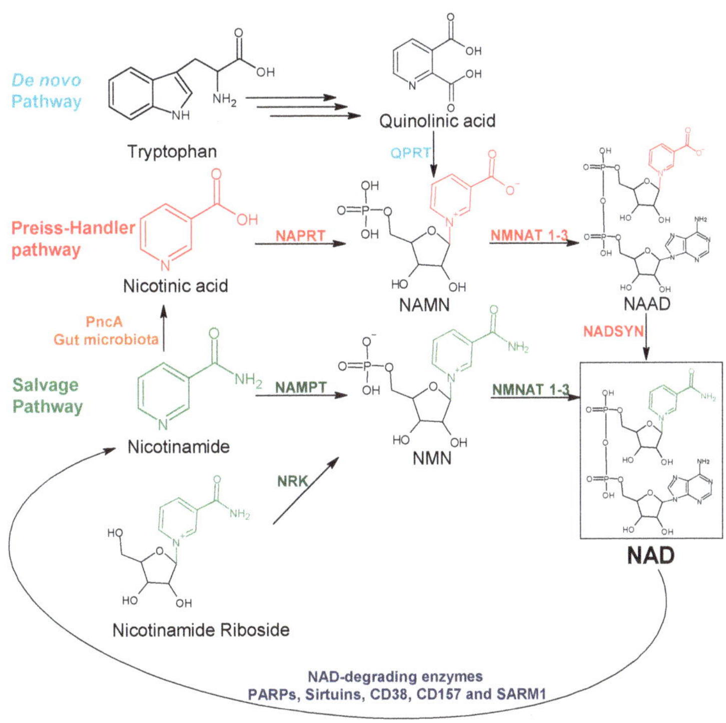

Figure 1. Schematic representation of the NAD–generating pathways in mammalian cells. NAMN, nicotinic acid mononucleotide; NMN, nicotinamide mononucleotide; NAAD, nicotinic acid adenine dinucleotide; NAD, nicotinamide adenine dinucleotide; QPRT, quinolinate phosphoribosyltransferase; NAPRT, nicotinic acid phosphoribosyltransferase; NAMPT, nicotinamide phosphoribosyltransferase; NRK, nicotinamide riboside kinase; NMNAT, nicotinamide mononucleotide adenylyltransferase; NADSYN, nicotinamide adenine dinucleotide synthetase; PncA, nicotinamidase; SARM1, sterile alpha and toll/interleukin receptor [TIR] motif-containing protein 1; PARPs, poly(ADP-ribose) polymerases.

In this context, the relevance of nicotinic acid phosphoribosyltransferase (NAPRT), one of the key enzymes from the PH pathway, as a viable antitumor target has gathered growing attention. NAPRT boosts and regulates NAD biosynthesis under certain conditions and in specific tissues [8,25]. Several studies indicated that the therapeutic activity of NAMPT inhibitors is largely dictated by the NAPRT expression status of the tumor cells [26–28]. Susceptibility to NAMPT inhibitors was particularly noted in epithelial-to-mesenchymal transition (EMT)-subtype gastric cancers, in isocitrate dehydrogenase 1 (*IDH1*)-mutant gliomas, and in protein phosphatase Mg^{2+}/Mn^{2+}-dependent 1D (*PPMD1*)-mutant gliomas as a result of the epigenetic loss of the *NAPRT* gene expression that frequently accompanies these cancer subtypes [29–31]. NAPRT was found to be amplified in a large subset of solid human cancers such as ovarian, pancreatic, and breast cancers [23,24]. Accordingly, we showed that several ovarian cancer cell lines became responsive to FK866 upon *NAPRT* knock-down both in vitro and in vivo (in mice ovarian cancer xenografts) [23]. We also demonstrated that NAPRT plays a central role in energy metabolism, DNA repair, and in protein synthesis in cancer cells via its ability to promote NAD production [23]. Based on the above findings, the development of NAPRT inhibitors holds promise for potential application as anti-cancer agents.

Very few NAPRT inhibitors have been reported so far. Early studies on human platelets discovered several compounds that displayed NAPRT inhibitory activity, including 2-hydroxynicotinic acid (2-HNA) and several non-steroidal anti-inflammatory drugs (NSAIDs) such as flufenamic acid, mefenamic acid, and phenylbutazone (Table 1) [32–34]. We demonstrated that 2-HNA was indeed able to sensitize NAPRT-expressing ovarian and pancreatic cancer cells to NAMPT inhibitors and recapitulated the effect of *NAPRT* silencing [23]. A series of endogenous NAPRT-inhibiting metabolic intermediates were also identified, with CoA being the most potent metabolite [35]. To our best knowledge, 2-HNA remains the only reported NAPRT inhibitor with proven anti-cancer activity, although its clinical use is limited by poor aqueous solubility. In this work, we performed a high-throughput molecular docking screen with the aim of identifying chemical scaffolds with inhibitory activity on the human NAPRT enzyme. Hence, we were able to identify and characterize as NAPRT inhibitors two compounds that show favorable drug-like properties and potency in the micromolar range.

Table 1. Known inhibitors of human NAPRT enzyme.

Compound	Reported Ki * (μM)	Compound	Reported Ki (μM)
Flufenamic acid	10	6-Chloronicotinic acid	560
Mefenamic acid	50	Isonicotinic acid	750
2-Pyrazinoic acid	75	3-Pyridylsulfonic acid	750
Phenylbutazone	100	Pyridine	780
Indomethacin	150	2-Aminonicotinic acid	820
Salicylic acid	160	Acetanilide	1000
2-Hydroxynicotinic acid	230	Aminopyrine	1000
2-Fluoronicotinic acid	280	Antipyrine	1000
Oxyphenbutazone	300	Picolinic acid	1160
Acetylsalicylic acid	500	3-Pyridylacetic acid	1280
Sulfinpyrazone	500	Benzoic acid	1900

* Ki, inhibition constant.

2. Results

2.1. Structure-Based Virtual High-Throughput Screening

2.1.1. Analysis of Available 3D Structures of NAPRT

At the time of this work, only one X-ray structure of human NAPRT was available in the public domain (PDB accession code 4YUB), in its ligand-free form [36]. This NAPRT structure was solved at a resolution of 2.9 Å, which usually allows for the unambiguous assignment of the main chain and side chains for the rigid parts of a protein, although a chance of an incorrectly placed side chains still exists. This potential issue, which, at least in principle, could affect the results of subsequent docking studies, was addressed by applying standard protein preparation procedures, such as restrained energy minimization (see methods section). The crystal structure of human NAPRT (Figure 2A) reveals that its monomer folds into 17 α-helices, 24 β-strands, and the connecting loops organized in two domains: a first domain characterized by an irregular α/β barrel and a second open-faced sandwich domain. The structural organization of human NAPRT is highly similar to that of bacterial NAPRT, e.g., the enzyme expressed by T. acidophilum or E. faecalis, for which the X-ray structures are also available (PDB codes 1YTD/1YTE/1YTK and 2FTF, respectively). Despite the low sequence identity (34%), many active site residues are conserved among the different species, suggesting a very similar mode of binding for substrates. The model also reveals the presence of an intimately associated dimer in the asymmetric unit; the two monomers are arranged head to tail with the N-terminal domain in one monomer contacting the α/β barrel in the other monomer (Figure 2A). When structurally compared, the structures of the two monomers showed similarity in terms of the overall protein conformation but slight differences in terms of the shape of the active site; in fact, some residues in the active site, such as Leu170, Arg171, Arg172, His213, and Tyr21, translated into a different conformation between the two protein chains, leading to a significant difference in active site shape and volume (Figure 2B).

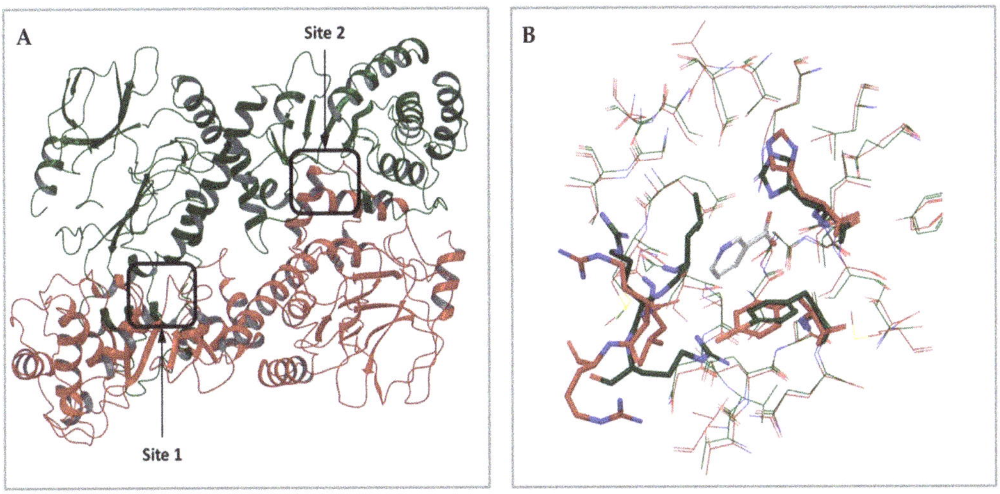

Figure 2. Analysis of the 3D structure of human NAPRT enzyme. (**A**) Overall oligomeric structure of human NAPRT. The enzyme has a dimeric structure and monomers A and B are colored in red and green, respectively. The two active sites are highlighted by black squares. (**B**) Structural superposition of the two active sites in the NAPRT dimer colored red and green, respectively; residues with relevant differences in conformation are drawn in thick tubes.

2.1.2. Virtual Screening Procedure

The crystal structure of human NAPRT was used as a template for our virtual screenings, which were aimed at identifying compounds with inhibitory activity on NAPRT. This structure was prepared with standard preparation procedures that include the correct assignment of bond orders, adding hydrogen, the optimization of protonation states of residues, and restrained energy minimization. Although some water molecules are present in the NAPRT crystal structure, none of these seem to be involved in stable interactions with active site residues; therefore, all water was removed. To maximize the probability of identifying active molecules, two different high-throughput virtual screening procedures were carried out: 1) a functional dimeric model of human NAPRT was taken as the docking target and 2) the single NAPRT monomer was taken as the docking target. For model 1), in order to tackle the differences in active site conformation (see Figure 2), both active sites identified in the functional dimeric model were used as targets, applying the so-called ensemble docking technique. This strategy allows the docking of a single ligand library against multiple rigid receptor conformations and the combining of the results.

In both models, the HTS Compound Collection from Life Chemicals (https://lifechemicals.com/screening-libraries/hts-compound-collection, accessed on 23 March 2016) consisting of 537,009 drug-like compounds, was docked into a docking grid of 18 Å centered on the active site residues, as shown in Figure 3. Docking results were ranked based on the score, and the first 500 hits were visually inspected to prioritize compounds that reproduced, at least in part, the putative binding mode of the NAPRT substrates. This evaluation led to a final list of 35 purchasable compounds to be tested in vitro as putative NAPRT inhibitors. In addition, from the same Life Chemicals compound collection, a set of 2-hydroxynicotinic acid (2-HNA) analogs (Figure 4) was manually selected, as 2-HNA is known to inhibit NAPRT in the micro/millimolar range of concentration. A brief description of the compounds selected for in vitro characterization can be found in Table 2.

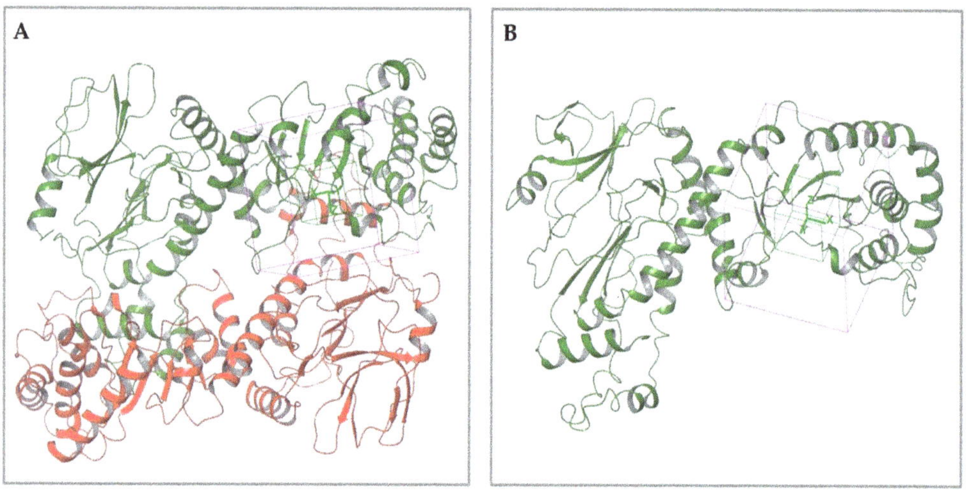

Figure 3. Representation of the NAPRT active site grids employed in docking-based virtual screenings. (**A**) Docking grid of the functional dimeric model of human NAPRT. (**B**) Docking grid of the human NAPRT monomer.

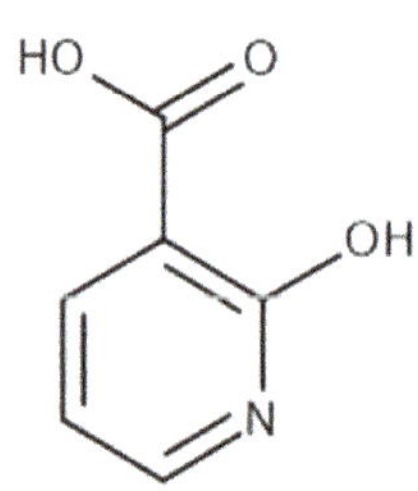

Figure 4. Chemical structure of 2-hydroxynicotinic acid. 2-Hydroxynicotinic acid is the chemical analog of nicotinic acid containing an additional hydroxyl group on carbon 2.

Table 2. Selected structurally diverse compounds and 2-HNA analogs for in vitro characterization as putative NAPRT inhibitors.

Compound ID	Structure	Vendor ID	M.W. *	Compound ID	Structure	Vendor ID	M.W.
1	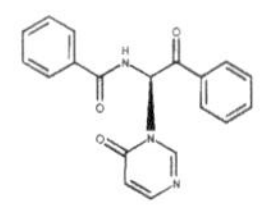	F0020-0171	333.3407	**26**	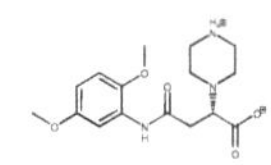	F2721-0386	337.3709
2	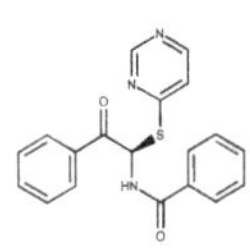	F0173-0133	349.4063	**27**	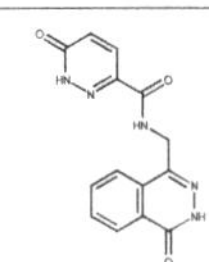	F2758-0213	297.2688
3	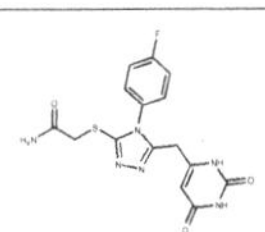	F0648-0699	376.3655	**28**	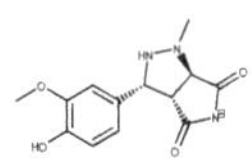	F3188-0088	277.2759
4	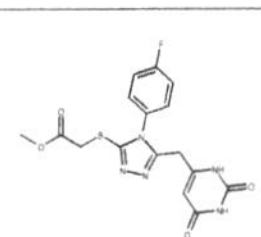	F0648-0785	391.3769	**29**	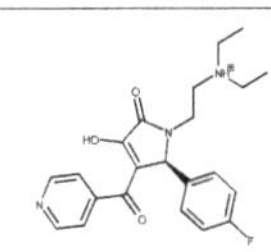	F3226-2226	397.4427
5	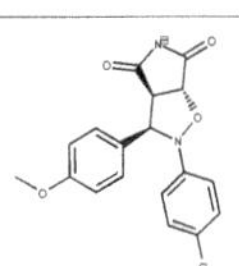	F1199-0146	358.7757	**30**	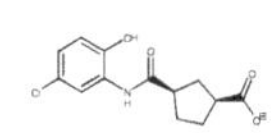	F3229-0191	283.7076
6	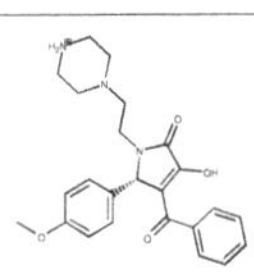	F1260-1693	421.4889	**31**	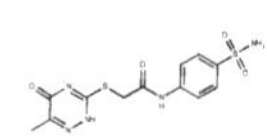	F3295-0007	355.3927
7	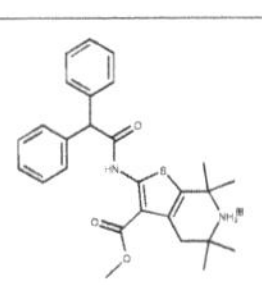	F1299-0156	462.6037	**32**	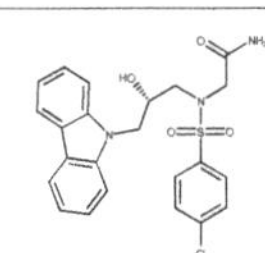	F3311-0032	471.9565
8	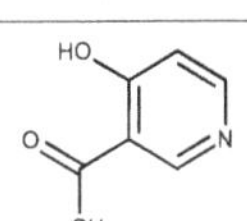	F1371-0219	139.1088	**33**	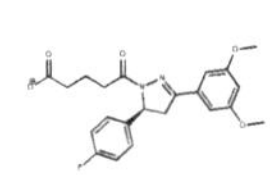	F3371-0859	414.4268
9	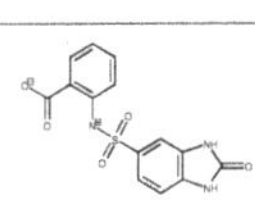	F1566-0988	333.3192	**34**	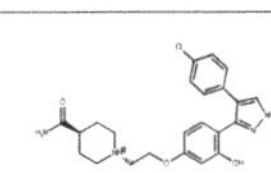	F3385-4161	440.9226

Table 2. *Cont.*

Compound ID	Structure	Vendor ID	M.W. *	Compound ID	Structure	Vendor ID	M.W.
10	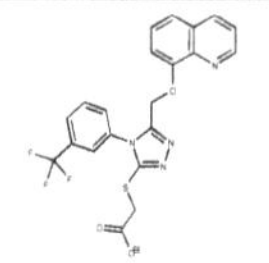	F1710-0049	460.429	**35**	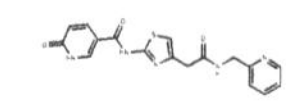	F5008-0290	369.3977
11	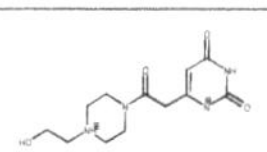	F1811-0048	282.2957	**36**	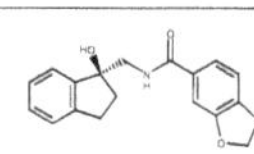	F5857-5354	311.3319
12	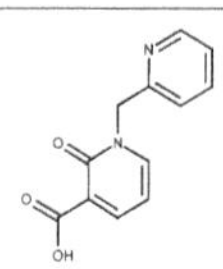	F1907-0958	230.2194	**37**	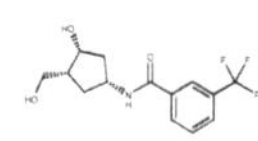	F6127-0104	303.2769
13	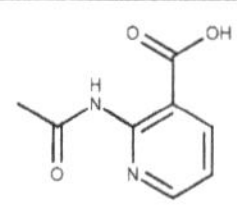	F1967s-1157	180.1607	**38**	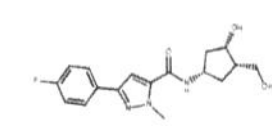	F6127-0210	333.3574
14	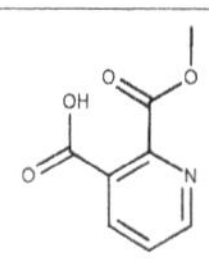	F2135-0162	181.1455	**39**	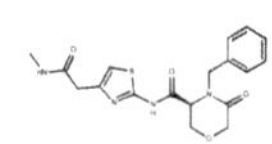	F6241-0336	388.4408
15	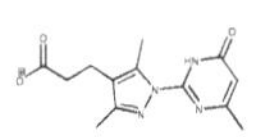	F2135-0875	276.2911	**40**	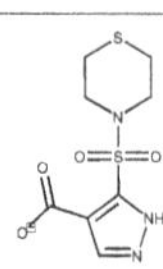	F6252-1248	277.3206
16	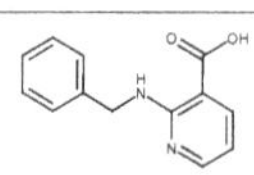	F2135-0897	228.2466	**41**	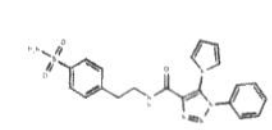	F6252-5764	436.4869
17	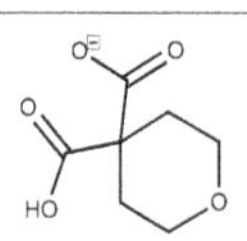	F2147-0724	174.1513	**42**	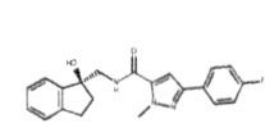	F6279-0434	365.4008
18	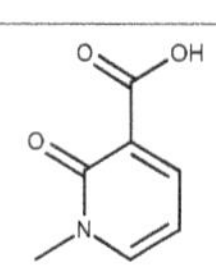	F2168-0001	153.1354	**43**	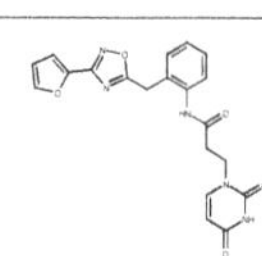	F6372-1828	407.3795
19	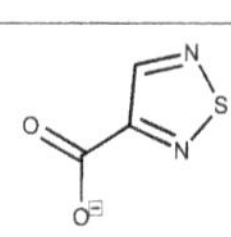	F2169-0490	130.1252	**44**	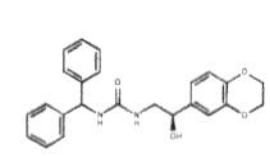	F6414-0992	404.4584

Table 2. *Cont.*

Compound ID	Structure	Vendor ID	M.W. *	Compound ID	Structure	Vendor ID	M.W.
20	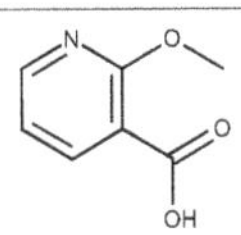	F2191-0003	153.1354	45	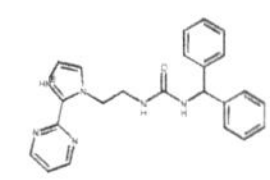	F6439-7266	398.4604
21	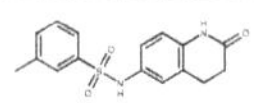	F2278-0232	316.3748	46	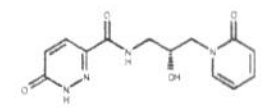	F6465-0031	290.2747
22	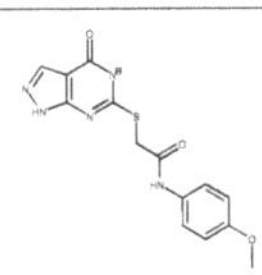	F2503-0045	331.3497	47	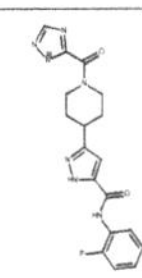	F6497-7775	383.3796
23	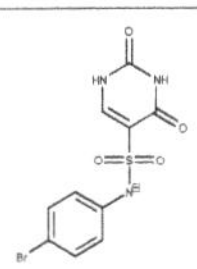	F2526-0046	346.1572	48	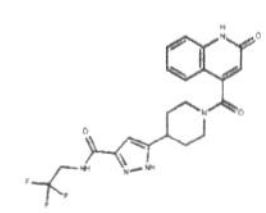	F6497-8054	447.4104
24	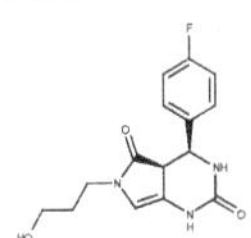	F2711-0182	305.3042	49	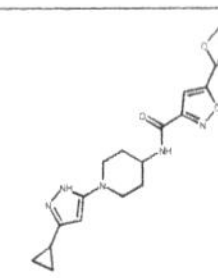	F6523-1712	367.4017
25	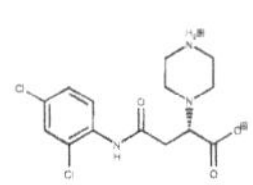	F2721-0331	346.2091	50	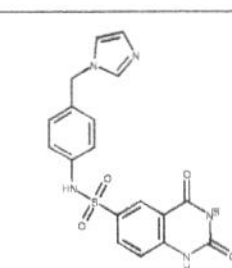	F9994-0201	397.4078

* M.W: Molecular weight.

2.2. Biological Annotation of the Selected Compounds

2.2.1. In Vitro Compound Screening

To rapidly screen the 50 selected compounds for their ability to inhibit NAPRT, we used their capacity to sensitize the NAPRT-proficient ovarian cancer cell line, OVCAR-5, to FK866 as a reading frame (since this cell line is normally resistant to the NAMPT inhibitor but becomes sensitized to it through either NAPRT silencing or inhibition). By itself, the addition of putative NAPRT inhibitors is postulated to show minimal anti-proliferative activity due to the ability of the cells to use the NAM that is present in the cell culture media to synthesize NAD [23]. OVCAR-5 cells were treated with the putative NAPRT inhibitors at 100 μM concentration with or without 100 nM FK866. As depicted in Figure 5A,B, five compounds out of the 50 that were tested (i.e., compounds **1**, **2**, **8**, **16**, and **19**) led to significant cancer cell growth inhibition when coupled with FK866 while being minimally active when used alone. The remaining compounds were discarded since they were either completely inactive or caused remarkable anti-proliferative activity without FK866 (as observed in Figure 5A with compound **5** and compound **7**). The complete inactivity of some compounds could be ascribed to their inability to bind NAPRT or to poor cell membrane permeability. The intrinsic anti-cancer effect of some of the compounds (i.e., without FK866) was considered to be indicative of non-specific toxicity that would possibly also affect healthy cells.

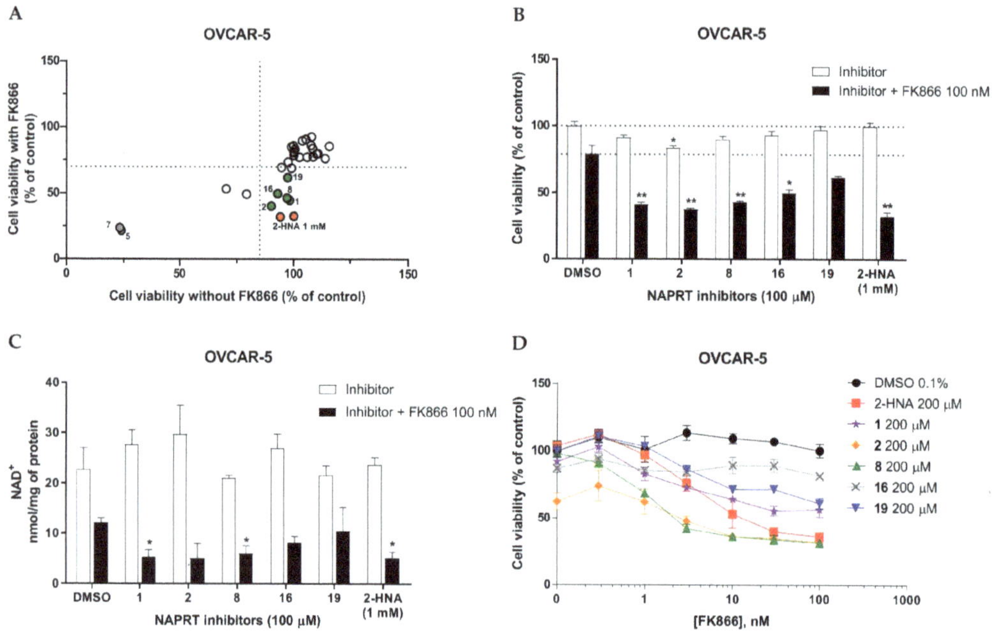

Figure 5. In vitro screening of the putative NAPRT inhibitors. (**A**) Graphical representation of the cell viability results obtained from screening our selected compounds in ovarian cancer cells. OVCAR-5 cells were plated in 96-well plates (2×10^3 cells/well) and left to adhere overnight. The following day, the culture media were replaced with new media containing the respective treatments (i.e., with or without 100 nM FK866 and the putative NAPRT inhibitors, all at 100 μM final concentration, except for 2-HNA, which was used at 1 mM). Each point is the mean of three experimental replicates normalized to the control. The green circles indicate the five most promising putative inhibitors, and the red circles represent 2-HNA as the control NAPRT inhibitor. (**B**) The viability results for the five most-promising NAPRT inhibitors from (**A**) are also represented in a bar graph. *, $p < 0.05$; **, $p < 0.01$ (**C**) OVCAR-5 cells were plated in 12-well plates (1×10^5 cells/well) and allowed to adhere overnight. The following day, the culture media were replaced with new media containing the respective treatments (i.e., with or without 100 nM FK866 and the putative NAPRT inhibitors, all at 100 μM final concentration, except for 2-HNA, which was used at 1 mM). After 24 h, intracellular NAD levels were measured. *, $p < 0.05$ (**D**) OVCAR-5 were plated in 96-well plates (2×10^3 cells/well) and allowed to adhere overnight. The following day, the culture media were replaced with new media that contain the respective treatments (i.e., with or without FK866 at increasing concentrations from 0.3 to 100 nM and the putative NAPRT inhibitors, added at 200 μM final concentration), and the plates were then incubated for 72 h. Afterwards, the cell viability was determined using the sulforhodamine B assay.

Afterwards, we aimed at assessing the downstream effects of inhibiting both enzymes in cancer cells, particularly in terms of intracellular NAD concentration. In line with our previous observations with NAPRT silencing, by themselves 2-HNA and the new 5 putative NAPRT inhibitors failed to reduce intracellular NAD levels [23]. However, 2-HNA and the new putative inhibitors did cooperate with the NAMPT inhibitor, FK866, to blunt intracellular NAD concentrations (Figure 5C). We next evaluated the ability of these compounds to sensitize OVCAR-5 cells to lower concentrations of FK866. Four out of the five compounds (i.e., compounds **1**, **2**, **8**, and **19**) were indeed able to sensitize OVCAR-5 cells when incubated (at 200 μM) with increasing concentrations of FK866 (Figure 5D). The degree of sensitization varied among the putative inhibitors, with compound **8** exhibiting the most potent sensitization effect. Notably, the sensitizing activity of compound **8** was even more pronounced than that of the classical NAPRT inhibitor, 2-HNA (Figure 5D). On the other hand, compound **16** was the only compound that completely failed to sensitize the ovarian cancer cells to FK866, and thus it was not further investigated.

To further confirm the observed sensitization effect of our putative inhibitors on the anti-tumor activity of NAMPT inhibitors, we extended our experiments in two additional NAPRT-expressing cancer cell lines (i.e., HCT116 and OVCAR-8). Consistent with our previous observations in OVCAR-5 cells, compound **8** and compound **19** also sensitized these other two cell models to FK866 when they were used at 100 μM concentration (Figure 6A–C). By contrast, compound **1** and compound **2**, when used at the same concentration, failed to sensitize these two cancer cell lines to FK866, with compound **2** even showing unspecific anti-proliferative activity in these models (Figure 6D,E). Since compound **8** (4-hydroxynicotinic acid) and 2-HNA are structural isomers, we decided to evaluate whether the remaining 2-HNA analogs [i.e., 5-hydroxynicotinic acid (5-HNA) and 6-hydroxynicotinic acid (6-HNA)] are also capable of inhibiting NAPRT. We tested this hypothesis in OVCAR-8 cells. Neither 5-HNA nor 6-HNA could recreate the effects of compounds **8** and 2-HNA in terms of cell sensitization to FK866 (Figure 6F). We hypothesize that this reflects the inability of these compounds to bind within the NAPRT enzymatic pocket. Overall, these findings highlight the specificity of compound **8**, since shifting the position of the -OH group from position 4 to position 5 or 6 entirely abolished their ability to sensitize cancer cells to FK866 (and thus, arguably, to inhibit NAPRT). Ultimately, these experiments indicate that the NAPRT-inhibitory activity of these hydroxylated analogs of nicotinic acid strictly relies on -OH substitution at position 2 or 4 of the pyridine ring.

In the PH pathway, NAPRT catalyzes the transfer of a phosphoribosyl group from phosphoribosyl pyrophosphate (PRPP) to its substrate NA, thereby yielding nicotinic acid mononucleotide (NAMN). The latter is converted into nicotinic acid adenine dinucleotide (NAAD) and, finally, amidated into NAD (Figure 1). In order to confirm that the ability of our new inhibitors to sensitize NAPRT-proficient cancer cells to FK866 is on-target, i.e., due to NAPRT obstruction, we supplemented HCT116 and OVCAR-8 cells with NA or NAMN (at 10 μM) while treating them with our putative NAPRT inhibitors, in the presence or absence of FK866. Both NA and NAMN fully rescued these cells from the marked anti-proliferative effect that was achieved by combining FK866 with 2-HNA, compound **8**, or compound **19** (Figure 6G,H). Taken together, these observations are in line with compound **8** and compound **19** being NAPRT inhibitors.

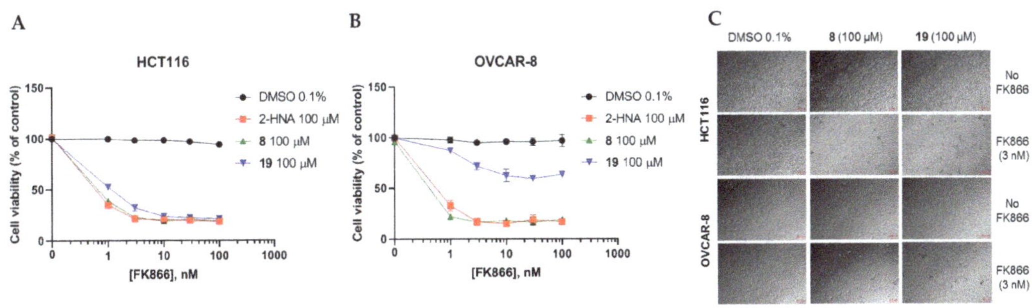

Figure 6. *Cont.*

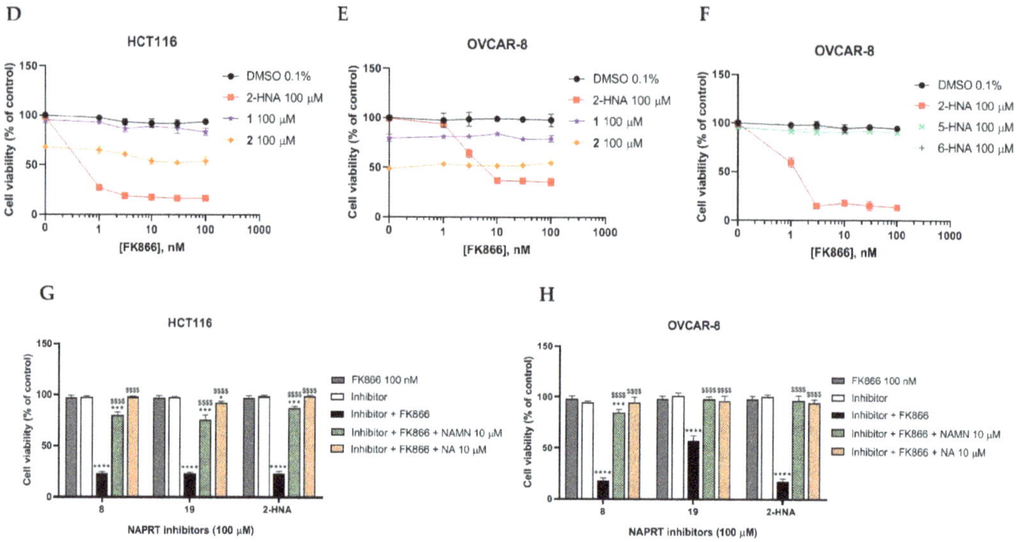

Figure 6. Compound 8 and compound 19 sensitize ovarian and colon cancer cells to FK866 via NAPRT inhibition. (**A–F**) HCT116 and OVCAR-8 were plated in 96-well plates (2×10^3 cells/well) and allowed to adhere overnight. The following day, culture media were replaced with new media containing the respective treatments (i.e., with or without FK866 at increasing concentrations from 1 to 100 nM and the putative NAPRT inhibitors, added at 100 μM final concentration), and the plates were then incubated for 72 h. Afterwards, the cells were imaged using light microscopy as in (**C**), and cell viability was determined using the sulforhodamine B assay. Data are mean ± SD of three experimental replicates. (**G,H**) The same experimental procedure was employed as in (**A–F**). Single concentrations of the NAPRT inhibitors (100 μM), 100 nM FK866, 10 μM NA, and 10 μM NAMN were added. Data are mean ± SD of 4 experimental replicates. One representative experiment is shown. *, $p < 0.05$; ***, $p < 0.001$; ****, $p < 0.0001$; $$$$, $p < 0.0001$. The * symbols refer to the statistical significance compared to the treatment with FK866 alone, whereas the $ symbols refer to the statistical significance compared to the combined treatment with FK866 and the NAPRT inhibitors.

2.2.2. Biochemical Activity on Recombinant Human NAPRT

Given these results in cancer cells, we evaluated the activity of our candidates on the recombinant human NAPRT protein. If the NAPRT enzyme is efficiently inhibited by our compounds, it is postulated to consume less NA compared to what is observed in the absence of NAPRT inhibitors. As expected, the chromatographic analysis revealed higher NA and lower NAMN amounts when compounds **1**, **2**, **8**, and **19** were added to the reaction mixture, in line with their on-target inhibitory activity (data not shown). We performed enzyme kinetic studies to determine the inhibition constant (K_i) of our putative inhibitors and decipher the fine mechanism underlying their binding to the NAPRT enzyme (Table 3 and Figure 7). Analysis of the kinetic data (V_{max} and K_m in the presence of the different inhibitors) suggests that compounds **1**, **2**, and **19** are un-competitive NAPRT inhibitors, with K_i of 2281, 89, and 295 μM, respectively. Compound **8** shows similar potency as compound **19** (K_i approximately equals 300 μM) and acts as a competitive NAPRT inhibitor, i.e., it competes with NA for the NAPRT catalytic site, as inferred by the fact that V_{max} was not affected and that K_m was increased, in the presence of compound **8** (Figure 7, Table 3). In addition, the fact that compound **8** is one of the structurally closest analogs of 2-HNA and NA also lends support to the proposed mechanism of action of this compound. Due to the low potency of compound **1** on the human NAPRT enzyme (K_i equals 2.3 mM), in addition to its limited anti-cancer activity in our cancer cell models, we decided to exclude this compound from further experiments. Despite the promising activity of compound **2** on the purified NAPRT protein and in OVCAR-5 cells at 100 μM concentration when coupled to FK866, a 200 μM concentration of the same compound showed remarkable activity in the absence of FK866 in the same cell line (Figure 5D). Similar activity without FK866 was also seen in OVCAR-8 and HCT116 cells at 100 μM (Figure 6D,E). Thus, in view of this intrinsic toxicity, compound **2** was also excluded from further testing. Nonetheless, the core structures of compound **1** and compound **2** could be a starting point for future compound optimization steps. Ultimately, we decided to focus on compound **8** and compound **19** for our subsequent analyses.

Table 3. Proposed mechanism of action for putative NAPRT inhibitors.

Compound ID	Vendor ID	Ki (μM)	V_{max}/K_m	Proposed Mechanism
1	F0020-0171	2281	$V_{max}\downarrow/K_m\downarrow$	Un-competetive
2	F0173-0133	88.99	$V_{max}\downarrow/K_m\downarrow$	Un-competetive
8	F1371-0219	307.5	$V_{max}=/K_m\uparrow$	Competitive
19	F2169-0490	295.1	$V_{max}\downarrow/K_m\downarrow$	Un-competitive

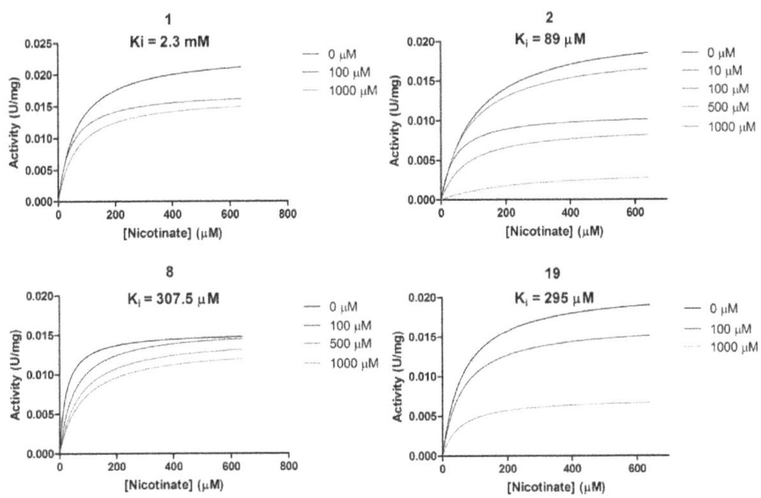

Figure 7. Analysis of NAPRT enzyme activity in the presence or absence of putative NAPRT inhibitors. Graphs represent Michaelis–Menten regression curves of NAPRT reactions performed in the presence of different concentrations of compounds **1**, **2**, **8**, and **19**. The concentration-dependent inhibiting effect on the NAPRT reaction is represented with different colors.

2.3. *In Silico Solubility Prediction and Pharmacokinetic Characterization*

In a previous study involving animal experiments, we were unable to dissolve 2-HNA in saline at the desired concentration for intraperitoneal injections, and, thus, we used its sodium salt as an alternative [23]. Poor water solubility is a major hurdle during the drug development process, especially when a drug is meant to be administered orally or parenterally [37]. It was estimated that approximately 40% of the new chemical entities demonstrate modest solubility in water [37]. Given the promising pharmacological results of our drug candidates compound **8** and compound **19**, we addressed their physicochemical and pharmacokinetic parameters. In order to predict their solubility, we made use of the SwissADME website, a publicly available online computational tool that characterizes physicochemical parameters, ADME properties, and the drug-likeliness of a molecule [38]. Compound **8** and compound **19** possess favorable drug-like properties since they don't violate Lipinski's rule of five. Based on 2 out of the 3 predictive models employed by the software to calculate water solubility, we found a 1.64- and a 2.25-fold improvement in the predicted molar solubility of compound **8** compared to 2-HNA (reported as Log(S) in Table 4). Likewise, the molar solubility of compound **19** was higher than that of 2-HNA according to all 3 estimating methods (Table 4). Moreover, compound **8** and compound **19** had high predicted GI absorption. Compound **8** had the same bioavailability score as 2-HNA and a higher bioavailability score than compound **19** (Table 4). Neither of the two chemical entities seemed to be able to cross the blood–brain barrier (BBB). Finally, neither compound **8** nor compound **19** was predicted to be a substrate of the efflux transporter P-glycoprotein (Pgp), which is frequently associated with cancer resistance against chemotherapeutics [39]. Collectively, these results indicate promising pharmacokinetic features for compound **8** and compound **19**.

Table 4. Predicted water solubility and additional pharmacokinetic properties of the two most promising putative NAPRT inhibitors.

Compound ID	Log S (ESOL)	Log *S* (Ali)	Log *S* (SILICOS-IT)	GI Absorption	Pgp Substrate	BBB Permeant	Bioavailability Score
2-HNA	−1.65	−1.97	−0.8	High	No	No	0.85
8	−1.44	−1.62	−0.8	High	No	No	0.85
19	−1.18	−1.77	−0.26	High	No	No	0.56

3. Discussion

We previously demonstrated that the prototypical NAPRT inhibitor, 2-HNA, synergizes with FK866 in the killing of NAPRT-expressing cancer cells [23]. Herein, we report on the identification of two additional chemical entities that function as NAPRT inhibitors with antineoplastic activity comparable to 2-HNA and with desirable drug-like features.

From a biological standpoint, our studies of cell growth, cellular NAD content, and the enzymatic activity of purified NAPRT unequivocally confirm that compound **8** and compound **19** indeed inhibit this enzyme. Cell-based assays clearly demonstrated that our best NAPRT inhibitor, compound **8** (4-hydroxynicotinic acid), but not 5-hydroxy or 6-hydroxynicotinic acid, exhibited marked anti-cancer activity when combined with FK866 while showing no significant growth inhibition when used alone. These results are consistent with our previous work, showing that per se NAPRT silencing or inhibition do strongly sensitize NAPRT-expressing cancer cells (such as OVCAR-5, OVCAR-8, as well as other cell lines) to NAMPT inhibitors but by themselves have minor anti-proliferative activity [23]. On the other hand, Chowdhry and colleagues showed that the inducible depletion of *NAPRT* caused the regression of OV4 xenografts (PH-amplified ovarian cancer), implying that NAPRT inhibitors might be effective as single agents in similar cancer models that highly depend on the PH pathway to survive [24]. These differences between our studies and the work by Chowdhry and coworkers could be explained by the different cell lines that were utilized.

Our enzyme kinetics analyses indicate that compound **8**, similar to 2-HNA, acts as a competitive NAPRT inhibitor that competes with NA for its enzymatic binding pocket, whereas compound **19** un-competitively inhibits NAPRT. Furthermore, the specificity of our inhibitors was testified by experiments demonstrating that the chemo-sensitizing activity of our inhibitors was abolished upon supplementing cancer cells with sufficient amounts of the substrate (NA) or the downstream product (NAMN) of the NAPRT enzyme. However, similar to the analyses conducted in the case of NAMPT inhibitors [40,41], additional crystallographic studies of the NAPRT enzyme in complex with one or more of our identified inhibitors are warranted to precisely disclose the binding mode of these compounds and describe their interactions within the enzymatic pocket.

Very recently, we demonstrated that gut microbiota caused leukemia cells to display resistance to FK866-induced cell death in vivo when mice were fed with NAM-rich diets through gut-microbiota-derived NA and the consequent activation of the PH pathway in cancer cells (since bacteria use their enzyme nicotinamidase to convert NAM to NA and thereby interconnect the salvage and the PH pathways) [42,43]. Accordingly, coupling FK866 therapy to our NAPRT inhibitors would presumably reverse the protective effect of the intestinal bacteria and restore the anti-tumor effect of NAMPT inhibitors in vivo. Since the in vitro anti-cancer activity observed upon combining NAMPT and NAPRT inhibitors was abrogated when NA was exogenously added in excess to the culture media, it could be argued that the in vivo activity of this combination therapy might be compromised when NA levels rise considerably in the body, as could happen in response to NA- or NAM-rich diets or to NA supplements (e.g., NA is used in gram doses in cases of dyslipidemia due to its lipid-modifying effects) [44]. Future studies should address whether the compounds we identified as NAPRT inhibitors actually show antitumor activity in vivo and whether conditions characterized by high circulating NA levels actually hamper their efficacy. Further improvements in the affinity of these NAPRT inhibitors will increase their therapeutic potential and also reduce the risk of reduced activity in the presence of high NA availability.

4. Materials and Methods

4.1. Reagents and Cell Lines

OVCAR-5 and OVCAR-8 cell lines were obtained from the NCI-60 panel. The HCT116 cell line was purchased from ATCC (LGC Standards S.r.l., Milan, Italy). The cells were maintained in RPMI-1640 cell culture medium supplemented with 10% heat-inactivated fetal bovine serum (FBS; Gibco, Waltham, MA, USA) and 1% antibiotics [penicillin (50 units/mL)/streptomycin (50 µg/mL) (Life Technologies Italia, Monza, Italy)]. Cells were incubated at 37 °C in a humidified atmosphere of 5% CO_2 and 95% air. FK866 was bought from the NIMH Chemical Synthesis and Drug Supply Program. All chemical compounds listed in Table 2 were obtained from Life Chemicals. 5-hydroxynicotinic acid and 6-hydroxynicotinic acid were purchased from Thermo Fisher Scientific. Stock solutions of all the putative NAPRT inhibitors were prepared by dissolving the compounds in DMSO at 100 mM. NA, 2-HNA, and NAMN were obtained from Sigma Aldrich S.r.l.

4.2. Sulforhodamine B (SRB)Assay

To evaluate the anti-proliferative activity of the putative NAPRT inhibitors in the presence or absence of FK866, the sulforhodamine B colorimetric assay was employed [45]. OVCAR-5, OVCAR-8, or HCT116 cells were seeded in 96-well plates (2×10^3 cells/well) and incubated overnight at 37 °C in a humidified atmosphere of 5% CO_2 and 95% air to allow cells to adhere. The day after, the old medium was removed from each well and replaced with a fresh culture medium containing the desired compounds at the indicated final concentrations in triplicate, and the plates were subsequently incubated at 37 °C in a humidified atmosphere of 5% CO_2 and 95% air for 72 h. Afterwards, cold 50% (w/v) trichloroacetic acid (TCA) was gently added to each well to fix the cells (final concentration, 10% TCA). The plates were incubated at 4 °C for 20 min, then washed four times with tap water and left to air-dry. Thereafter, a SRB solution (0.057% w/v in 1% acetic acid) was added to stain the fixed cells, and the plates were shaken for 10 min at room temperature. After staining, the SRB solution was removed, and the plates were washed four times with 1% (v/v) acetic acid and left to air-dry. To solubilize the protein-bound dye, 100 µL of 10 mM trizma base was next added, and the plates were shaken for 10 min at room temperature. Finally, the absorbance was measured at a wavelength of 515 nM by an automated plate reader (Tecan Infinite® 200 PRO instrument).

4.3. Intracellular NAD Levels Measurements

To assess whether the antitumor activity of the newly identified putative NAPRT inhibitors, when combined with FK866, was due to their ability to decrease intracellular NAD levels, we performed NAD measurement as follows: 1×10^5 OVCAR-5 cells were plated in each well of a 12-well plate and left to adhere overnight. The day after, cells were treated with combinations of NAPRT inhibitors and FK866 and incubated at 37 °C in a humidified atmosphere of 5% CO_2 and 95% air for 24 h. The NAPRT inhibitors were used at 100 μM, except for 2-HNA, which was used at 1 mM, and FK866 at 100 nM. After 24 h, cells were lysed with 0.6 M perchloric acid (PCA) at 4 °C and manually detached by a scrapper. The cell lysates were subsequently collected, transferred to new tubes, and diluted in 100 mM Na_2HPO_4 at pH 8. To determine the amount of NAD^+, we utilized a sensitive cyclic assay that takes advantage of the enzymatic activity of alcohol dehydrogenase [46]. Briefly, 100 μL of the diluted samples were pipetted into a white 96-well plate, followed by the addition of 100 μL of the cycling reaction mixture (100 mM Na_2HPO_4, 90 U/mL alcohol dehydrogenase, 10 μM flavinmononucleotide, 2% ethanol, 130 mU/mL diaphorase, 2.5 μg/mL resazurin, and 10 mM nicotinamide). Fluorescence increase was measured every 60 s over 30 min using a fluorescence plate reader (544 nm excitation, 590 nm emission). The NAD content was calculated from a standard curve and normalized against the total protein content that was previously quantified for every test sample using the standard Bradford colorimetric assay (Bio-Rad).

4.4. Recombinant Human NAPRT Production and Purification

The coding sequence for human NAPRT was cloned in a pET23a vector to insert an N-terminal His-Tag. The recombinant protein was produced in BL-21 (DE3) *E. coli* cells as follows: a starting culture of 5 mL was grown overnight at 37 °C in Luria–Bertani medium supplemented with 100 μg/mL ampicillin. The day after, the culture was diluted at 1:100 in a fresh medium and incubated at 25 °C with the addition of 1 mM NA. When 0.3-0.4 OD_{600} was reached, protein expression was induced with 1 mM IPTG, and bacteria growth was continued overnight at 20 °C. The day after, bacteria were harvested by mild centrifugation (5000 rpm, 10 min) in a Beckman Coulter J6-HC centrifuge and resuspended in 1/50 of the original volume with an equilibration buffer composed of 100 mM K_2HPO_4, pH 7.4, 300 mM KCl, and 5 mM imidazole. The cell suspension was sonicated for 10 min at 10 s intervals to disrupt the bacteria cells, and the crude extract was clarified by centrifugation (6000 rpm, 15 min). The recombinant human NAPRT was purified by His-tag affinity chromatography as follows: the supernatant was batch-mixed for 1 h at 4 °C with a HisPur Cobalt resin (Thermo Fisher Scientific, Pittsburg, PA, USA) on a rotating mixer, then the resin was packed into a chromatographic column. The flow-through and the subsequent 10 mM imidazole wash buffer were discarded. The recombinant protein was finally eluted three times with 1 mL of an equilibration buffer containing 150 mM imidazole. The three elutions were merged, and the protein was concentrated with a Protein Concentrator 10K (Pierce-Thermo Fisher Scientific, Pittsburg, PA, USA). The concentrated protein was dialyzed in a SnakeSkin™ dialysis tubing 10K (Thermo Fisher Scientific) overnight at 4 °C against 50 mM Tris/HCl, pH 7.4, 10 mM KCl, and 1 mM DTT to remove the imidazole of the elution buffer. The dialyzed protein was quantified by spectrophotometer absorbance at 280 nm and stored at 4 °C after the addition of 500 μM PRPP to stabilize the protein structure.

4.5. Enzymatic Activity Assays and K_i Calculation

To determine if the effects of the putative NAPRT inhibitors on cell viability and NAD content were indeed caused by an inhibition of NAPRT enzymatic activity, we set enzymatic reactions with the recombinant human NAPRT, analyzed NAMN formation by HPLC, and the K_i was calculated. For each NAPRT inhibitor, reactions with variable NA concentration (between 10 and 640 μM) and variable inhibitor concentration (between 0 and 1000 μM) were performed at 37 °C for a time in which the amount of the product NAMN did not exceed the 10% of the total NA amount. The reactions were blocked by heating samples at 85 °C for 3 min, and the protein was removed by centrifugation. The clarified reactions were analyzed by HPLC with an XTerra MS C18 Column, 125Å, 5 μm, 4.6 mm × 150 mm (Waters) in 100 mM phosphate buffer pH 5 with a gradient of methanol from 0 to 30%. The initial velocities (V_0) were calculated and inserted in the Michaelis–

Menten equation, and subsequently, the K_i of each NAPRT inhibitor was calculated with GraphPad Prism 8.

4.6. In Silico Screening of the Life Chemicals HTS Compound Collection

In order to identify new drug-like small molecules with inhibitory properties towards human NAPRT, we performed a high-throughput docking virtual screening of the Life Chemicals HTS Compound Collection into the crystal structure of human NAPRT (PDB code: 4YUB). The protein structure was processed with the Protein Preparation Wizard (Schrodinger Maestro v. 2017-4), and 18 Å grids encompassing the catalytic pocket of human NAPRT were generated for the functional human NAPRT dimer and its monomer. Ligands were standardly prepared with Gypsum-DL and docked into the two generated models of human NAPRT with AutoDock Vina [47,48]. The top-500 ranked binding poses from each virtual screening were rigorously evaluated to prioritize compounds that displayed favorable interactions with key catalytic residues and suitable fitting in the NAPRT active site. Selected docking hit compounds were purchased from commercial sources and tested in vitro as putative NAPRT inhibitors.

4.7. Statistics

Statistical analyses were carried out with GraphPad Prism software v. 8 (GraphPad Software). All two-group comparisons were performed using an unpaired *t*-test. *p*-values less than 0.05 were considered statistically significant.

5. Conclusions

In summary, we took advantage of in silico drug design techniques to identify two small molecules that selectively inhibit NAPRT. The hit rate observed with our virtual screening procedure is essentially consistent with the hit rate obtained in our previous in silico screenings, including work that led us to discover the first selective SIRT6 inhibitors [49]. These studies further underscore the advantage of virtual screening approaches for drug discovery when compared to the traditional high-throughput screening procedures that are time- and resource-intensive and that typically achieve lower hit rates [50]. Our best candidates, compound **8** and compound **19**, were able to restore the sensitivity of NAPRT-expressing cancer cells to NAMPT inhibitors through NAPRT inhibition. Similar to 2-HNA, they showed anti-cancer activity in the micromolar range. Although a substantial improvement in the potency of NAPRT inhibitors has not been achieved yet, the structural backbones of these two inhibitors lend themselves to future optimization efforts. Lastly, computational analysis supported desirable drug-like and pharmacokinetic features of these agents. Altogether, our study lays the background for further studies of these new NAPRT inhibitors, including in vivo testing in mouse tumor models and further drug optimization steps.

Author Contributions: Conceptualization, A.N. (Alessio Nencioni), M.S.G. and F.P.; conduct of the experiments, M.S.G., F.P., S.R., N.B., I.C., A.K. and M.D.P.; writing—original draft preparation, M.S.G.; writing—review and editing, M.S.G., F.P., A.N. (Alessio Nencioni), A.N. (Aimable Nahimana), S.B., J.F., A.D.R., A.K., F.M., M.C. and M.A.D.; funding acquisition, A.N. (Alessio Nencioni). All authors have read and agreed to the published version of the manuscript.

Funding: This work was supported in part by the Associazione Italiana per la Ricerca sul Cancro (AIRC; IG#17736 and #22098) to A. Nencioni and (AIRC; IG#19172) to A. Del Rio, the 5 × 1000 Funds to the IRCCS Ospedale Policlinico San Martino to A. Nencioni, the BC161452P1 grant of the Breast Cancer Research Program (U.S. Department of Defense) to A. Nencioni and the Italian Ministry of Health (PE-2016-02362694 and PE-2016-02363073). This work received funding from the European Union's Horizon 2020 research and innovation programme under the Marie Skłodowska-Curie grant agreement no. 813284.

Institutional Review Board Statement: Not applicable.

Informed Consent Statement: Not applicable.

Data Availability Statement: Data is contained within the article.

Conflicts of Interest: The authors declare no conflict of interest.

References

1. Chiarugi, A.; Dölle, C.; Felici, R.; Ziegler, M. The NAD Metabolome—A Key Determinant of Cancer Cell Biology. *Nat. Rev. Cancer* **2012**, *12*, 741–752. [CrossRef] [PubMed]
2. Cantó, C.; Menzies, K.J.; Auwerx, J. NAD+ Metabolism and the Control of Energy Homeostasis: A Balancing Act between Mitochondria and the Nucleus. *Cell Metab.* **2015**, *22*, 31–53. [CrossRef] [PubMed]
3. Xie, N.; Zhang, L.; Gao, W.; Huang, C.; Huber, P.E.; Zhou, X.; Li, C.; Shen, G.; Zou, B. NAD + Metabolism: Pathophysiologic Mechanisms and Therapeutic Potential. *Signal Transduct. Target. Ther.* **2020**, *5*, 1–37. [CrossRef]
4. Navas, L.E.; Carnero, A. NAD^+ Metabolism, Stemness, the Immune Response, and Cancer. *Signal Transduct. Target. Ther.* **2021**, *6*, 1–20. [CrossRef]
5. Covarrubias, A.J.; Perrone, R.; Grozio, A.; Verdin, E. NAD^+ Metabolism and Its Roles in Cellular Processes during Ageing. *Nat. Rev. Mol. Cell Biol.* **2021**, *22*, 119–141. [CrossRef]
6. Preiss, J.; Handler, P. Biosynthesis of Diphosphopyridine Nucleotide I. Identification of intermediates. *J. Biol. Chem.* **1958**, *233*, 488–492. [CrossRef]
7. Preiss, J.; Handler, P. Biosynthesis of Diphosphopyridine Nucleotide. II. Enzymatic Aspects. *J. Biol. Chem.* **1958**, *233*, 493–500. [CrossRef]
8. Hara, N.; Yamada, K.; Shibata, T.; Osago, H.; Hashimoto, T.; Tsuchiya, M. Elevation of Cellular NAD Levels by Nicotinic Acid and Involvement of Nicotinic Acid Phosphoribosyltransferase in Human Cells. *J. Biol. Chem.* **2007**, *282*, 24574–24582. [CrossRef]
9. Liu, L.; Su, X.; Quinn, W.J.; Hui, S.; Krukenberg, K.; Frederick, D.W.; Redpath, P.; Zhan, L.; Chellappa, K.; White, E.; et al. Quantitative Analysis of NAD Synthesis-Breakdown Fluxes. *Cell Metab.* **2018**, *27*, 1067–1080.e5. [CrossRef]
10. Bogan, K.L.; Brenner, C. Nicotinic Acid, Nicotinamide, and Nicotinamide Riboside: A Molecular Evaluation of NAD^+ Precursor Vitamins in Human Nutrition. *Annu. Rev. Nutr.* **2008**, *28*, 115–130. [CrossRef]
11. Bieganowski, P.; Brenner, C. Discoveries of Nicotinamide Riboside as a Nutrient and Conserved NRK Genes Establish a Preiss-Handler Independent Route to NAD+ in Fungi and Humans. *Cell* **2004**, *117*, 495–502. [CrossRef]
12. Yang, Y.; Zhang, N.; Zhang, G.; Sauve, A.A. NRH Salvage and Conversion to NAD + Requires NRH Kinase Activity by Adenosine Kinase. *Nat. Metab.* **2020**, *2*, 364–379. [CrossRef] [PubMed]
13. Giroud-Gerbetant, J.; Joffraud, M.; Giner, M.P.; Cercillieux, A.; Bartova, S.; Makarov, M.V.; Zapata-Pérez, R.; Sánchez-García, J.L.; Houtkooper, R.H.; Migaud, M.E.; et al. A Reduced Form of Nicotinamide Riboside Defines a New Path for NAD+ Biosynthesis and Acts as an Orally Bioavailable NAD^+ Precursor. *Mol. Metab.* **2019**, *30*, 192–202. [CrossRef] [PubMed]
14. Hanahan, D.; Weinberg, R.A. Hallmarks of Cancer: The Next Generation. *Cell* **2011**, *144*, 646–674. [CrossRef] [PubMed]
15. Hasmann, M.; Schemainda, I. FK866, a Highly Specific Noncompetitive Inhibitor of Nicotinamide Phosphoribosyltransferase, Represents a Novel Mechanism for Induction of Tumor Cell Apoptosis. *Cancer Res.* **2003**, *63*, 7436–7442. [PubMed]
16. Hjarnaa, P.J.; Jonsson, E.; Latini, S.; Dhar, S.; Larsson, R.; Bramm, E.; Skov, T.; Binderup, L. CHS 828, a Novel Pyridyl Cyanoguanidine with Potent Antitumor Activity in Vitro and in Vivo. *Cancer Res.* **1999**, *59*, 5751–5757.
17. Ghanem, M.S.; Monacelli, F.; Nencioni, A. Advances in NAD-Lowering Agents for Cancer Treatment. *Nutrients* **2021**, *13*, 1665. [CrossRef]
18. Yaku, K.; Okabe, K.; Hikosaka, K.; Nakagawa, T. NAD Metabolism in Cancer Therapeutics. *Front. Oncol.* **2018**, *8*, 622. [CrossRef]
19. Goldinger, S.M.; Gobbi Bischof, S.; Fink-Puches, R.; Klemke, C.-D.; Dréno, B.; Bagot, M.; Dummer, R. Efficacy and Safety of APO866 in Patients With Refractory or Relapsed Cutaneous T-Cell Lymphoma: A Phase 2 Clinical Trial. *JAMA Dermatol.* **2016**, *152*, 837–839. [CrossRef]
20. Hovstadius, P.; Larsson, R.; Jonsson, E.; Skov, T.; Kissmeyer, A.-M.; Krasilnikoff, K.; Bergh, J.; Karlsson, M.O.; Lönnebo, A.; Ahlgren, J. A Phase I Study of CHS 828 in Patients with Solid Tumor Malignancy. *Clin. Cancer Res. Off. J. Am. Assoc. Cancer Res.* **2002**, *8*, 2843–2850.
21. von Heideman, A.; Berglund, A.; Larsson, R.; Nygren, P. Safety and Efficacy of NAD Depleting Cancer Drugs: Results of a Phase I Clinical Trial of CHS 828 and Overview of Published Data. *Cancer Chemother. Pharmacol.* **2010**, *65*, 1165–1172. [CrossRef] [PubMed]
22. Holen, K.; Saltz, L.B.; Hollywood, E.; Burk, K.; Hanauske, A.-R. The Pharmacokinetics, Toxicities, and Biologic Effects of FK866, a Nicotinamide Adenine Dinucleotide Biosynthesis Inhibitor. *Invest. New Drugs* **2008**, *26*, 45–51. [CrossRef] [PubMed]
23. Piacente, F.; Caffa, I.; Ravera, S.; Sociali, G.; Passalacqua, M.; Vellone, V.G.; Becherini, P.; Reverberi, D.; Monacelli, F.; Ballestrero, A.; et al. Nicotinic Acid Phosphoribosyltransferase Regulates Cancer Cell Metabolism, Susceptibility to NAMPT Inhibitors, and DNA Repair. *Cancer Res.* **2017**, *77*, 3857–3869. [CrossRef]
24. Chowdhry, S.; Zanca, C.; Rajkumar, U.; Koga, T.; Diao, Y.; Raviram, R.; Liu, F.; Turner, K.; Yang, H.; Brunk, E.; et al. NAD Metabolic Dependency in Cancer Is Shaped by Gene Amplification and Enhancer Remodelling. *Nature* **2019**, *569*, 570–575. [CrossRef] [PubMed]
25. Rongvaux, A.; Andris, F.; Van Gool, F.; Leo, O. Reconstructing Eukaryotic NAD Metabolism. *BioEssays News Rev. Mol. Cell. Dev. Biol.* **2003**, *25*, 683–690. [CrossRef]

26. Shames, D.S.; Elkins, K.; Walter, K.; Holcomb, T.; Du, P.; Mohl, D.; Xiao, Y.; Pham, T.; Haverty, P.M.; Liederer, B.; et al. Loss of NAPRT1 Expression by Tumor-Specific Promoter Methylation Provides a Novel Predictive Biomarker for NAMPT Inhibitors. *Clin. Cancer Res.* **2013**, *19*, 6912–6923. [CrossRef]
27. Watson, M.; Roulston, A.; Bélec, L.; Billot, X.; Marcellus, R.; Bédard, D.; Bernier, C.; Branchaud, S.; Chan, H.; Dairi, K.; et al. The Small Molecule GMX1778 Is a Potent Inhibitor of NAD^+ Biosynthesis: Strategy for Enhanced Therapy in Nicotinic Acid Phosphoribosyltransferase 1-Deficient Tumors. *Mol. Cell. Biol.* **2009**, *29*, 5872–5888. [CrossRef]
28. Peterse, E.F.P.; van den Akker, B.E.W.M.; Niessen, B.; Oosting, J.; Suijker, J.; de Jong, Y.; Danen, E.H.J.; Cleton-Jansen, A.-M.; Bovée, J.V.M.G. NAD Synthesis Pathway Interference Is a Viable Therapeutic Strategy for Chondrosarcoma. *Mol. Cancer Res.* **2017**, *15*, 1714–1721. [CrossRef]
29. Lee, J.; Kim, H.; Lee, J.E.; Shin, S.-J.; Oh, S.; Kwon, G.; Kim, H.; Choi, Y.Y.; White, M.A.; Paik, S.; et al. Selective Cytotoxicity of the NAMPT Inhibitor FK866 Toward Gastric Cancer Cells With Markers of the Epithelial-Mesenchymal Transition, Due to Loss of NAPRT. *Gastroenterology* **2018**, *155*, 799–814.e13. [CrossRef]
30. Tateishi, K.; Wakimoto, H.; Iafrate, A.J.; Tanaka, S.; Loebel, F.; Lelic, N.; Wiederschain, D.; Bedel, O.; Deng, G.; Zhang, B.; et al. Extreme Vulnerability of IDH1 Mutant Cancers to NAD^+ Depletion. *Cancer Cell* **2015**, *28*, 773–784. [CrossRef]
31. Fons, N.R.; Sundaram, R.K.; Breuer, G.A.; Peng, S.; McLean, R.L.; Kalathil, A.N.; Schmidt, M.S.; Carvalho, D.M.; Mackay, A.; Jones, C.; et al. PPM1D Mutations Silence NAPRT Gene Expression and Confer NAMPT Inhibitor Sensitivity in Glioma. *Nat. Commun.* **2019**, *10*, 3790. [CrossRef]
32. Gaut, Z.N.; Solomon, H.M. Inhibition of Nicotinate Phosphoribosyltransferase in Human Platelet Lysate by Nicotinic Acid Analogs. *Biochem. Pharmacol.* **1971**, *20*, 2903–2906. [CrossRef]
33. Gaut, Z.N.; Solomon, H.M. Uptake and Metabolism of Nicotinic Acid by Human Blood Platelets. Effects of Structure Analogs and Metabolic Inhibitors. *Biochim. Biophys. Acta* **1970**, *201*, 316–322. [CrossRef]
34. Gaut, Z.N.; Solomon, H.M. Inhibition of Nicotinate Phosphoribosyl Transferase by Nonsteroidal Anti-Inflammatory Drugs: A Possible Mechanism of Action. *J. Pharm. Sci.* **1971**, *60*, 1887–1888. [CrossRef] [PubMed]
35. Galassi, L.; Di Stefano, M.; Brunetti, L.; Orsomando, G.; Amici, A.; Ruggieri, S.; Magni, G. Characterization of Human Nicotinate Phosphoribosyltransferase: Kinetic Studies, Structure Prediction and Functional Analysis by Site-Directed Mutagenesis. *Biochimie* **2012**, *94*, 300–309. [CrossRef] [PubMed]
36. Marletta, A.S.; Massarotti, A.; Orsomando, G.; Magni, G.; Rizzi, M.; Garavaglia, S. Crystal Structure of Human Nicotinic Acid Phosphoribosyltransferase. *FEBS Open Bio* **2015**, *5*, 419–428. [CrossRef]
37. Savjani, K.T.; Gajjar, A.K.; Savjani, J.K. Drug Solubility: Importance and Enhancement Techniques. *ISRN Pharm.* **2012**, *2012*, 195727. [CrossRef]
38. Daina, A.; Michielin, O.; Zoete, V. SwissADME: A Free Web Tool to Evaluate Pharmacokinetics, Drug-Likeness and Medicinal Chemistry Friendliness of Small Molecules. *Sci. Rep.* **2017**, *7*, 42717. [CrossRef]
39. Nanayakkara, A.K.; Follit, C.A.; Chen, G.; Williams, N.S.; Vogel, P.D.; Wise, J.G. Targeted Inhibitors of P-Glycoprotein Increase Chemotherapeutic-Induced Mortality of Multidrug Resistant Tumor Cells. *Sci. Rep.* **2018**, *8*, 967. [CrossRef]
40. Khan, J.A.; Tao, X.; Tong, L. Molecular Basis for the Inhibition of Human NMPRTase, a Novel Target for Anticancer Agents. *Nat. Struct. Mol. Biol.* **2006**, *13*, 582–588. [CrossRef]
41. Kang, G.B.; Bae, M.-H.; Kim, M.-K.; Im, I.; Kim, Y.-C.; Eom, S.H. Crystal Structure of Rattus Norvegicus Visfatin/PBEF/Nampt in Complex with an FK866-Based Inhibitor. *Mol. Cells* **2009**, *27*, 667–671. [CrossRef] [PubMed]
42. ElMokh, O.; Matsumoto, S.; Biniecka, P.; Bellotti, A.; Schaeuble, K.; Piacente, F.; Gallart-Ayala, H.; Ivanisevic, J.; Stamenkovic, I.; Nencioni, A.; et al. Gut Microbiota Severely Hampers the Efficacy of NAD-Lowering Therapy in Leukemia. *Cell Death Dis.* **2022**, *13*, 320. [CrossRef] [PubMed]
43. Shats, I.; Williams, J.G.; Liu, J.; Makarov, M.V.; Wu, X.; Lih, F.B.; Deterding, L.J.; Lim, C.; Xu, X.; Randall, T.A.; et al. Bacteria Boost Mammalian Host NAD Metabolism by Engaging the Deamidated Biosynthesis Pathway. *Cell Metab.* **2020**, *31*, 564–579.e7. [CrossRef] [PubMed]
44. Gille, A.; Bodor, E.T.; Ahmed, K.; Offermanns, S. Nicotinic Acid: Pharmacological Effects and Mechanisms of Action. *Annu. Rev. Pharmacol. Toxicol.* **2008**, *48*, 79–106. [CrossRef]
45. Vichai, V.; Kirtikara, K. Sulforhodamine B Colorimetric Assay for Cytotoxicity Screening. *Nat. Protoc.* **2006**, *1*, 1112–1116. [CrossRef]
46. Graeff, R.; Lee, H.C. A Novel Cycling Assay for Cellular CADP-Ribose with Nanomolar Sensitivity. *Biochem. J.* **2002**, *361*, 379. [CrossRef]
47. Trott, O.; Olson, A.J. AutoDock Vina: Improving the Speed and Accuracy of Docking with a New Scoring Function, Efficient Optimization, and Multithreading. *J. Comput. Chem.* **2010**, *31*, 455–461. [CrossRef]
48. Ropp, P.J.; Spiegel, J.O.; Walker, J.L.; Green, H.; Morales, G.A.; Milliken, K.A.; Ringe, J.J.; Durrant, J.D. Gypsum-DL: An Open-Source Program for Preparing Small-Molecule Libraries for Structure-Based Virtual Screening. *J. Cheminformatics* **2019**, *11*, 34. [CrossRef]
49. Parenti, M.D.; Grozio, A.; Bauer, I.; Galeno, L.; Damonte, P.; Millo, E.; Sociali, G.; Franceschi, C.; Ballestrero, A.; Bruzzone, S.; et al. Discovery of Novel and Selective SIRT6 Inhibitors. *J. Med. Chem.* **2014**, *57*, 4796–4804. [CrossRef]
50. Sliwoski, G.; Kothiwale, S.; Meiler, J.; Lowe, E.W. Computational Methods in Drug Discovery. *Pharmacol. Rev.* **2014**, *66*, 334–395. [CrossRef]

pharmaceuticals

Article

Physiologically Based Pharmacokinetic (PBPK) Modeling to Predict PET Image Quality of Three Generations EGFR TKI in Advanced-Stage NSCLC Patients

I. H. Bartelink [1,*,†], E. A. van de Stadt [2,†], A. F. Leeuwerik [1], V. L. J. L. Thijssen [3], J. R. I. Hupsel [1], J. F. van den Nieuwendijk [1], I. Bahce [2], M. Yaqub [4] and N. H. Hendrikse [4]

1 Department of Clinical Pharmacology and Pharmacy, Amsterdam UMC Location Vrije Universiteit Amsterdam, Boelelaan 1117, 1081 HV Amsterdam, The Netherlands; a.f.leeuwerik@gmail.com (A.F.L.); j.r.i.hupsel@amsterdamumc.nl (J.R.I.H.); j.f.vandennieuwendijk@amsterdamumc.nl (J.F.v.d.N.)
2 Department of Pulmonary Medicine, Amsterdam UMC Location Vrije Universiteit Amsterdam, Boelelaan 1117, 1081 HV Amsterdam, The Netherlands; e.vandestadt@amsterdamumc.nl (E.A.v.d.S.); i.bahce@amsterdamumc.nl (I.B.)
3 Department of Radiation Oncology, Amsterdam UMC Location Amsterdam Medical Center, Meibergdreef 9, 1105 AZ Amsterdam, The Netherlands; v.thijssen@amsterdamumc.nl
4 Department of Radiology and Nuclear Medicine, Amsterdam UMC Location Vrije Universiteit Amsterdam, Boelelaan 1117, 1081 HV Amsterdam, The Netherlands; maqsood.yaqub@amsterdamumc.nl (M.Y.); nh.hendrikse@amsterdamumc.nl (N.H.H.)
* Correspondence: i.bartelink@amsterdamumc.nl
† These authors contributed equally to this work.

Abstract: Introduction: Epidermal growth factor receptor (EGFR) mutated NSCLC is best treated using an EGFR tyrosine kinase inhibitor (TKI). The presence and accessibility of EGFR overexpression and mutation in NSCLC can be determined using radiolabeled EGFR TKI PET/CT. However, recent research has shown a significant difference between image qualities (i.e., tumor-to-lung contrast) in three generation EGFR TKIs: ^{11}C-erlotinib, ^{18}F-afatinib and ^{11}C-osimertinib. In this research we aim to develop a physiological pharmacokinetic (PBPK)-model to predict tumor-to-lung contrast and as a secondary outcome the uptake of healthy tissue of the three tracers. Methods: Relevant physicochemical and drug specific properties (e.g., pKa, lipophilicity, target binding) for each TKI were collected and applied in established base PBPK models. Key hallmarks of NSCLC include: immune tumor deprivation, unaltered tumor perfusion and an acidic tumor environment. Model accuracy was demonstrated by calculating the prediction error (PE) between predicted tissue-to-blood ratios (TBR) and measured PET-image-derived TBR. Sensitivity analysis was performed by excluding each key component and comparing the PE with the final mechanistical PBPK model predictions. Results: The developed PBPK models were able to predict tumor-to-lung contrast for all EGFR-TKIs within threefold of observed PET image ratios (PE tumor-to-lung ratio of −90%, +44% and −6.3% for erlotinib, afatinib and osimertinib, respectively). Furthermore, the models depicted agreeable whole-body distribution, showing high tissue distribution for osimertinib and afatinib and low tissue distribution at high blood concentrations for erlotinib (mean PE, of −10.5%, range −158%–+190%, for all tissues). Conclusion: The developed PBPK models adequately predicted the image quality of afatinib and osimertinib and erlotinib. Some deviations in predicted whole-body TBR lead to new hypotheses, such as increased affinity for mutated EGFR and active influx transport (erlotinib into excreting tissues) or active efflux (afatinib from brain), which is currently unaccounted for. In the future, PBPK models may be used to predict the image quality of new tracers.

Keywords: NSCLC; EGFR TKI; PBPK modeling; PET/CT

check for updates

Citation: Bartelink, I.H.; van de Stadt, E.A.; Leeuwerik, A.F.; Thijssen, V.L.J.L.; Hupsel, J.R.I.; van den Nieuwendijk, J.F.; Bahce, I.; Yaqub, M.; Hendrikse, N.H. Physiologically Based Pharmacokinetic (PBPK) Modeling to Predict PET Image Quality of Three Generations EGFR TKI in Advanced-Stage NSCLC Patients. *Pharmaceuticals* **2022**, *15*, 796. https://doi.org/10.3390/ph15070796

Academic Editors: Marialuigia Fantacuzzi and Mariangela Agamennone

Received: 30 May 2022
Accepted: 21 June 2022
Published: 27 June 2022

1. Introduction

Lung cancer is one of the most prevalent cancer types with over 2 million new cases each year worldwide [1–3]. Chemotherapy has long been the sole treatment of metastatic nonsmall cell lung cancer (NSCLC), but over the past decade, targeted therapies against specific oncogenic driver pathways have been developed [4]. One of these oncogenic driver pathways is the epidermal growth factor receptor (EGFR) pathway [5]. An activating mutation in the kinase domain of the EGF receptor can lead to activation of the receptor. EGFR tyrosine kinase inhibitors (TKIs) block this activating pathway [5,6]. In recent years, three generations of EGFR-TKIs have been developed. First-generation EGFR inhibitors such as erlotinib bind reversibly to EGFR harboring mutations and to a lesser extent to wild-type EGFR [5–7]. To overcome resistance, second-generation TKIs such as afatinib were developed, which bind irreversibly to EGFR [6]. To achieve more selective binding to T790M mutant EGFR, third-generation TKIs such as osimertinib were developed [8]. These three generations of TKIs vary in potency and binding characteristics [9–13]. Treatment using EGFR-TKIs has shown better response rates and longer durations of response and has become the standard of care in patients where an activating EGFR mutation is present [1,3,14].

In recent years, research has been conducted using positron emission tomography (PET) to assess EGFR mutational status and to explore PET drug uptake as a predictive biomarker for response to EGFR-TKI treatment. Three generations of EGFR-directed PET tracers were developed by our group and the tracer uptake was studied in patients exposed to ^{11}C-erlotinib, ^{18}F-afatinib and ^{11}C-osimertinib [15–20]. Such PET tracers can be used to predict whole-body and tumor drug uptake and thereby guide EGFR-TKI drug treatment. In a recent study, we compared published data from NSCLC patient scans of the three generations radiolabeled EGFR TKIs [21]. Three tracers were analyzed: ^{11}C-erlotinib, ^{18}F-afatinib and ^{11}C-osimertinib. Tracer uptake was quantified using tumor-to-blood ratio (TBR). Previous research has shown that TBR is an adequate measure for quantification of tracer uptake in ^{11}C-erlotinib and ^{18}F-afatinib, and could also be applied for ^{11}C-osimertinib [15,20,21]. Furthermore, tumor-to-lung contrast was used to assess PET image quality of each tracer. ^{11}C-osimertinib showed a negative contrast: tumor-tissue uptake was 20% lower than surrounding lung tissue. In contrast, ^{18}F-afatinib showed a better contrast, 96% higher in tumor tissue than in surrounding tissue. ^{11}C-erlotinib image quality was deemed superior, with a tumor-to-lung contrast value of 178% [21].

Based on our analyses [21], we hypothesized that the physicochemical drug properties may explain some of the variability in penetration of the tracers in different tissues. Important physicochemical properties are lipophilicity (log P) and basicity, defined by the pKa (negative log of the acid dissociation constant or Ka value). Lipophilicity affects protein binding (e.g., albumin (ALB)) and transport and binding to neutral lipids and phospholipids (NL/NP) in the cell membranes. A strong basic drug (a compound with a high pKa), is highly protonated at physiological (plasma) pH levels, whereas weak bases (pKa < 7) are mostly unprotonated at physiological pH levels (Figure 1A). Protonation, whether an H^+ atom is added to a base, affects transport over negatively charged acidic phospholipid membranes (Figure 1C). Unprotonated lipophilic drugs will bind to albumin (Figure 1D) and membrane NL/PL (Figure 1E,F).

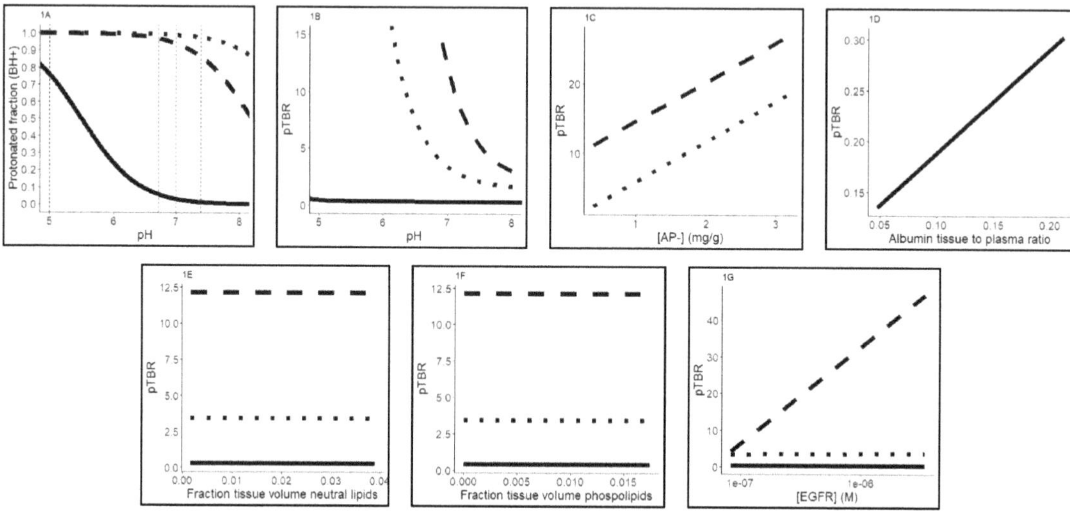

Figure 1. Schematic representation of the change in drug distribution in relation to relevant physiological values in human tissues. Uninterrupted lines represent erlotinib, dotted lines represent osimertinib and dashed lines represent afatinib. (**A**) pH versus protonation (Equation (A)). (**B**) pH of the intracellular water versus predicted TBR, tumor-to-blood ratio (Equation (B)). (**C**) AP-, acidic phospholipids versus predicted TBR (Equation (C)). (**D**) Albumin tissue-to-plasma ratio versus predicted TBR (Equation (D)). (**E**, **F**) fraction of neutral lipids and phospholipids versus predicted TBR (Equation (E)). (**G**) EGFR tissue concentration versus predicted TBR. pTBR: predicted tumor to blood ratio, BH: protonated base, AP-: acidic phospholipids, EGFR: epidermal growth factor receptor, Fiw: fraction intracellular water, B:P: blood to plasma partition coefficient of compound, Fu: fraction unbound drug of compound, pHiw: pH of intracellular water, pHp: pH of plasma, Fnp: fraction of neutral phospholipids, Fnl: fraction of neutral lipids, pKa: basicity, Ka: association constant.

Equations:

$$Fraction\ BH^{+} = \frac{1}{1+10^{pKa-pH}} \tag{1}$$

$$pTBR_{pHiw}\ or\ contribution\ pH = \left(\frac{1+10^{pKa-pHiw}}{1+10^{pKa-pHp}} * fiw\right) * \frac{Fu}{B:P} \tag{2}$$

$$pTBR_{AP-}\ or\ AP\text{−}binding = \left(\frac{Ka * [AP-] * 10^{pKa-pHiw}}{1+10^{pKa-pHp}}\right) * \frac{Fu}{B:P} \tag{3}$$

$$pTBR_{ALB}\ or\ Albumin\ binding = \left(Ka, ALB * \frac{[ALB, tissue]}{[ALB, plasma]}\right) * \frac{Fu}{B:P} \tag{4}$$

$$pTBR_{Lipids}\ or\ Lipid\ binding = \left(\frac{P * Fnl + (0.3P+0.7) * Fnp}{1+10^{pKa-pHp}}\right) * \frac{Fu}{B:P} \tag{5}$$

$$pTBR_{EGFR}\ or\ EGFR\ binding = \left(\frac{\frac{[EGFR]}{Kd} * \left(1+10^{pKa-pHiw}\right)}{1+10^{pKa-pHp}} * fiw\right) * \frac{Fu}{B:P} \tag{6}$$

Because of the differences in basicity, we hypothesize that lysosomal sequestration, or the trapping of the tracer in the lysosome, plays a role in the differences in tracer uptake. Lysosomes are acidic membrane-bound organelles which are capable of digesting biomolecules [22]. Since the pH level of the lysosome is lower (i.e., the lysosome is more acidic) than the cytosol, protonation will occur here to a greater extent than in the neutral environment of the cytosol (Figure 1A). When protonated transport through the lysosomal membrane decreases, it leads to "trapping" of the protonated base in the lysosome. Lysosomal sequestration is a well-known cellular distribution pattern and may cause intrinsic resistance to strong basic EGFR-TKIs due to decreased cytosolic concentrations and thereby less availability for EGFR binding [23,24].

In addition to EGFR abundance, other factors in the microenvironment of the tumor may predict the tumor-to-lung contrast. These compound- and tissue-specific properties may influence the whole-body distribution resulting in different tumor-to-lung contrast of the three compounds.

Biodistribution of small molecules, such as EGFR-TKIs, related to compound- and tissue-specific properties can be described using physiologically based pharmacokinetic models (PBPK). A physiologically based pharmacokinetic model (PBPK model) is a mathematical, mechanistical description of a physiological system where compartments are used to represent various tissues in the NSCLC population. Each compartment corresponds with physiological volumes of each tissue/organ [25]. Although PBPK models are frequently used in pharmacology research, no prior research has been done using PBPK models in relationship to PET tracers. An advantage of such mechanistic models is its predictive potential. In this study, we developed a PBPK model reflecting the essential features of tissue distribution of EGFR-TKIs with the primary aim to predict the image quality by predicting the right tumor-to-lung contrast. As a secondary aim, the model was used to predict the whole-body distribution. When fully validated, mechanistic PBPK models can be applied to predict tumor drug uptake in a wide array of diseases and structurally diverse compounds, and can be used to predict the image quality of new tracers—drugs with large tumor/tissue contrast.

2. Results

2.1. Components of the Mechanistical PBPK Model

The mechanistical PBPK model consists of the base model for each TKI, lysosomal sequestration for the strong bases (afatinib and osimertinib), tumor immune deprivation, unaltered tumor perfusion, EGFR target binding and a more acidic extracellular water (i.e., water located just outside the tumor cells in the tumor microenvironment) in tumor tissue. Each component added to the literature based on physicochemical base models was analyzed during the sensitivity analyses (below). The model, including unaltered perfusion but without a tumor vascularization coefficient, optimally predicted the tumor-to-lung contrast of all three EGFRs. See Supplemental Materials S-IV for full analysis of the mechanistical model including vascularization versus the final model excluding this component.

2.2. PBPK Model Validation Using PET Data

The tumor-to-lung contrast was predicted using the mechanistic PBPK model to visualize differences in image quality observed in PET imaging, as shown in Figure 2A. The predicted TBR by using the final PBPK model are also shown in Figure 2B. Both observed and predicted TBR values showed high uptake of osimertinib and afatinib in lung and tumor tissue (TBR > 1) and high blood concentrations compared to tissue concentrations for erlotinib (TBR < 1, Figure 2A,B). The tumor-to-lung contrast for each EGFR TKI was predicted within the established boundaries, i.e., within threefold of the observed value. Furthermore, the model correctly predicts a tumor-to-lung contrast of >1 for erlotinib and afatinib, and <1 for osimertinib (Table 1).

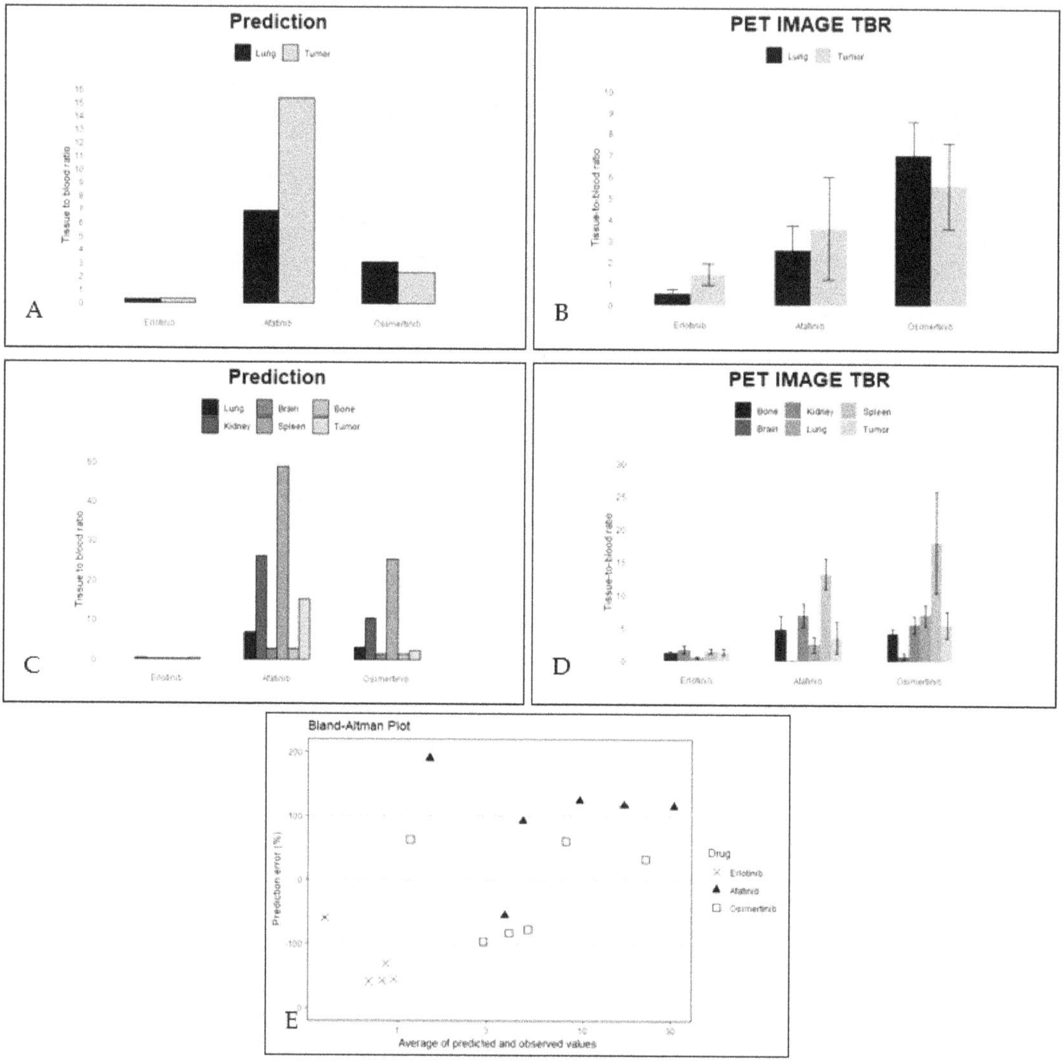

Figure 2. The association between PET-image-derived TBR and model-predicted TBR. (**A**) PET-image-derived TBR (left) vs. (**B**) model-predicted TBR of lung and tumor. (**C**) PET-image-derived TBR (left) vs. (**D**) model-predicted TBR of all tissues of interest. For patient data, standard deviations are given. (**E**) Bland–Altman plot showing accuracy of the model to predict tissue uptake. The solid black line represents the mean and the dashed lines a factor 3 of both sides of zero. Percentage of predictions falling within 3-fold: erlotinib 16.6%, afatinib 33.3% and osimertinib 100%.

Table 1. Contribution of the different components in the final mechanistic PBPK models to the predicted TBR in the lung and the tumor.

	Erlotinib	Afatinib	Osimertinib
		Lung	
EGFR binding	0.21%	16.73%	0.14%
Lysosomal trapping	n.a.	49.19%	59.16%
NL/NP	1.12%	0.01%	0.00%
Albumin	73.01%	n.a.	n.a.
AP-	n.a.	32.63%	39.74%
IW	14.54%	1.08%	0.72%
EW	0.36%	0.36%	0.23%
		Tumor	
EGFR binding	1.89%	72.17%	1.85%
Lysosomal trapping	n.a.	11.99%	42.79%
NL/NP	1.05%	0.00%	0.00%
Albumin	72.32%	n.a.	n.a.
AP-	n.a.	14.63%	52.91%
IW	13.71%	0.48%	0.96%
EW	11.02%	0.73%	1.49%

NL/NP neutral (phospho)lipids AP- acidic phospholipids IW intracellular water EW extracellular water.

Table 1 describes the contribution of different components in the mechanistic PBPK model. Erlotinib binds extensively to albumin in tissue, whereas osimertinib and afatinib predominantly bind to AP- in the cellular membranes and are sequestered in the lysosome. Substantial decrease in lysosomal sequestration was predicted for both strong basic EGFR-TKIs in the less lysosome-rich tumor in comparison with lung tissue. Tumor EGFR binding of afatinib was predicted to be extensive for and comprised 72.17% of all tissue binding, whereas for erlotinib and osimertinib the model showed that only a minor fraction of the tissue fraction bound to EGFR (1.89 and 1.85%, respectively). The whole-body distribution of the three EGFR-TKIs was described by the mechanistic PBPK models. The predicted TBR values by the mechanistic PBPK models are shown in Figure 2D compared to the measured TBR (obtained during PET imaging, Figure 2C).

As secondary outcome, the mechanistical PBPK model is able to predict the overall body distribution of the TKIs, with extensive distribution to most tissues for osimertinib and afatinib and limited tissue penetration for erlotinib. The observed and predicted TBR values of osimertinib and afatinib in most tissues were >1 and for erlotinib <1 (Figure 2C,D). The TBR predicted by the mechanistic PBPK model correlated strongly with the PET imaging data (r2: 0.593 with $p < 0.0003$; $\alpha = 0.01$) with a mean PE of −10.5% (CI95% of the data: −126.0 to 104.9, Figure 2E). For afatinib, 33.3% of tissues were predicted within a factor of 3 of the observed value. The data point falling outside this range represented brain uptake of afatinib, and was predicted to be 189.7% times higher than observed, and tumor predictions above the 3-fold limit of the observed mean value (Table 2). However, for osimertinib, 100% of tissues were predicted to be within 3-fold of the observed tissue uptake of PET imaging. For erlotinib, the PBPK model predicted the TBR compared to the observed ratio less accurately: only 16.6% of TBR was predicted within 3-fold of the observed values. The lung was predicted accurately with a PE of −58.8%. However, the predicted TBR of spleen, kidney, bone and tumor were underestimated (Table 2).

Table 2. PET-image-derived tissue-to-blood ratios compared to predicted TBR in all tissues of interest. SD is given in brackets.

	Erlotinib			Afatinib			Osimertinib		
	Predicted	Observed	Prediction Error (%)	Predicted	Observed	Prediction Error (%)	Predicted	Observed	Prediction Error (%)
Brain	0.13	n.a.	n.a.	2.85	0.08 (0.03)	189.7	1.52	0.79 (0.5)	62.7
Lung	0.28	0.51 (0.2)	−58.8	6.89	2.54 (1.2)	92.4	3.11	7.01 (1.6)	−77.1
Spleen	0.17	1.46 (0.4)	−157.8	48.72	13.23 (2.3)	114.6	25.33	18.09 (7.7)	33.3
Kidney	0.21	1.69 (0.6)	−155.6	26.20	6.93 (1.8)	116.3	10.48	5.61 (2.0)	60.6
Bone	0.14	1.23 (0.2)	−158.3	2.72	4.81 (2.0)	−55.3	1.48	4.24 (0.7)	−96.6
Tumor	0.30	1.42 (0.5)	−131.1	15.36	3.60 (2.4)	124.1	2.33	5.60 (2.0)	−82.4

2.3. Sensitivity Analysis

The influence of including only a pH-driven approach for lysosomal sequestration (excluding lysosomal membrane-binding distribution [26] (Supplemental Materials S-III, Equation (S7)) in the PBPK model was simulated. The correct tumor-to-lung contrast was not simulated by using this simplified approach for lysosomal sequestration, indicated by the increase in PE for both afatinib and osimertinib (Afatinib PE 45% to 93%, Osimertinib: −6.3% to +27%; see Supplemental Materials S-IV, Table S1). Although the PE decreased for lung TBR in afatinib, tumor TBR remained relatively similar, leading to a worse outcome when tumor-to-lung contrast was simulated.

The influence of EGFR on the mechanistic PBPK model was researched by simulation of the PBPK model without EGFR binding. This PBPK model without EGFR was not able to capture the right tumor-to-lung contrast for mainly afatinib (PE 44% to −62%). Contrary to the observed contrast, without EGFR target binding a higher uptake in lung than in tumor was predicted for afatinib. (Supplemental Materials S-IV, Table S1) The PBPK model including EGFR was able to capture the image quality by predicting the right predictive values in 33.3% of tissues for afatinib, 100% (osimertinib) and 16.6% (erlotinib), within 3-fold of the observed values (Supplemental Materials S-IV, Table S1).

Immune deprivation in the tumor tissue may lead to less macrophages and type II cells in the tumor core, influencing distribution to the tumor. In the final model, this is corrected by adding the fraction F_{cell} to healthy lung tissue, but not to tumor tissue, simulating immune deprivation in tumor tissue. To analyze whether this difference in immune cell presence plays a role in determining drug distribution, we added the same fraction of immune cells to the tumor tissue as well. This model showed a worse outcome when compared to the mechanistic model as described above (Supplemental Materials S-IV), leading to a decrease in accuracy predicting afatinib tumor tissue (PE 124% to 130%) and a decrease in accuracy predicting tumor-to-lung contrast for both afatinib (PE 45% to 53%)) and osimertinib (PE −6.3% to 24%, Supplemental Materials S-IV, Table S1). Since erlotinib is a weak base, immune deprivation was not simulated.

Using the mechanistic PBPK model, we hypothesized that not just perfusion but also vascularization of the tumor determines tumor drug penetration. Histological analysis of the healthy lung tissue samples and adenocarcinoma samples yielded a vasculature coefficient of 0.36, indicating that tumor tissue shows approximately 2.8 times less vessels per mm^2 tissue than lung tissue. We assumed that all three EGFR TKIs were perfusion independent [15,20]. The influence of the variability in vascularization between tumor and nontumorous lung tissue was tested by including this vasculature reflection coefficient. The prediction of lung uptake decreased by including this parameter for all TKI, compared to the final model, presuming unaltered tumor perfusion (Supplemental Materials S-IV, Table S1, PE −90 to −152%, 45% to −56% and −6.3 to −99% for erlotinib, afatinib and osimertinib, respectively). Therefore, only the perfusion coefficient and not the vasculature coefficient was retained in the final mechanistic PBPK model.

The extracellular water of the tumor microenvironment might become more acidic, leading to a pH shift from 7.4 to 6.7. To determine the influence of this pH change, a simulation was used. Therefore, the model was simulated including the pH change and without. The results showed that taking into account the more acidic extracellular water resulted in more precise prediction. Excluding the more acidic tumor microenvironment mainly affected erlotinib (PE −90 to −93%). The effect on afatinib and osimertinib was minimal (PE 44 to 44% and PE −6.3 to −7.5%, respectively).

3. Discussion

The PBPK model in this study was developed to capture essential features of tissue distribution by extending previously published physicochemical base models with EGFR binding and lysosomal sequestration and tumor immune deprivation at unrestricted tumor perfusion [8,9,19]. The developed mechanistical PBPK model was able to capture the right whole-body distribution with a high tissue distribution for osimertinib and afatinib and low tissue distribution and high whole blood concentrations for erlotinib. Furthermore, the model captured the right tumor-to-lung contrast for all EGFR-TKIs and therefore was able to predict image quality.

^{11}C-erlotinib reached relatively high concentrations in the blood compared to tissues (TBR < 1), compared to ^{18}F-afatinib and ^{11}C-osimertinib. This relates to a small whole-body volume of drug distribution, which is similar to the volume of distribution estimated at therapeutic dose levels (erlotinib 232 L, afatinib 2370 L and osimertinib 918 L) [12,27,28]. We have shown that our PBPK model based on physicochemical drug properties of the TKIs predicted these differences in distribution profiles. Furthermore, the negative tumor-to-lung contrast as seen by ^{11}C-osimertinib is also predicted by the PBPK model including these parameters.

For osimertinib, we hypothesized that a lower lysosomal volume in tumors, assuming an immune suppressive microenvironment, would lead to a decreased cellular concentration of osimertinib compared to lung tissue. Indeed, this resulted in a correctly predicted tumor-to-lung contrast for osimertinib. Since the decrease in lysosomal sequestration mainly impacts the tumor uptake (Table 2), we showed that the low tumor-to-lung contrast for osimertinib may be explained by immune deprivation and subsequent decrease in lysosomal volume in tumor tissue. The same phenomenon was observed for other TKIs such as nintedanib where increased lysosomal number and lysosomal size decreases sensitivity toward these drugs [29]. This hypothesis is further strengthened by the sensitivity analysis where immune deprivation is excluded from the model. This model was less accurate in predicting uptake, indicating that immune cells play a significant role in tissue uptake.

For afatinib, the predicted decrease in lysosomal sequestration in tumor compared to lung was accompanied by a relative high percentage of EGFR binding (Table 2). For all three compounds, the tumor-to-lung contrast was predicted adequately after accounting for EGFR binding in the model. For afatinib, EGFR binding had the highest influence on tumor distribution in the mechanistical PBPK model due to its low dissociation constant (K_D) [9,10]: EGFR binding showed the highest contribution to the overall tissue uptake. As shown in the sensitivity analyses, when EGFR binding is removed, tumor-to-lung contrast was highly underpredicted. However, tumor-tracer uptake of erlotinib and osimertinib was underpredicted by our model and for most tissues, erlotinib predictions fell outside of the 3-fold range. We hypothesize that the variation in EGFR abundance and target affinity among patients' tumors relates to results in high variability in tumor-tracer uptake. Erlotinib and osimertinib EGFR binding may be underpredicted as affinity for wild-type EGFR was applied. Previous research from our group provided the framework for EGFR binding in tissue by demonstrating the ability of PET/CT to distinguish between wild-type and mutated EGFR [17,18,20]. Therefore, future studies should include EGFR-binding affinities for mutated and wild-type receptors, specifically for drugs with differences in affinity between wild-type and mutation.

The distribution of drugs into tissues with high drug transporter abundancy, e.g., brain, kidney and spleen, was less accurate. For erlotinib, only lung was predicted correctly. The underpredictions of the other tissues of interest are best explained by the effect of influx transporters. Erlotinib is a substrate for the influx transporters OAT3 and OCT2 [23]. The influence of the influx transporter OAT on tissue distribution of erlotinib was investigated in a ^{11}C-erlotinib PET imaging study in rats. The OAT influx transporter was inhibited by rifampicin and decreasing erlotinib exposure was measured in the kidneys and liver, but the exposure in lung was unaffected [30]. In contrast, the overprediction of the TBR for the brain of afatinib may be caused by drug efflux by MDR1 and the BCRP [31]. These drug efflux transporters are highly abundant in the blood–brain barrier (BBB). Similar to the observations in our PET study, a preclinical permeability study showed a low brain-to-blood ratio of 0.31 for afatinib [31]. Further studies of brain uptake can be performed after extension of the PBPK model to include the central nervous system (CNS) physiologically and brain tumor compartments, as described by Hirasawa et al. [32]. In future studies CNS physiology and both influx and efflux transport processes should be studied and research into how to optimally implement these processes is needed.

Another possible explanation for underprediction of erlotinib is that albumin binding may not be the only process affecting tissue distribution of weak bases to be accounted for. Multiple lipophilic-basic drugs bind with a high affinity to the immune-activated protein alpha-1-acid glycoprotein (AAG) [33]. Prior studies show that in NSCLC patients' plasma, AAG levels are increased, but little is known of AAG in the extracellular water of tissues in cancer. In lysosomal-rich tissues such as lung, AAG levels may potentially be higher compared to the immune-suppressive microenvironment of the tumor, leading to differences in tissue distribution. As the role of AAG in plasma binding and drug transport of weak bases has been established, further research on the role of AAG for tissue distribution is needed [34]. Finally, the binding of TKIs to the lysosomal membrane had an important influence on the TBR (Assmus vs. Schmitt). The degree of binding to the membrane was based on the compound logP. However, Pearce et al. (2018) showed in their research that the usage of the actual membrane affinity (instead of logP) is more accurate to predict the KPU [35]. They also presented a critical analysis on how to predict a membrane affinity, in case it is not available.

Sensitivity analysis demonstrates the need for inclusion of the lysosomal membrane, since the correct tumor-to-lung contrast was not captured for all compounds when only a pH-driven approach was included in the PBPK model (Schmitt vs. Assmus, [26,36]). The high impact of lysosomal sequestration (Table 1) after microdosed PET may be due to the unsaturated lysosomes. Fluoxetine, a basic lipophilic compound with comparable physicochemical properties (log P = 3.2, pKa = 9.8) to afatinib and osimertinib, shows that at prolonged exposure of therapeutic doses, lysosomal saturation curve occurs [37]. When extrapolating the results to therapeutic PK, potential saturation of lysosomes needs to be accounted for. Additionally, further research into nonlinear processes of drug binding and sequestration may improve the model predictions [24,38], when comparing microdose and therapeutic dose PK.

In the sensitivity analysis, vascularization-driven drug penetration was studied as a hallmark of NSCLC tumors [39,40]. Our results show that vascularization does not drive drug penetration since the model performs worse when this component was added. Previous results have shown that tracer uptake is perfusion-independent. As our research shows that vasculature is reduced in NSCLC, this may result in hypoxia in some regions with reduced O_2 perfusion within cancers and a decreased capacity to deliver nutrients or remove metabolic waste from rapidly proliferating cells and increased utilization of glucose and the production of lactate [41]. Prior studies show that while the intracellular water pH is highly balanced, extracellular pH in the cells in the tumor microenvironment decreases, resulting in increased uptake of strong basic compounds [41]. The sensitivity analysis showed that the more acidic extracellular water in tumor tissue has only a minor influence on the tumor drug uptake. However, adding this change resulted in a more precise prediction for all three TKIs.

In this study, we extended previously published physicochemical base models with EGFR binding, lysosomal sequestration and more acidic extracellular tumor water. These base models were validated with in vivo data obtained at steady state therapeutic drug concentrations. The effect of each component was studied using sensitivity analyses, but uncertainty of the parameter was not included. Uncertainty evaluation could help predict variability in drug penetration between individuals and is a topic for further research. The PET imaging data used in this study was obtained 1–2 h after a single microdose. Therefore, it should be noted that discrepancies between predicted and observed TBR may occur due to lower drug exposure and nonsteady state. Another study limitation is that drug metabolism and elimination were not taken into account. The half-life of the three researched EGFR-TKIs was comparable and greatly exceeded the 1–2 h scan period (erlotinib = 36 h; afatinib = 45 h; osimertinib = 49 h) [12,27,28], making the assumption of absence of elimination reasonable. Prior research of midazolam, a compound with similar metabolic profile to erlotinib, suggests that metabolism at microdose level is not different from metabolism at therapeutic dose level [42]. However, when these models are applied for predicting the whole-body distribution and target uptake of new tracers with shorter half-lives, inclusion of metabolism and elimination might be needed.

To our knowledge, this is the first PBPK study that addresses these environmental differences. Prior studies either did not account for variability in tumor vs. normal tissue or used an in vitro based partitioning coefficient (e.g., 0.73 in colon cancer [43]). However, other hallmarks of cancer such as the collagen matrix and its effect on the penetration of drugs should be studied.

Future Perspective

If drug properties can be used to predict differences in image quality, it may be possible to predict tracers with an optimal image quality: drugs with large tumor/tissue contrast. Before applying these predictions in drug development, prospective validation of the predictive value of the PBKP model using new tracers is needed. Furthermore, to predict TBR in future studies more precisely, both active and passive influx and efflux transport needs to be included in the mechanistic PBPK model. Therefore, further in vitro research into binding affinities for transporters and transporter tissue concentrations is needed.

In order to study differences in tumor-to-lung contrast and whole-body distribution between microdose and therapeutic dose, the mechanistic PBPK model needs to be extended in a concentration-dependent manner. The injected dose of ^{11}C-erlotinib corresponded with 2.2 µg (±0.46) erlotinib. When compared to the regular therapeutic dose of 150 mg, this is a >10,000-fold difference [20]. With this difference in dosing, lysosomal sequestration, albumin, lipoprotein, AP^- and EGFR binding and EGFR target binding will become saturated, and by including these nonlinear processes, the influence of different doses on TBR can be assessed.

The impact of mutational status on tumor-to-lung contrast and whole-body distribution can be investigated by the use of affinity constants for EGFR wild-type and mutated EGFR. First, the activating EGFR mutation needs to be identified in order to use the right affinity value, thereby increasing accuracy of the prediction of tumor uptake. When such models are fully validated, and combined with optimized individual imaging-based uptake measurements, they may predict individualized dosing regimens intended to optimize drug exposure at the site of disease, thereby improving drug efficacy.

4. Materials and Methods

4.1. Overview

Our goal was to develop a PBPK model that captures essential features of tissue distribution of the three EGFR TKI. First, the base model that best describes the researched drugs was identified based on physicochemical properties. Key components (identified in prior research [21]) were added to the base model: EGFR target binding, lysosomal sequestration of strong bases (osimertinib and afatinib), tumor immune deprivation and unaltered perfusion. Last, we validated our mechanistical PBPK model using data from prior PET research by our group and evaluated whether reduced tumor vascularity could influence drug penetration. All equations and detailed explanation can be found in Supplemental Materials S-I, and an overview of included parameters can be found in Table 3 and schematic overview in Figure 3.

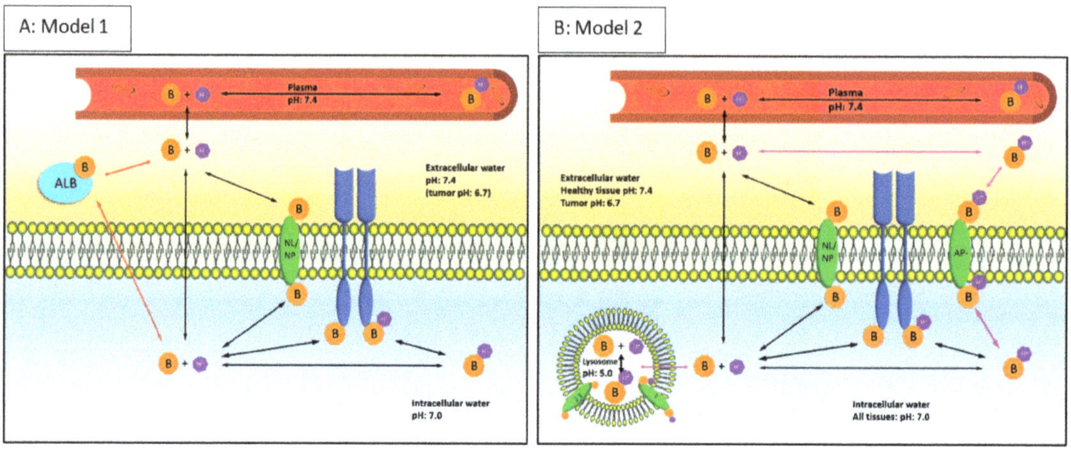

Figure 3. Schematic overview of the final mechanistic PBPK models for weak basic or acidic EGFR-TKIs (**A**: Model 1) and strong basic EGFR-TKIs (**B**: Model 2). The plasma compartment is depicted in red, extracellular space in yellow and intracellular space in blue. Orange hexagons "B" (base) depict the drug; purple hexagons depict H+ atoms. When depicted together, bases are protonated. When depicted separately, the base is unprotonated. Blue receptors depict epidermal growth factor receptor (EGFR), green ovals over the cell membranes are neutral (phospho)lipids (NL/NP) and acidic phospholipids (AP-). Light blue ALB is a representation of albumin. Black arrows depict processes that are included in both models, red/purple arrows depict processes specific for each model. pH values for each compartment are given. Equations for model 1 and 2 below and the model structure are further explained in the Supplemental Materials S-II,III. B = basic unprotonated drug, BH+ = protonated drug, AP- = acidic phospholipids, NL = neutral lipids PL = phospholipids, ALB = albumin, EGFR = epidermal growth factor receptor.

Model 1 full equation:

$$pTBR = \left(F_{vasc/perf} * (Kpu(1) + EGFR\ binding)\right) * \frac{Fu}{B:P}$$

Model 2 full equation:

$$pTBR = \left(F_{vasc/perf} * \left(Kpu(2) + EGFR\ binding + \left(\frac{1+10^{pKa-pHiw}}{1+10^{pKa-pHp}}\right) * Kpu, lys * F_{lys} * F_{cell\ type}\right)\right) * \frac{Fu}{B:P}$$

Table 3. Tissue- and compound-specific input parameters. Tissue-specific parameters were adapted from Table 1 in Rodgers et al., 2005, Table 1 in Rodgers et al., 2006 and Table 1 in Schmitt et al., 2021, EGFR concentrations from Table 3 in Glassman et al., 2016, and lung-specific parameters from Table 1 in Assmuss et al., 2017.

Tissue-Specific Input Parameters

	F_{nl}	F_{np}	F_{ew}	F_{iw}	F_{lys} [2]	Tissue Concentration of AP- (mg/g) [2]	Albumin Tissue to Plasma Ratio [3]	EGFR (nM)
Blood cells	1.7×10^{-3}	0.0029	n.a.	0.60	n.a.	0.50	n.a.	n.a.
Bone	0.017	0.0017	0.1	0.35	n.d.	0.67	0.10	n.a.
Brain	0.039	0.0015	0.16	0.61	0.014	0.40	0.048	n.a.
Kidney	0.039 [1]	0.012 [1]	0.27	0.47	0.017	2.44 [1]	0.13	177
Lung [3] Tumor	0.0088 [1]	0.0030 [1]	0.34	0.43	0.015 0.01	0.57 [1]	0.21	31.1 299
Spleen	0.021 [1]	0.017 [1]	0.21	0.53	0.053	3.18	0.097	54.6
Plasma [4]	n.a.	n.a.	n.a.	n.a.	n.a.	n.a.	n.a.	n.a.

Lung-specific parameters [2]

	F_{nl}	F_{np}	F_{ew}	F_{iw}	pH_{ew}	F_{lys}	pH lysosome	$F_{cell\ type}$
-Alveolar macrophages -Type II cells -Residual cells	0.0088 [1]	0.0030 [1]	0.34	0.45	7.4	0.078 0.03 0.01	4.75 5.1 5.1	0.041 0.083 0.88

Tumor-specific input parameters

	F_{nl}	F_{np}	F_{ew}	F_{iw}	pH_{ew}	F_{lys}	pH lysosome	$F_{cell\ type}$
Residual cells	0.008	0.0030	0.34	0.45	6.7	0.01	5.1	1

Compound-specific parameters

	Erlotinib	Afatinib	Osimertinib	References
Log P	3.3	3.6	3.2	Colclough et al. (2021) [31]
pKa	5.5	8.2	9.0	Colclough et al. (2021) [31]
B:P ratio [5]	0.95	1.27	0.79	Van de Stadt et al. (2021) [21]
Kd EGFR (nM)	2164	2	155	Joly-Tonetti et al. (2021) [44]
$F_{unbound}$ [6]	0.088	0.095	0.017	Colclough et al. (2021) [31]

[1] Translation factor from rats to human [45,46]. [2] Input parameter only used in model 1. [3] Input parameter only used in model 2; Lung pHew: 7.22; pHp 7.4; pHiw 7.0; pHlys: 5.3. [4] Hematocrit (H): 0.45. [5] Blood-to-plasma concentration ratio. [6] Unprotonated fraction [31]. F_{iw}, F_{ew}, F_{nl} and F_{np} reflect tissue-specific fractional tissue volumes of the cellular components intracellular water, extracellular water, neutral lipids and neutral phospholipids. Flys, pHlys and Fcell reflect lysosomal volume fraction, lysosomal pH and the fraction of various cell types in tissue. Fvasc: 0.36, Supplemental Materials S-V, and Fperf: 1 reflect the vascular and perfusion coefficient in the tumor compared to the surrounding lung tissue.

4.2. Scan Data

All PET scans were performed in advanced stage, EGFR mutated NSCLC patients. No patients were treated with the treatment analog of the PET tracer prior to scanning (e.g., a patient undergoing ^{11}C-erlotinib scanning was treatment-naïve for erlotinib). The PET data used in this research are derived from static, 40–60 min post-tracer dose whole-body PET/CT scans. All regions of interest were delineated by the same experienced researcher in a standardized manner using in-house-developed software. For all tracers, spleen, kidney, tumor, lung (contralateral from tumor site) and vertebra were included. For afatinib and osimertinib, brain was additionally delineated, but erlotinib scans did not include brain tissue in field of view. The full-scan protocol and evaluation of the biodistribution is currently under submission.

4.3. PBPK Model: Base Model Selection

The biodistribution of basic lipophilic drugs such as the three researched EGFR-TKIs has been described extensively by well-established PBPK models [45,46]. The PBPK models from Rodgers et al. provide the most accurate tissue distribution predictions [45–47]. Choice of model is dependent on compound properties, most importantly basicity. Relevant physicochemical properties used in PBPK modeling of the three EGFR TKIs, erlotinib, afatinib and osimertinib, can be found in Table 3. Since erlotinib is a weak base and osimertinib and afatinib are strong basic drugs, two base models are used. Model 1 is applicable for predicting tissue uptake of the weak basic drugs and was used for erlotinib [45]. Model 2 can be used for afatinib and osimertinib predictions since this model applies to strong basic compounds [46].

The following assumptions were made in all models, as validated by Rogers et al.: drug transport into tissues only occurs passively; conditions are nonsaturating; the drug is at steady state and well-stirred in all tissues of interest; metabolism and drug clearance are negligible at the time of scanning (at <0.05 of the biological half-life); the tissue PET scans did not contain a significant number of blood vessels (in the PBPK model, only the concentration outside of the blood perfusing the tissue was calculated).

In the PBPK model, we focused on tissues in the PET field of view that are large enough to allow adequate PET data analysis (regions of interest > 1.5 cm). The liver was excluded since all EGFR TKI included in this study are metabolized in the liver, leading to unreliable uptake assessments.

4.4. Base Models: Physicochemical Drug Distribution

In the physicochemical model, tissue-to-blood ratios (predicted TBR, pTBR) are predicted based on distribution to albumin (ALB), neutral lipids and phospholipids (NL/NP), acidic phospholipids (AP-) and to cellular spaces such as the extra- and intracellular water (EW/IW). The described physicochemical base models predict pTBR at steady state by inclusion of drug-specific physicochemical properties and tissue composition (Table 3). If available, drug-specific properties were adapted from the PET imaging data, such as the blood-to-plasma ratio. Physicochemical properties including pKa values were retrieved from the same in vitro research publication to prevent insecurities and enable comparison of the outcomes [31], and can be found in Table 3. All formulas used in the base models and subsequent additions can be found in Supplemental Materials S-II,III.

Base model 1, the model reflecting weak bases, predicts the pTBR by calculation of the pH-driven distribution to cellular components (Figure 3B). Tissue-specific fractional tissue volumes of cellular components, including intracellular water, extracellular water, neutral lipids and neutral phospholipids are reflected by, respectively, F_{iw}, F_{ew}, F_{nl} and F_{np}. By use of the pH values of the cellular components intracellular water, neutral lipids and neutral phospholipids pH_{iw}, pH_{nl}, pH_{np} relative to the pH of plasma (pH_p), the fraction of unprotonated drug available for diffusion to these cellular parts is predicted. The pH values of the cellular components are shown in Figure 3. The octanol/water partition coefficient (P) is included for binding affinity of the unprotonated drug to neutral lipids and phospholipids in the cell membrane. Since a weak base such as erlotinib is highly (99%) unprotonated in plasma, albumin binding in the extracellular water is a predominant process of tissue distribution. The albumin binding was predicted based on the multiplication of the association constant (Ka) for albumin (Figure 1D) with the tissue specific albumin tissue-to-blood ratio [46]. The formula for calculation of Ka and the base model equations can be found in Supplemental Materials S-II (base model 1, Equations (S1)–(S4)). A schematic overview of base model 1 is depicted in Figure 3A.

In model 2, the model reflecting strong bases, the pTBR contains the same elements for the distribution to neutral (phospho) lipids, intracellular and extracellular water. In contrast to weak bases, afatinib and osimertinib are strong basic drugs (pKa > 7) and are mostly protonated (respectively, 98% and 86%) at physiological pH levels [31], as shown in Figure 3A. This protonation leads to binding to acidic phospholipids (AP-) (Figure 1C). Distribution to acidic phospholipids was predicted using association constant Ka, Supplemental Materials S-II Equation (S4) and tissue-specific concentration [AP-]. Model 2 equations can be found in Supplemental Materials S-III (base model 2, Equations (S5)–(S10)). A schematic overview of model 2 is depicted in Figure 3B.

4.5. Extension of the Physicochemical Base Models with EGFR Target Binding

Only nonspecific binding is described by the physicochemical base models. Intracellularly, TKIs bind with high affinity to EGFR [6,9,44]. Differences in affinity of EGFR-TKIs for their target may influence tissue binding and is therefore an essential feature for tissue distribution of TKIs. By adding EGFR binding to base models 1 and 2, target binding was included in the PBPK model. Tissue-specific EGFR concentrations ([EGFR]) and drug-specific dissociation constants (Kd) for wild-type EGFR are shown in Table 3. For two tissues of interest, bone and brain, which lacked relevant literature data, we assumed EGFR was not present.

4.6. Extension of the Physicochemical-EGFR Models with Lysosomal Sequestration (Mechanistical PBPK Model)

Because of the protonated status in an environment with physiological pH, the lysosomal trapping was added to the physicochemical base model for strong bases (model 2) only [26,36]. To estimate the binding, the same composition was assumed for the lysosomal membrane as for the outer membrane of the cell. Since immune cells, mostly consist of a higher lysosomal volume and a lower lysosomal pH than normal tissue cells, tissue-specific cell types were included to predict the TBR [48].

4.7. Including Hallmarks of NSCLC

Four of the hallmarks of NSCLC tumors are a potential immune-suppressive microenvironment and erratic (and potential inadequate) neovascularization and perfusion caused by changes in the microenvironment [39,40,49] resulting in decrease in pH of the tumor microenvironment [41]. We hypothesized that either of these hallmarks could predict a decreased cellular concentration of the TKIs, even at a high affinity and higher expression of EGFR in the tumor.

The impact of the acidic tumor environment was added. Based on prior research, the pH extracellular water in tumor cells was set at 6.7, while intracellular pH was not altered compared to surrounding lung tissue (Table 3, Supplemental Materials S-III Equation (S12)) [41].

The impact of the lysosomal volume of different cell types on tissue uptake in tumor compared to normal lung was researched. Lung-tissue uptake was simulated by use of a physiological composition including the different immune cells: 4.1% alveolar macrophages, 8.3% type II cells and 87.6% residual cells (Table 3) [36]. To reflect an immune-suppressive microenvironment, tumor-tissue uptake with input parameters concerning only residual lung cells was applied (Table 3, Supplemental Materials S-III Equations (S6) and (S10)).

As a final step in the modeling, we hypothesized that the number of vessels drives drug penetration. The vascular coefficient was calculated by dividing the microvessel density (MVD) of normal lung tissue (4 samples), obtained from the Human Protein Atlas, by the MVD of 8 samples of adenocarcinoma NSCLC patients. MVD was calculated per surface area of CD31+ vessels and tissues. A full description of this analysis can be found in Supplemental Materials S-V. Since tumor uptake of ^{11}C-erlotinib and ^{18}F-afatinib has previously been shown to be independent of tumor perfusion, we assumed that all three EGFR TKIs were perfusion independent (50).

4.8. Simulation of Tumor-to-Lung Contrast and Tissue Distribution

For all EGFR-TKIs, the tumor-to-lung contrast was estimated by dividing the uptake in tumor by the uptake in lung (contrast = pTBR tumor/pTBR lung). This contrast was subsequently validated with the PET imaging data. Furthermore, tissue distribution was assessed by predicting the TBR of the lung, tumor, spleen, kidney, brain and bone, and compared to PET imaging tissue uptake data. The functions of the systemic level are as described in the Supplemental Materials S-II,III.

4.9. Software and Statistics

R software (version 4.0.3; R Foundation for Statistical Computing, Vienna, Austria) was used for simulations of the PBPK models and graphical visualization of the predictions and PET observations. PET-TBR data were used to validate the developed PBPK models [21].

The accuracy of mechanistic PBPK model predicted tumor-to-lung contrast and the TBR was assessed by determination of the percentage of tissues falling within 3-fold of the observed data, as is done in the referenced research by Rodger et al. [45,46]. This was researched by calculating prediction errors (PE) and mean prediction error of the mechanistical PBPK model and of subsequent sensitivity analyses. The strength of correlation between the predicted and the PET image-observed TBR was assessed by the Pearson correlation coefficient and with a two-sample t-test significance of the correlation:

$$PE = \left(\frac{PRED - OBS}{mean\ (PRED + OBS)}\right) * 100\% \quad Pearson\ R = \frac{1}{n-1}\sum\left(\frac{x-\overline{x}}{s}\right)\left(\frac{y-\overline{y}}{s}\right)$$

Sensitivity analyses were performed by researching the impact of extension with EGFR binding, use of a different lysosomal extension models on tissue-to-blood ratios and the effect of tumor immune deprivation on all tissues of interest. The effect of the aberrant tumor vasculature was determined by comparing the results after inclusion of a vascular versus the (unaltered) perfusion coefficient. When in sensitivity analyses the removal of the extension of the base model showed a significant decrease in predictivity (the mean PE decrease of more than 10% in tumor-to-lung contrast of all three models), the component was retained in the final mechanistic PBPK model.

5. Conclusions

Our mechanistic PBPK model consisting of a base model—EGFR binding, lysosomal sequestration, tumor immune deprivation, a changed tumor microenvironment, unaltered tumor perfusion and dependent on physicochemical properties of the relevant drug—was able to accurately predict tumor-to-lung contrast. We therefore conclude that our mechanistical PBPK model accurately predicts image quality for EGFR expressing NSCLC tumors, while further study of distribution for drugs into tissues with high drug transporter abundancy and the effect of EGFR mutation on drug penetration is needed.

Supplementary Materials: The following are available online at https://www.mdpi.com/article/10.3390/ph15070796/s1, S-I: Physiologic equations used to describe the physicochemical parameters, S-II: Components of the mechanistical PBPK-model for weak bases (model 1), S-III: Components of the mechanistical PBPK-model for strong bases (model 2), S-IV: Sensitivity analyses, S-V: Extension Rodgers' base model, S-VI: PET scan data.

Author Contributions: All authors contributed substantially to conceptualizing, designing and writing this manuscript. All authors have read and agreed to the published version of the manuscript.

Funding: This research received no external funding.

Institutional Review Board Statement: Data used in this study was obtained from research approved by either the Medical Ethics Committee of the VU University Medical Center or the Medical Ethics Committee of the Netherlands Cancer Institute-Antoni van Leeuwenhoek.

Informed Consent Statement: Informed consent was obtained from all subjects involved in the study. All procedures performed in studies involving human participants were in accordance with the ethical standards of the institutional and/or national research committee and with the 1964 Helsinki declaration and its later amendments or comparable ethical standards.

Data Availability Statement: Data is contained within the article and supplementary material.

Acknowledgments: We would like to thank R.J. Honeywell for his contribution in the design of this research.

Conflicts of Interest: The authors declare no conflict of interest.

References

1. Siegel, R.L.; Miller, K.D.; Fuchs, H.E.; Jemal, A. Cancer Statistics, 2021. *CA Cancer J. Clin.* **2021**, *71*, 7–33. [CrossRef] [PubMed]
2. Siegel, R.L.; Miller, K.D.; Jemal, A. Cancer statistics, 2020. *CA Cancer J. Clin.* **2020**, *70*, 7–30. [CrossRef]
3. Sung, H.; Ferlay, J.; Siegel, R.L.; Laversanne, M.; Soerjomataram, I.; Jemal, A.; Bray, F. Global Cancer Statistics 2020: GLOBOCAN Estimates of Incidence and Mortality Worldwide for 36 Cancers in 185 Countries. *CA Cancer J. Clin.* **2021**, *71*, 209–249. [CrossRef] [PubMed]

4. Faehling, M.; Schwenk, B.; Kramberg, S.; Eckert, R.; Volckmar, A.L.; Stenzinger, A.; Strater, J. Oncogenic driver mutations, treatment, and EGFR-TKI resistance in a Caucasian population with non-small cell lung cancer: Survival in clinical practice. *Oncotarget* **2017**, *8*, 77897–77914. [CrossRef] [PubMed]
5. Red Brewer, M.; Yun, C.H.; Lai, D.; Lemmon, M.A.; Eck, M.J.; Pao, W. Mechanism for activation of mutated epidermal growth factor receptors in lung cancer. *Proc. Natl. Acad. Sci. USA* **2013**, *110*, E3595–E3604. [CrossRef]
6. Eck, M.J.; Yun, C.H. Structural and mechanistic underpinnings of the differential drug sensitivity of EGFR mutations in non-small cell lung cancer. *Biochim. Biophys. Acta* **2010**, *1804*, 559–566. [CrossRef]
7. Yun, C.H.; Boggon, T.J.; Li, Y.; Woo, M.S.; Greulich, H.; Meyerson, M.; Eck, M.J. Structures of lung cancer-derived EGFR mutants and inhibitor complexes: Mechanism of activation and insights into differential inhibitor sensitivity. *Cancer Cell* **2007**, *11*, 217–227. [CrossRef] [PubMed]
8. Yi, L.; Fan, J.; Qian, R.; Luo, P.; Zhang, J. Efficacy and safety of osimertinib in treating EGFR-mutated advanced NSCLC: A meta-analysis. *Int. J. Cancer* **2019**, *145*, 284–294. [CrossRef]
9. Hirano, T.; Yasuda, H.; Tani, T.; Hamamoto, J.; Oashi, A.; Ishioka, K.; Arai, D.; Nukaga, S.; Miyawaki, M.; Kawada, I.; et al. In vitro modeling to determine mutation specificity of EGFR tyrosine kinase inhibitors against clinically relevant EGFR mutants in non-small-cell lung cancer. *Oncotarget* **2015**, *6*, 38789–38803. [CrossRef]
10. Wind, S.; Schnell, D.; Ebner, T.; Freiwald, M.; Stopfer, P. Clinical Pharmacokinetics and Pharmacodynamics of Afatinib. *Clin. Pharm.* **2017**, *56*, 235–250. [CrossRef]
11. Freiwald, M.; Schmid, U.; Fleury, A.; Wind, S.; Stopfer, P.; Staab, A. Population pharmacokinetics of afatinib, an irreversible ErbB family blocker, in patients with various solid tumors. *Cancer Chemother Pharm.* **2014**, *73*, 759–770. [CrossRef] [PubMed]
12. EMA Assessment Report. Giotrif (INN-Afatinib) European Public Assessment Report. Available online: http://www.ema.europa.eu/docs/en_GB/document_library/EPAR_Public_assessment_report/human/004124/WC500202024.pdf (accessed on 8 February 2022).
13. Nelson, V.; Ziehr, J.; Agulnik, M.; Johnson, M. Afatinib: Emerging next-generation tyrosine kinase inhibitor for NSCLC. *Onco Targets* **2013**, *6*, 135–143. [CrossRef]
14. Planchard, D.; Popat, S.; Kerr, K.; Novello, S.; Smit, E.F.; Faivre-Finn, C.; Mok, T.S.; Reck, M.; Van Schil, P.E.; Hellmann, M.D.; et al. Metastatic non-small cell lung cancer: ESMO Clinical Practice Guidelines for diagnosis, treatment and follow-up. *Ann. Oncol.* **2018**, *29*, iv192–iv237. [CrossRef] [PubMed]
15. van de Stadt, E.A.; Yaqub, M.; Lammertsma, A.A.; Poot, A.J.; Schober, P.R.; Schuit, R.C.; Smit, E.F.; Bahce, I.; Hendrikse, N.H. Quantification of [(18)F]afatinib using PET/CT in NSCLC patients: A feasibility study. *EJNMMI Res.* **2020**, *10*, 97. [CrossRef]
16. van de Stadt, E.A.; Yaqub, M.; Lammertsma, A.A.; Poot, A.J.; Schuit, R.C.; Remmelzwaal, S.; Schwarte, L.A.; Smit, E.F.; Hendrikse, H.; Bahce, I. Identifying advanced stage NSCLC patients who benefit from afatinib therapy using (18)F-afatinib PET/CT imaging. *Lung Cancer* **2021**, *155*, 156–162. [CrossRef]
17. Bahce, I.; Smit, E.F.; Lubberink, M.; van der Veldt, A.A.; Yaqub, M.; Windhorst, A.D.; Schuit, R.C.; Thunnissen, E.; Heideman, D.A.; Postmus, P.E.; et al. Development of [(11)C]erlotinib positron emission tomography for in vivo evaluation of EGF receptor mutational status. *Clin. Cancer Res.* **2013**, *19*, 183–193. [CrossRef]
18. Bahce, I.; Yaqub, M.; Errami, H.; Schuit, R.C.; Schober, P.; Thunnissen, E.; Windhorst, A.D.; Lammertsma, A.A.; Smit, E.F.; Hendrikse, N.H. Effects of erlotinib therapy on [(11)C]erlotinib uptake in EGFR mutated, advanced NSCLC. *EJNMMI Res.* **2016**, *6*, 10. [CrossRef] [PubMed]
19. Bahce, I.; Yaqub, M.; Smit, E.F.; Lammertsma, A.A.; van Dongen, G.A.; Hendrikse, N.H. Personalizing NSCLC therapy by characterizing tumors using TKI-PET and immuno-PET. *Lung Cancer* **2017**, *107*, 1–13. [CrossRef]
20. Yaqub, M.; Bahce, I.; Voorhoeve, C.; Schuit, R.C.; Windhorst, A.D.; Hoekstra, O.S.; Boellaard, R.; Hendrikse, N.H.; Smit, E.F.; Lammertsma, A.A. Quantitative and Simplified Analysis of 11C-Erlotinib Studies. *J. Nucl. Med.* **2016**, *57*, 861–866. [CrossRef]
21. van de Stadt, E.A. Relationship between biodistribution and tracer kinetics of 11C-erlotinib, 18F-afatinib and 11C-osimertinib and image quality evaluation using pharmacokinetic/pharmacodynamic analysis in advanced stage non-small cell lung cancer patients. *Diagnostics* **2022**, *12*, 883. [CrossRef]
22. Hallifax, D.; Houston, J.B. Saturable uptake of lipophilic amine drugs into isolated hepatocytes: Mechanisms and consequences for quantitative clearance prediction. *Drug Metab. Dispos.* **2007**, *35*, 1325–1332. [CrossRef] [PubMed]
23. de Klerk, D.J.; Honeywell, R.J.; Jansen, G.; Peters, G.J. Transporter and Lysosomal Mediated (Multi)drug Resistance to Tyrosine Kinase Inhibitors and Potential Strategies to Overcome Resistance. *Cancers* **2018**, *10*, 503. [CrossRef] [PubMed]
24. Krchniakova, M.; Skoda, J.; Neradil, J.; Chlapek, P.; Veselska, R. Repurposing Tyrosine Kinase Inhibitors to Overcome Multidrug Resistance in Cancer: A Focus on Transporters and Lysosomal Sequestration. *Int. J. Mol. Sci.* **2020**, *21*, 3157. [CrossRef] [PubMed]
25. Aarons, L. Physiologically based pharmacokinetic modelling: A sound mechanistic basis is needed. *Br. J. Clin. Pharm.* **2005**, *60*, 581–583. [CrossRef]
26. Schmitt, M.V.; Reichel, A.; Liu, X.; Fricker, G.; Lienau, P. Extension of the Mechanistic Tissue Distribution Model of Rodgers and Rowland by Systematic Incorporation of Lysosomal Trapping: Impact on Unbound Partition Coefficient and Volume of Distribution Predictions in the Rat. *Drug Metab. Dispos.* **2021**, *49*, 53–61. [CrossRef]
27. EMA Assessment Report. Tarceva (INN-Erlotinib) Summary of Product Characteristics. Available online: https://www.ema.europa.eu/en/documents/product-information/tarceva-epar-product-information_en.pdf (accessed on 8 February 2022).

28. EMA Assessment Report. Tagrisso (INN-Osimertinib) European Public Assessment Report. Available online: http://www.ema.europa.eu/en/documents/assessment-report/tagrisso-epar-public-assessment-report_en.pdf (accessed on 8 February 2022).
29. Englinger, B.; Kallus, S.; Senkiv, J.; Heilos, D.; Gabler, L.; van Schoonhoven, S.; Terenzi, A.; Moser, P.; Pirker, C.; Timelthaler, G.; et al. Intrinsic fluorescence of the clinically approved multikinase inhibitor nintedanib reveals lysosomal sequestration as resistance mechanism in FGFR-driven lung cancer. *J. Exp. Clin. Cancer Res.* **2017**, *36*, 122. [CrossRef]
30. Amor, D.; Goutal, S.; Marie, S.; Caille, F.; Bauer, M.; Langer, O.; Auvity, S.; Tournier, N. Impact of rifampicin-inhibitable transport on the liver distribution and tissue kinetics of erlotinib assessed with PET imaging in rats. *EJNMMI Res.* **2018**, *8*, 81. [CrossRef]
31. Colclough, N.; Chen, K.; Johnstrom, P.; Strittmatter, N.; Yan, Y.; Wrigley, G.L.; Schou, M.; Goodwin, R.; Varnas, K.; Adua, S.J.; et al. Preclinical Comparison of the Blood-brain barrier Permeability of Osimertinib with Other EGFR TKIs. *Clin. Cancer Res.* **2021**, *27*, 189–201. [CrossRef]
32. Hirasawa, M.; Saleh, M.A.A.; de Lange, E.C.M. The Extension of the LeiCNS-PK3.0 Model in Combination with the "Handshake" Approach to Understand Brain Tumor Pathophysiology. *Pharm. Res.* **2022**, 1–9. [CrossRef]
33. Zsila, F.; Fitos, I.; Bencze, G.; Keri, G.; Orfi, L. Determination of human serum alpha1-acid glycoprotein and albumin binding of various marketed and preclinical kinase inhibitors. *Curr. Med. Chem.* **2009**, *16*, 1964–1977. [CrossRef]
34. Huang, Z.; Ung, T. Effect of alpha-1-acid glycoprotein binding on pharmacokinetics and pharmacodynamics. *Curr. Drug Metab.* **2013**, *14*, 226–238. [PubMed]
35. Pearce, O.M.T.; Delaine-Smith, R.M.; Maniati, E.; Nichols, S.; Wang, J.; Bohm, S.; Rajeeve, V.; Ullah, D.; Chakravarty, P.; Jones, R.R.; et al. Deconstruction of a Metastatic Tumor Microenvironment Reveals a Common Matrix Response in Human Cancers. *Cancer Discov.* **2018**, *8*, 304–319. [CrossRef] [PubMed]
36. Assmus, F.; Houston, J.B.; Galetin, A. Incorporation of lysosomal sequestration in the mechanistic model for prediction of tissue distribution of basic drugs. *Eur. J. Pharm. Sci.* **2017**, *109*, 419–430. [CrossRef] [PubMed]
37. Schmitt, M.V.; Lienau, P.; Fricker, G.; Reichel, A. Quantitation of Lysosomal Trapping of Basic Lipophilic Compounds Using In Vitro Assays and In Silico Predictions Based on the Determination of the Full pH Profile of the Endo-/Lysosomal System in Rat Hepatocytes. *Drug Metab. Dispos.* **2019**, *47*, 49–57. [CrossRef] [PubMed]
38. Ling, J.; Johnson, K.A.; Miao, Z.; Rakhit, A.; Pantze, M.P.; Hamilton, M.; Lum, B.L.; Prakash, C. Metabolism and excretion of erlotinib, a small molecule inhibitor of epidermal growth factor receptor tyrosine kinase, in healthy male volunteers. *Drug Metab. Dispos.* **2006**, *34*, 420–426. [CrossRef]
39. Folkman, J. Tumor angiogenesis: Therapeutic implications. *N. Engl. J. Med.* **1971**, *285*, 1182–1186. [CrossRef]
40. Herbst, R.S.; Onn, A.; Sandler, A. Angiogenesis and lung cancer: Prognostic and therapeutic implications. *J. Clin. Oncol.* **2005**, *23*, 3243–3256. [CrossRef]
41. Ward, C.; Meehan., J.; Gray, M.E.; Murray, A.F.; Argyle, D.J.; Kunkler, I.H.; Langdon, S.P. The impact of tumour pH on cancer progression: Strategies for clinical intervention. *Explor. Target Antitumor.* **2020**, *1*, 71–100. [CrossRef]
42. Williams, J.A.; Ring, B.J.; Cantrell, V.E.; Jones, D.R.; Eckstein, J.; Ruterbories, K.; Hamman, M.A.; Hall, S.D.; Wrighton, S.A. Comparative metabolic capabilities of CYP3A4, CYP3A5, and CYP3A7. *Drug Metab. Dispos.* **2002**, *30*, 883–891. [CrossRef]
43. Abuqayyas, L.; Balthasar, J.P. Application of PBPK modeling to predict monoclonal antibody disposition in plasma and tissues in mouse models of human colorectal cancer. *J. Pharm. Pharm.* **2012**, *39*, 683–710. [CrossRef]
44. Joly-Tonetti, N.; Ondet, T.; Monshouwer, M.; Stamatas, G.N. EGFR inhibitors switch keratinocytes from a proliferative to a differentiative phenotype affecting epidermal development and barrier function. *BMC Cancer* **2021**, *21*, 5. [CrossRef] [PubMed]
45. Rodgers, T.; Leahy, D.; Rowland, M. Physiologically based pharmacokinetic modeling 1: Predicting the tissue distribution of moderate-to-strong bases. *J. Pharm. Sci.* **2005**, *94*, 1259–1276. [CrossRef] [PubMed]
46. Rodgers, T.; Rowland, M. Physiologically based pharmacokinetic modelling 2: Predicting the tissue distribution of acids, very weak bases, neutrals and zwitterions. *J. Pharm. Sci.* **2006**, *95*, 1238–1257. [CrossRef]
47. Graham, H.; Walker, M.; Jones, O.; Yates, J.; Galetin, A.; Aarons, L. Comparison of in-vivo and in-silico methods used for prediction of tissue: Plasma partition coefficients in rat. *J. Pharm. Pharm.* **2012**, *64*, 383–396. [CrossRef] [PubMed]
48. de Araujo, M.E.G.; Liebscher, G.; Hess, M.W.; Huber, L.A. Lysosomal size matters. *Traffic* **2020**, *21*, 60–75. [CrossRef]
49. Liu, L.; Wang, C.; Li, S.; Bai, H.; Wang, J. Tumor immune microenvironment in epidermal growth factor receptor-mutated non-small cell lung cancer before and after epidermal growth factor receptor tyrosine kinase inhibitor treatment: A narrative review. *Transl. Lung Cancer Res.* **2021**, *10*, 3823–3839. [CrossRef]

pharmaceuticals

Review

Advances of Artificial Intelligence in Anti-Cancer Drug Design: A Review of the Past Decade

Liuying Wang, Yongzhen Song, Hesong Wang, Xuan Zhang, Meng Wang, Jia He, Shuang Li, Liuchao Zhang, Kang Li * and Lei Cao *

Department of Biostatistics, School of Public Health, Harbin Medical University, Harbin 150081, China
* Correspondence: likang@ems.hrbmu.edu.cn (K.L.); caolei@hrbmu.edu.cn (L.C.)

Abstract: Anti-cancer drug design has been acknowledged as a complicated, expensive, time-consuming, and challenging task. How to reduce the research costs and speed up the development process of anti-cancer drug designs has become a challenging and urgent question for the pharmaceutical industry. Computer-aided drug design methods have played a major role in the development of cancer treatments for over three decades. Recently, artificial intelligence has emerged as a powerful and promising technology for faster, cheaper, and more effective anti-cancer drug designs. This study is a narrative review that reviews a wide range of applications of artificial intelligence-based methods in anti-cancer drug design. We further clarify the fundamental principles of these methods, along with their advantages and disadvantages. Furthermore, we collate a large number of databases, including the omics database, the epigenomics database, the chemical compound database, and drug databases. Other researchers can consider them and adapt them to their own requirements.

Keywords: artificial intelligence; machine learning; neoplasms; drug design; databases

check for updates

Citation: Wang, L.; Song, Y.; Wang, H.; Zhang, X.; Wang, M.; He, J.; Li, S.; Zhang, L.; Li, K.; Cao, L. Advances of Artificial Intelligence in Anti-Cancer Drug Design: A Review of the Past Decade. *Pharmaceuticals* **2023**, *16*, 253. https://doi.org/10.3390/ph16020253

Academic Editors: Marialuigia Fantacuzzi, Mariangela Agamennone and Osvaldo Andrade Santos-Filho

Received: 29 December 2022
Revised: 25 January 2023
Accepted: 6 February 2023
Published: 7 February 2023

1. Introduction

In recent years, many companies have ramped up their R&D (research and development) efforts for anti-cancer drugs [1]. There is a growing number of large and long-term clinical trials providing a possible therapeutic opportunity for more cancer patients [2,3]. Recently, the American Cancer Society announced that the three-year survival rate for lung cancer from 2014 to 2021 was raised from 21% to almost 31% [4]. The efficacy of targeted therapies and immunotherapeutics has been investigated in a variety of solid tumors [5]. Thus, a greater investment in targeted therapies and immunotherapeutics to realize the benefits of precision medicine will benefit the long-term survival of cancer patients [6–8].

The anti-cancer drug design and discovery workflow comprises target recognition, hit exploration, hit-to-lead development, lead optimization, preclinical drug candidate identification, and preclinical and clinical research [9–11]. Despite the improvements in tumor biotechnology and the advances in cancer mechanism research, the development of novel and effective anti-cancer drugs from scratch remains an arduous, expensive, and time-consuming process [12] that will require close multidisciplinary collaborations, including medicinal chemistry, computational chemistry, biology, pharmacology, and clinical research [13]. Statistically, it can take more than 10–17 years and almost 2.8 billion dollars to bring a new drug into clinical practice [14,15]. Apart from that, only 10% of the tested compounds in clinical trials reach the market [16].

It is especially difficult to design anti-cancer drugs due to challenges such as undruggable targets [17], chemoresistance in oncology [18], tumor heterogeneity [19], and metastasis [20]. The conventional drug design approaches may seem poorly effective. With so many challenges still to be faced, the treatment effects among cancer patients are actually suboptimal. Thus, more effective anti-cancer drug design strategies are urgently needed. They will reduce the cost of drug development and the time required for clinical trials. They can also help increase the global life expectancy and improve human health [4].

Computer-aided drug design (CADD) is a method that began in the early 1980s [21]. The use of computer-aided methods to guide drug screening is emerging as an important component in the practice of drug design [22–25]. This approach enabled medicinal chemists to calculate the interactions between a ligand and receptors and to design and optimize lead compounds by computer simulation [26]. The typical role of CADD in drug design is to screen out large compound libraries into smaller clusters of predicted active compounds based on computational chemistry. It can greatly speed up the process of anti-drug design and save a huge amount of time and money [27].

In the context of the rapid development of computer hardware and artificial intelligence techniques, researchers in academia and the pharmaceutical industry are turning to artificial intelligence to improve drug design processes [28]. Artificial intelligence (AI) refers to the simulation of human intelligence in machines that are programmed to think and act like humans [15]. A common presumption about artificial intelligence is that its goal is to build machines with a similar capacity for "understanding" [29]. Artificial intelligence is now used in many applications for cancer research, such as image classification of abnormal cancer cells [30], prediction of target protein structures [31], and prediction of drug–protein interactions [32]. These studies demonstrate that artificial intelligence techniques have the power to revolutionize anti-cancer drug design processes. Some applications using artificial intelligence in anti-cancer drug design processes are illustrated in Figure 1. This paper reviewed some of the advances in anti-cancer drug design based on artificial intelligence, presented some of the most classic examples, and clarified the fundamental principles of these methods.

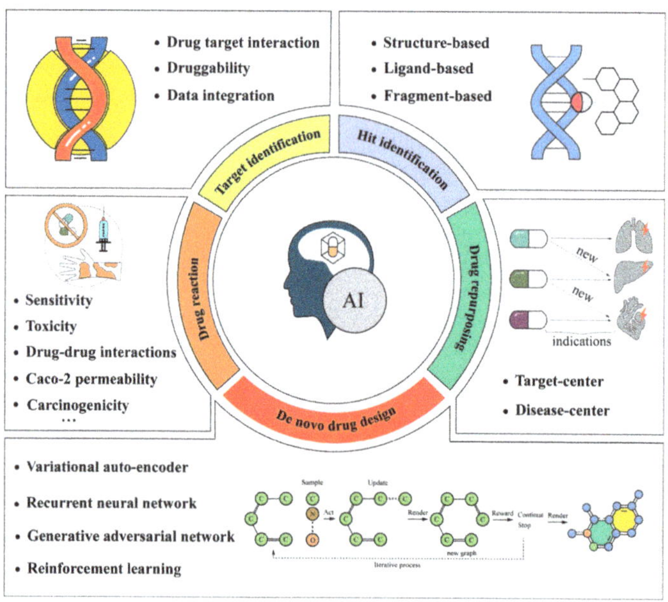

Figure 1. Some applications of artificial intelligence in anti-cancer drug design. The bottom (de novo drug design) is usually implemented using the deep learning-based models listed above. Recently, reinforcement learning has been used often. The above workflow example of a graphical chemical structure with an O–C–O connection is an iterative chemical graph generation process [33].

2. Method

The present study is a narrative review of the literature. We performed searches in the US National Library of Medicine (PubMed) to find original articles. The search strategy used in PubMed is shown in Table 1. We mainly focused on the articles and reviews published in the past decade. The last search of the present narrative review was performed on 10 December 2022.

Table 1. Search strategies used in the US National Library of Medicine (PubMed), according to selected descriptors.

Strategy	Descriptors Used
#1	("cancer" [Title/Abstract] AND "artificial intelligence" [Title/Abstract] AND "drug" [Title/Abstract]) AND (y_10[Filter])
#2	("cancer" [Title/Abstract] AND "drug discovery" [Title/Abstract] AND "AI" [Title/Abstract]) AND (y_10[Filter])
#3	("cancer" [Title/Abstract] AND "drug design" [Title/Abstract] AND "machine learning" [Title/Abstract]) AND (y_10[Filter])
#4	("database" [Title/Abstract] AND "drug" [Title/Abstract] AND "artificial intelligence" [Title/Abstract]) AND (y_10[Filter])

3. Artificial Intelligence in Anti-Cancer Drug Target Identification

The identification of drug–target interactions (DTIs) is the initial step in anti-cancer drug design. The strength of drug–target binding is often described by binding affinity constants, including indicators such as a dissociation constant (Kd), an inhibition constant (Ki), and a half-maximal inhibitory concentration (IC50) [34]. Since the experimental determination of DTIs is a time-consuming and expensive process, its computational prediction is of great interest. Accurate and effective DTI predictions can greatly aid drug development and accelerate lead or hit compound discovery.

3.1. Artificial Intelligence Efficiently Elevates the Prediction Accuracy of DTI

Traditionally, the computational methods for DTI predictions have included molecular docking simulation and machine learning-based methods. However, these studies would be expensive, time-consuming, and difficult to conduct without knowing the 3D structures of the drug targets. Peng et al. developed a novel end-to-end learning framework based on heterogeneous graph convolutional networks (EEG)-DTI for DTI predictions. A graph convolutional network-based model was used to learn the low-dimensional feature representations of drugs and targets and predict the DTI based on the learned features. It achieved a promising DTI prediction performance even when the 3D structures of the drug targets were not used [35]. To further improve the prediction performance, Shao et al. considered the DTI prediction as a link prediction problem and proposed an end-to-end model based on the heterogeneous graph with attention mechanism (DTI-HETA), which outperformed the state-of-the-art models [36]. Meanwhile, to address the explanation problem of deep learning, Yang et al. proposed a drug–target interaction prediction method based on mutual learning mechanisms without 3D structural information and with explanation [37].

3.2. Artificial Intelligence Could Integrate Data from Multiple Sources to Help with Anti-Drug Target Identification

Drug target identification is a key step in drug development. However, most previous studies were confined to a single data type and did not integrate multiple data types. Thus, they were vulnerable to data-specific noise and needed to be improved in terms of practicality and accuracy [38]. Recently, there has been a growing number of methods within similarity-based or data-driven frameworks that attempt to use artificial intelligence to improve the predictive power by integrating multiple different data types. Madhukar et al. developed a Bayesian-based machine learning method (BANDIT), which achieved approximately 90% target prediction accuracy on more than 2000 small molecules by integrating six types of data, including growth inhibition data, gene expression data, adverse reaction data, chemical structure data, and drug data [39]. Olayan et al. proposed a method named DDR to investigate how to predict drug–target interactions more efficiently by using data from different sources, which included eight drug similarity networks and eight target similarity networks. The drug similarity networks included the following: gene expression similarity, disease-based similarity, drug side effect-based similarity, chemical structure fingerprint-based similarity, etc. The target similarity networks included the following: gene ontology-based similarity, protein sequence-based similarity, etc. [40]. The above studies illustrated that integrating data from multiple sources through artificial intelligence could increase the biological explanation of drug target prediction and prediction accuracy.

3.3. *Artificial Intelligence Could Help Predict the Druggability of Anti-Cancer Drug Targets*

The selection of drug targets is also a very critical step in the cancer drug design process, and it has a great impact on the success rate of later clinical trials. Therefore, many related methods were developed. Raies et al. proposed a prediction model called DrugnomeAI to address the problem of targeted drug synthesis. The stochastic semi-supervised machine learning framework was used to develop DrugnomeAI for predicting the druggability of drug targets in the human exome. It also demonstrated how the application of DrugnomeAI can predict the druggability of drug targets in oncology diseases [41]. In recent years, an increasing number of studies have identified synthetic lethality (SL) as a promising approach for the discovery of anticancer drug targets [42]. However, the wet experimental screening for SL has problems, including high costs, batch effects, and off-target results. Wang et al. designed a new model based on a graph neural network (GNN) called KG4SL. It incorporates knowledge graph (KG) messaging into a graph neural network prediction. The experimental results demonstrated a significant beneficial effect of incorporating KG into the GNN for SL predictions [43]. The Table 2 below lists some of the methods for anti-cancer drug target identification based on artificial intelligence that have been developed in recent years.

Table 2. Methods for anti-cancer drug target identification based on artificial intelligence.

Model	Data Source	Code	References
EEG-DTI	Luo dataset [44], Yamanishi dataset [45]	https://github.com/MedicineBiology-AI/EEG-DTI (5 July 2022)	[35]
DTI-HETA	Yamanishi dataset	https://github.com/ZhangyuXM/DTI-HETA (13 October 2022)	[36]
ML-DTI	Metz dataset, KIBA dataset, Davis dataset, [46–48], Drugbank	https://github.com/guaguabujianle/ML-DTI.git (19 June 2021)	[37]
DDR	Yamanishi dataset, KEGG BRITE, BRENDA, SuperTarget, DrugBank	https://bitbucket.org/RSO24/ddr/ (22 November 2017)	[40]
DrugnomeAI	TCRD, StringDB, CTDbase, InterPro, OMIM	https://github.com/astrazeneca-cgr-publications/DrugnomeAI-release (4 November 2022)	[41]
KG4SL	SynLethDB	https://github.com/JieZheng-ShanghaiTech/KG4SL (12 September 2021)	[43]

4. Artificial Intelligence in the Screening of Anti-Cancer Drug Hit Compounds

After the identification of therapeutic targets for anti-cancer drugs, we need to screen for anti-cancer drug hit compounds, which are molecules with initial activities against a specific target or linkage of action [49]. The discovery of computer-aided hit compounds is mainly through high-throughput screening. High-throughput screening can be performed in the following two ways: structure-based screening and ligand-based screening [50]. Fragment-based screening methods are also effective for the discovery of hit compounds, as shown in recent studies [51]. High-throughput screening techniques have been highly successful in many R&D projects, but the efficiency of screening compounds by the millions has reached a bottleneck, and the cost is also significant [52]. With the proliferation of GPUs, increased computer power, and the rapid development of artificial intelligence technologies, more virtual hit compound screening tools have been developed to enrich the drug design toolkit.

4.1. Structure-Based Screening of Hit Compounds Using Artificial Intelligence

Structure-based virtual screening uses docking and scoring to select molecules that have good binding affinity with a target protein [53]. This strategy is an important tool for anti-cancer drug design, but many of the current docking procedures are time-consuming and pose challenges for large-scale virtual screening. Lu et al. accelerated the evaluation process through structure screening with the help of deep learning models. They constructed a deep learning model to predict molecular docking scoring [54]. Yasuo et al. used artificial intelligence to propose a new structure-based virtual screening method for hit compounds, called SIEVE-Score, which provided substantial improvements over other state-of-the-art virtual screening methods [55].

4.2. Ligand-Based Screening of Hit Compounds Using Artificial Intelligence

Ligand-based screening is based on taking small molecules with known activities and searching for structures with similar physical or chemical characteristics in a compound library as candidates. Krasoulis et al. proposed an end-to-end method called DENVIS, a scalable and novel algorithm for high-throughput screening using graphical neural networks with atomic and surface protein pocket features. By conducting experiments on two benchmark databases, DENVIS was much faster than other models [56]. This method was not only advantageous in terms of speed and had an impressive success rate, but it was also easy to use. Generally, most of these methods could only receive one representative molecular structure as a search template [57], which may result in data waste. To address this problem, Hutter developed a cumulative molecular fingerprinting algorithm that can take all structure data into account in the calculation, effectively improving the utilization of experimental data and achieving an organic combination of molecular fingerprinting and experimental data. It inherited the speed advantage of the former method with higher information utilization [58].

4.3. Fragment-Based Screening of Hit Compounds Using Artificial Intelligence

In recent years, the rise of emerging technologies such as high-throughput screening (HTS) and combinatorial chemistry (CC) has led to the gradual systematization of drug discovery from the randomized screening of known drugs [59]. These methods can significantly increase the speed of drug discovery and shorten the process of new drug development, but the high cost of screening has also increased the research burden on small drug development companies and research institutions. Therefore, many researchers are focusing on fragment-based drug design (FBDD) [60]. Compared with the traditional screening methods, FBDD starts with small molecular fragments, which greatly reduces the size of the required screening compound library, circumvents the undesirable ADMET properties of molecules, and enhances the diversity of the designed structures [61]. In addition, FBDD has potential advantages for the drug design of difficult targets and has gradually developed into a mainstream drug design method in small drug development companies and research units [62]. To ligate fragments rationally, it is necessary to know where the fragments bind in a pocket. Currently, the main computational prediction methods are molecular docking, functional group mapping, and molecular structure splitting and reconstruction. These methods are more or less limited by computational costs and manual judgement and cannot fully utilize the structural data of protein–ligand complexes. To solve this problem, Didier Rognan's group proposed the method POEM, which is based on the recognition and matching of the pocket environment in which the fragments are located [63]. Another challenge of FBDD is linking fragments to generate interest libraries of compounds for specific drug targets. To address this issue, Yang et al. proposed a model based on automatic fragment linking with deep conditional transformer neural networks called SyntaLinker [64]. Caburet et al. screened the activity of NDM-1 β-lactamase inhibitors using the FBDD method. They finally found 37 fragments for pharmacophore establishment, which was proven to be accurate and efficient. Table 3 lists all of these methods [65].

Table 3. Artificial intelligence-based screening methods of anti-cancer drug hit compounds.

Model	Data Source	Code	References
SECSE	PDB	http://github.com/KeenThera/SECSE (15 July 2022)	[54]
SIEVE-Score	ChEMBL, ZINC	https://github.com/sekijima-lab/SIEVE-Score (15 November 2019)	[55]
DENVIS	PDB, DUD-E, LIT-PCBA	https://github.com/deeplab-ai/denvis (3 October 2022)	[56]
DMMFP	ChEMBL, DUD-E, ZINC	https://github.com/michahutter/multimolecule_fingerprints (28 April 2022)	[58]
POEM	ChEMBL, ZINC, PDB	https://github.com/kimeguida/POEM (30 December 2022)	[63]
SyntaLinker	ChEMBL	https://github.com/YuYaoYang2333/SyntaLinker (18 June 2021)	[64]

5. Artificial Intelligence in De Novo Anti-Cancer Drug Design

The chemical space of drug-like molecules is extremely vast; the number is estimated to be 10^{23}~10^{60} [66]. Therefore, it is nearly impossible to completely mine the entire chemical space using computational methods. In this context, finding specific lead compounds in the vast chemical space is a major challenge. With the rapid development of computational power and experimental techniques, high-throughput screening (HTS) and virtual screening (VS) methods can effectively evaluate molecules in large compound libraries with a wide variety of filters [67,68].

However, both traditional HTS and vs. methods that are based on molecular docking can only screen the known compound library to find molecules that satisfy specific properties [69]. De novo drug design and virtual screening are very similar in the sense that they both search for molecules that meet specific requirements in the chemical space. However, their processes are very different. Instead, de novo drug design is a molecule generation method that generates and optimizes a molecule by ultimately using artificial intelligence [70]. Molecular generation methods include variational auto-encoders (VAEs), the recurrent neural network (RNN), the generative adversarial network (GAN), and deep reinforcement learning (DRL) [71].

5.1. Application of Variational Auto-Encoder to De Novo Design of Anticancer Drugs

The variational auto-encoder (VAE) is an important type of generative model that was proposed by Diederik P. Kingma and Max Welling in 2013 [72]. Born et al. constructed a hybrid VAE model to generate candidate molecules with anti-cancer drug properties. The model was able to generate molecules with strong inhibitory effects against specific diseases. The generated molecules were similar to existing drugs in terms of structure, synthesizability, and solubility [73]. Hong et al. proposed a molecular structure tree generation model in which the molecules were generated by gradually adding substructures [74]. The proposed model was based on a VAE architecture, which used an encoder to map molecules into the latent vector space and then built an autoregressive generative model as a decoder to generate new molecules from a Gaussian distribution. It showed that the model can generate efficient and new molecules and that the optimized model can effectively improve the properties of the molecules. Samanta et al. proposed the NEVAE method, which solved the problems of current methods. For instance, existing models can only generate molecules with the same number of atoms but fail to utilize a large number of macromolecules in the training process, limiting the diversity of the generated molecules. In addition, they cannot provide the spatial coordinates of the generated atoms [75].

5.2. Application of the Recurrent Neural Network to De Novo Design of Anti-Cancer Drugs

The recurrent neural network (RNN) model uses basic units, such as atoms or fragments of molecules, as the basic vocabulary and generates molecules in a temporal order. The output probability of the next atom character generated by the RNN model depends on the previous generated atom. The RNN-based model has been widely used to process time-series-related data, such as language, text, video, etc. [76]. Grisoni et al. proposed a new bidirectional RNN molecular generation model, or BIMODAL, that can be used for SMILES generation and data enhancement [77]. The model performed bidirectional molecular design by alternate learning, and the model was compared with other bidirectional RNNs. BIMODAL was promising in terms of molecular novelty, backbone diversity, and chemical and biological relevance of the generated molecules and was superior to the state-of-the-art methods [78,79]. To address problems such as the poor performance of DL on small training datasets, Krishnan et al. designed a de novo drug design method based on RNN generative models and migration learning to generate molecules with not only the desired drug-like properties but also target specificity [80]. In addition, Moret et al. combined the RNN generation model with three optimization methods, namely data augmentation, temperature sampling, and transfer learning. This method can generate new molecules with the desired properties with a small amount of data [81].

5.3. Application of Generative Adversarial Network to De Novo Design of Anti-Cancer Drugs

The generative adversarial network (GAN) is an unsupervised learning method proposed by Goodfellow in 2014. It consists of the following two networks: the generative network G, which is used to fit the data distribution, and the discriminative network D, which is used to determine whether the input is "real" or not. In the training process, the generative network G tries to "cheat" D by accepting random noise to imitate the real images in the training set, while D tries to distinguish the real data from the output of the generative network as much as possible, thus forming a game process between the two networks. Ideally, the game results in a generative model that can be "faked" [82]. Maziarka et al. proposed the Mol-Cycle GAN method. Mol-Cycle GAN is a conditional generative adversarial network-based method for de novo drug design and synthesis optimization of molecules through a generative model. It can solve the problem of difficult-to-synthesize compounds given a starting molecule. It can also generate molecules with similar structures and desired properties [83]. ABbbasi et al. proposed a feedback-based GAN framework that implemented an optimization strategy by connecting an encoder–decoder, a GAN, and a predictor depth model with a feedback loop. The results showed that molecules with high binding affinity can be generated by the GAN optimization model [84].

5.4. Application of Deep Reinforcement Learning to De Novo Design of Anti-Cancer Drugs

Even though a variety of drug generation models have been developed, they all focus on the following two points: molecular representation and optimization strategies [71]. Deep reinforcement learning (DRL) is an artificial intelligence technique that combines the perceptual capabilities of deep learning with the decision-making capabilities of reinforcement learning to solve decision-making problems in high dimensional and state spaces [85]. A novel computational strategy, called ReLeaSE, was proposed by Tropsha for designing molecules with desired properties from scratch. ReLeaSE was built on deep learning (DL) and reinforcement learning (RL) methods by integrating two deep neural networks (generative and predictive), which were trained to generate novel libraries of molecules with specified properties [86]. Goel et al. combined RNN and reinforcement learning to propose a molecule generation model named MoleGuLAR that can perform multi-objective optimization of molecules in terms of drug-like properties and binding affinity. In particular, they proposed a new alternating reward strategy where the reward function changes dynamically as different molecules are generated, allowing the model to alternately explore different chemical intervals and sample more reasonable molecules [87]. Table 4 shows some of these methods.

Table 4. Methods for de novo anti-cancer drug design through artificial intelligence.

Model	Data Source	Code	References
PaccMannRL	TCGA, ChEMBL, GDSC, CCLE	https://github.com/PaccMann/ (10 February 2022)	[73]
ACGT	QM9, ZINC	https://github.com/gicsaw/ARAE_SMILES (14 October 2022)	[74]
NEVAE	QM9, ZINC	https://github.com/Networks-Learning/nevae (22 November 2019)	[75]
BIMODAL	ChEMBL	https://github.com/ETHmodlab/BIMODAL (3 June 2020)	[77]
Mol-CycleGAN	ChEMBL, ZINC	https://github.com/ardigen/mol-cycle-gan (6 February 2019)	[83]
GAN-Drug-Generator	ChEMBL, ZINC	https://github.com/larngroup/GAN-Drug-Generator (13 April 2022)	[84]
ReLeaSE	ChEMBL, ZINC	https://github.com/isayev/ReLeaSE (9 December 2021)	[86]
MoleGuLAR	ChEMBL, ZINC	https://github.com/devalab/MoleGuLAR (21 October 2021)	[87]

6. Artificial Intelligence in Anti-Cancer Drug Repurposing

Effective identification of new indications from approved or established clinical drugs plays a critical role in drug discovery. Such a process is also known as drug repositioning. Despite tremendous efforts in academic and pharmacological research worldwide, current anti-cancer therapies have achieved success in only a few tumor types. The application of drug repositioning in tumor therapies is a hot topic in current research. In theory, repurposing is faster, safer, easier, and less expensive than the known barriers to developing new molecular entities. Opportunities for drug repurposing are often based on incidental observations or time-consuming preclinical drug screens that are not usually hypothesis-driven. Indeed, the widespread use of histology technologies, improved electronic medical record systems, improved data storage, data meaning, machine learning algorithms, and computational modeling have provided unprecedented knowledge of the biological mechanisms of cancer and drug modes of action, providing broad availability of both disease-related and drug-related data. Drug repositioning strategies are often categorized as "target-center" and "disease-center" methods for predicting unknown drug–target and drug–disease interactions.

6.1. Artificial Intelligence in Anti-Drug Repositioning Based on the Interaction between a Drug and a Target

Many artificial intelligence-based methods have been used to predict drug–target relationships, as described above. At present, predicting drug–target relationships is one of the main approaches for drug repurposing. To achieve personalized drug repurposing using genomic information, Cheng et al. developed a genome-wide localization system network algorithm (GPSnet) [88]. This method uses patient-specific DNA and RNA sequencing profiles of specific targets to obtain disease modules for repurposing drugs. They validated that the approved arrhythmia and heart failure drug Ouabain specifically targets the HIF1α/LEO1-mediated cellular metabolic pathways in lung adenocarcinomas, showing potential anti-tumor activities. Wang et al. proposed a deep learning framework through kernel-based data integration, known as DeepDRK [89]. The model was trained on over

20,000 pairs of pan-cancer cell line anti-cancer drug pairs. These pairs were characterized by using kernel-based similarity matrices that integrate multi-source and multi-omics data, including genomics, transcriptomics, epigenomics, chemical properties of compounds, and known drug–target interactions. They provided a computational approach to predict cancer cell responses to drugs by integrating pharmacogenomic data, offering an alternative approach to repurposing drugs in cancer precision therapy.

6.2. Artificial Intelligence in Anti-Drug Repositioning Based on the Interaction between Drugs and Diseases

Predicting drug–disease interactions is essential for disease-centric drug repurposing. The current identification of drug–disease interactions is mainly based on similarity and network, respectively. For the similarity-based methods, Zhang et al. proposed a multiscale drug–disease topology learning framework (MTRD). By learning the representative properties of drug–disease, this method explored a new therapeutic effect of existing drugs based on the relevant similarity and association information of drug–disease node pairs. [90]. Jarada et al. proposed a novel framework based on deep learning, known as SNF-NN, to predict new drug–disease interactions using drug-related similarity information, disease-related similarity information, and known drug–disease interactions [91]. Luo et al. proposed a new computational method named MBiRW [92], which uses combined similarity measurements and a birandom walk (BiRW) algorithm to identify potential new indications for known drugs. This method was based on the assumption that similar drugs are usually associated with similar diseases. Moreover, Sadeghi et al. proposed a new model named DR-HGNN for drug repositioning using multiple labeling of heterogeneous graph neural networks [93]. Doshi et al. proposed a graph neural network-based drug repositioning model called GDRnet [94], which was able to efficiently screen the database for existing drugs and predict their unknown therapeutic effects. Table 5 shows some of the methods mentioned above.

Table 5. Methods for anti-cancer drug repurposing based on artificial intelligence.

Model	Data Source	Code	References
GPSnet	DrugBank, TTD, PharmGKB, ChEMBL, BindingDB, UniProt, TCGA	https://github.com/ChengF-Lab/GPSnet (16 December 2018)	[88]
DeepDRK	CTRP, GDSC, TCGA, DrugBank, KEGG	https://github.com/wangyc82/DeepDRK (16 January 2021)	[89]
MBiRW	Drugbank, OMIM	http://github.com//bioinfomaticsCSU/MBiRW (19 December 2016)	[92]
DR-HGNN	Drugbank, CTD, SIDER	https://github.com/sshaghayeghs/DR_HGNN (26 April 2022)	[93]
GDRnet	Drugbank, Hetionet, GNBR, STRING, IntAct, DGIdb	https://github.com/siddhant-doshi/GDRnet (27 December 2021)	[94]

7. Artificial Intelligence-Assisted Accurate Prediction of Anti-Cancer Drug Reactions

Drug reactions are related to their ADMET properties, which may influence drug sensitivity, drug toxicity, and drug–drug interactions [95,96]. The accurate prediction of drug reactions can effectively increase the success rate of clinical trials and improve patient outcomes. With the rapid development of artificial intelligence technologies, more and more related studies are being proposed at the drug design stage using artificial intelligence techniques.

7.1. Artificial Intelligence Aids in Predicting the ADMET Properties of Anti-Cancer Drugs

To explore drug reactions, the ADMET properties should be accurately predicted first. Several ADMET properties, including Caco-2 permeability, carcinogenicity, blood–brain barrier permeability, and plasma protein binding, are included in previous studies. For instance, Selvaraj et al. reviewed the applications of various machine learning models, such as SVM regression and partial least squares (PLSs), for the prediction of the Caco-2 permeability coefficient [97]. Li et al. proposed a DeepCarc model to predict the carcinogenicity of small molecules using deep learning-based model-level representations [98]. Vatansever et al. reviewed the current state-of-the-art methods in AI-guided central nervous system (CNS) drug discovery, focusing on the blood–brain barrier permeability prediction [99]. To predict the plasma protein binding of a drug, Mulpuru et al. built a prediction model of a fraction of unbound drug in human plasma using a chemical fingerprint and a freely available AutoML framework [100].

7.2. *Artificial Intelligence Aids in Predicting Anti-Cancer Drug Sensitivity*

Anti-cancer drug sensitivity predictions are important in guiding the enrollment of those patients who may benefit from specific treatments. Chawla et al. developed a deep neural network named Precily, which uses gene expression data to predict drug sensitivity for cancer therapy. The model combines the structural properties of drugs with the pathway specificity of gene expression as features to train the model [101]. Eliseo Papa et al. built a recommendation system based on the BIKG knowledge graph to predict drug sensitivity and identified effective patient subgroups early in clinical trials [102]. Gerdes et al. proposed a model called DRUML, which uses omics data to rank over 400 drugs based on their anti-tumor cell proliferation efficacy. The results showed that DRUML can accurately rank anti-cancer drugs based on their efficacy [103].

7.3. *Artificial Intelligence Aids in Predicting Toxicity of Anti-Cancer Drugs*

Drug toxicity is a central issue to be considered in the drugs development process. Recently, Wang et al. proposed a machine learning classifier that combines chemical structure (CS) and gene expression (GE) features. In addition, they prioritized the adverse effects of approved drugs and preclinical small-molecule compounds. The results showed that integrating GE data with drug CSs can significantly improve the predictability of adverse effects [104]. However, most of the current studies only predict the occurrence of adverse drug reactions, not their intensity or frequency. To address this issue, Zhao et al. designed a novel graphical attention model for predicting drug side effect frequency from multi-view data. The computational results showed the best performance on the benchmark dataset, illustrating effectiveness in predicting the frequency of drug side effects [105].

7.4. *Artificial Intelligence Can Predict Drug–Drug Interactions*

Zhu et al. proposed a unified multi-attribute discriminative representation learning (MADRL) model for DDI predictions. MADRL uses a generative adversarial network (GAN) to capture intra-attribute specificity information of DDI attributes and uses them for DDI predictions. The effectiveness of the MADRL algorithm was validated on a publicly available dataset [106]. Most methods for predicting drug–drug interactions only predict whether there is an interaction between two drugs, but it is more relevant to investigate the hidden mechanisms behind DDIs. Therefore, Zhang et al. proposed a deep learning method (DDIMDL) that used multiple drug features to predict the types of drug–drug interaction events and explored their hidden mechanisms [107]. To further increase the model's accuracy and biological explanation, Chen et al. developed 3DGT-DDI, which consists of a 3D graph neural network and a pre-trained textual attention module. The innovation of the method is that it utilizes a 3D molecular graph structure and location information to enhance the prediction ability of DDIs. The experiments showed that the prediction performance of 3DGT-DDI outperformed other baseline models [108]. Table 6 table shows some of the methods mentioned above.

Table 6. Methods for prediction of cancer drug reactions based on artificial intelligence.

Model	Data Source	Code	References
DeepCarc	CPDB, Pubchem, Drugbank	https://github.com/TingLi2016/DeepCarc (6 July 2022)	[98]
Precily	CCLE, MSigDB, GDSC, Pubchem	https://github.com/SmritiChawla/Precily (26 August 2022)	[101]
DRUML	PharmacoDB, DepMap portal, PRIDE dataset, DrugBank, ChEMBL	https://github.com/CutillasLab/DRUMLR (24 March 2022)	[103]
MGPred	SIDER, STITCH, DrugBank, PubChem	https://github.com/zhc940702/MGPred (6 May 2018)	[105]
DLADE	cTAKES, EHR and PubMed article	https://github.com/qinxiao (7 October 2022)	[109]
MADRL	KEEG, SIDER, CTD, DrugBank	https://github.com/AdverseDDI/MADRL (18 January 2022)	[106]
DDIMDL	DrugBank, KEGG	https://github.com/YifanDengWHU/DDIMDL (1 May 2021)	[107]
3DGT-DDI	DrugBank, DDI extraction 2013	https://github.com/hehh77/3DGT-DDI (21 February 2022)	[108]

8. Data Sources of Artificial Intelligence to Anti-Cancer Drug Designs

A large number of artificial intelligence-based algorithms, including deep learning, have become powerful tools in AI-assisted anti-cancer drug design [110,111]. Scientists are developing algorithms that can learn and analyze large amounts of data with superhuman efficiency to speed up the anti-cancer drug design process [112]. However, artificial intelligence is not universal and requires large amounts of reliable data or training experiences [113]. Nowadays, there are some specific databases for artificial intelligence-based anti-cancer drug design. They are listed in Table 7.

Table 7. Different data sources for anti-cancer drug design.

Database	Website
BindingDB	https://www.bindingdb.org/bind (24 December 2022)
BRENDA	https://www.brenda-enzymes.org/ (1 February 2023)
CCLE	https://sites.broadinstitute.org/ccle/ (23 December 2019)
chEMBL	https://www.ebi.ac.uk/chembldb (12 July 2022)
CPDB	https://www.nlm.nih.gov/databases/download/cpdb.html (12 October 2022)
CPTAC	https://proteomics.cancer.gov/programs/cptac (7 February 2023)
CTDbase	http://ctdbase.org (1 February 2023)
CTRP	https://portals.broadinstitute.org/ctrp.v2.1/ (7 February 2023)
DepMap	https://depmap.org/portal/ (14 December 2022)
DGIdb	www.dgidb.org. (21 October 2020)
Drugbank	https://www.drugbank.com/ (7 February 2023)
DUD-E	http://dude.docking.org/ (14 July 2012)
GDSC	https://www.cancerrxgene.org/ (July 2022)
GEO	https://www.ncbi.nlm.nih.gov/geo/ (7 February 2023)
HCA	https://data.humancellatlas.org/ (7 February 2023)
Hetionet	https://het.io/ (7 February 2023)
IntAct	https://www.ebi.ac.uk/intact/ (December 2021)
InterPro	https://www.ebi.ac.uk/interpro (November 2022)
JingleBells	http://jinglebells.bgu.ac.il/ (7 February 2023)
KEGG	https://www.genome.jp/kegg/ (1 January 2023)
LIT-PCBA	https://drugdesign.unistra.fr/LIT-PCBA/ (7 February 2023)
MSigDB	https://www.gsea-msigdb.org/gsea/msigdb/ (August 2022)
OMIM	https://www.omim.org. (5 February 2023)

Table 7. Cont.

Database	Website
Open Targets	https://www.opentargets.org/ (19 January 2023)
PDB	https://www.rcsb.org/docs/general-help/organization-of-3d-structures-in-the-protein-data-bank (31 August 2022)
PharmacoDB	https://pharmacodb.ca/ (7 February 2023)
PharmGKB	https://www.pharmgkb.org/ (7 February 2023)
portal	https://help.hcltechsw.com/digital-experience/9.5/plan/db_domains.html (7 February 2023)
PubChem	https://pubchem.ncbi.nlm.nih.gov/ (7 February 2023)
QM9	https://paperswithcode.com/dataset/qm9 (7 February 2023)
reactome	https://reactome.org/ (7 December 2022)
repoDB	https://repodb.net/ (7 February 2023)
scRNASeqDB	https://bioinfo.uth.edu/scrnaseqdb/ (7 February 2023)
SEER	https://seer.cancer.gov/ (27 October 2022)
SIDER	http://sideeffects.embl.de (7 February 2023)
STITCH	http://stitch.embl.de/ (7 February 2023)
STRING	https://string-db.org/ (7 February 2023)
SuperTarget	http://insilico.charite.de/supertarget/ (7 February 2023)
SynLethDB	http://synlethdb.sist.shanghaitech.edu.cn/v2/#/ (14 October 2022)
TCGA	https://portal.gdc.cancer.gov/ (10 January 2023)
TCRD	http://juniper.health.unm.edu/tcrd (7 February 2023)
TTD	https://db.idrblab.net/ttd/ (29 September 2021)
UniProt	https://www.uniprot.org/ (7 February 2023)
ZINC	http://zinc15.docking.org/ (7 February 2023)

9. Successful Cases Applying AI in Anti-Cancer Drug Design

To depict how AI facilitates the development of anticancer drugs, we list some of the anticancer drugs that have successfully entered human phase 2/3 clinical trials in the last 5 years in Table 8. For instance, Recursion identified REC-2282 as a potential candidate for the treatment of diseases caused by mutations in the NF2 gene through its proprietary AI-driven drug discovery platform, Recursion OS. REC-2282 is a permeable, orally bioavailable, small-molecule HDAC inhibitor that is being developed for the treatment of meningiomas with mutations in the NF2 gene. This molecule appears to be well tolerated, including in patients that have been administering it over several years, and different from other HDAC inhibitors in that it may reduce cardiotoxicity. It was granted both orphan drug status and fast-track status by the U.S. FDA [114]. Relay Therapeutics developed the FGFR2-specific inhibitor RLY-4008 by analyzing the dynamic balance of protein conformations through an artificial intelligence platform. Preclinical studies have shown that RLY-4008 exhibits high selectivity for FGFR2 targets in cancer cell lines, shrinking tumors with minimal impact on other targets [115]. Breg developed a new drug, BPM 31510, through an artificial intelligence platform that is currently in clinical testing. The drug restructures the metabolism of cancer cells so that patients do not have to undergo chemotherapy, allowing cancer cells to die naturally [116]. EXS-21546 is an AI-designed A2A receptor antagonist. Some tumors produce high levels of adenosine, which binds to and activates the A2A receptors on immune cells, thereby inhibiting the anti-tumor activity of the immune system [117]. PHI-101 is an orally available, selective checkpoint kinase 2 (Chk2) inhibitor designed by an AI-driven drug discovery platform [118].

Table 8. Some of the AI-designed anti-cancer drugs that have successfully entered human phase 2/3 clinical trials in the last 5 years.

Name	Chemical Structure	Company	Therapeutic Area	Target/Function	Phase
REC-2282	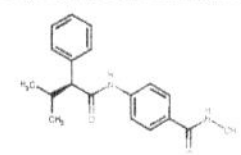	Recursion	Neurofibromatosis Type 2	HDAC	Phase 2/3
RLY-4008	not disclosed	Relay Therapeutics	Solid tumor	FGFR	Phase 2
BPM31510	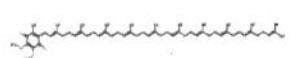	Berg	Solid tumor	Protein cbcl2 modulators	Phase 2
EXS-21546	not disclosed	Exscientia	Solid tumor	A2aR	Phase 1
PHI-101	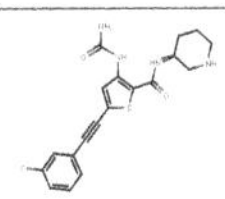	Pharos iBio	Ovarian cancer Breast cancer	Flt3 tyrosine kinase inhibitor	Phase 1

10. Conclusions and Prospects of Future Challenges

This review focuses on work that has been performed in the past decade on anti-cancer drug design based on artificial intelligence. Compared to other reviews, our study collated a large number of databases and source codes. It will offer some guidelines for other researchers to apply to their own research. This means our review has great practicality.

Artificial intelligence (AI) has strong logical reasoning and independent learning abilities that can simulate the thinking process of a human brain. AI technologies, such as machine learning, can profoundly optimize the existing anti-cancer drug research paradigm. In recent years, AI has already made unique contributions to the development and treatment of anti-cancer drugs. Artificial intelligence can accelerate the discovery of new drug molecules and the synthesis of more desirable drug molecules. This process may greatly accelerate the development of anti-cancer drugs. It is believed that artificial intelligence will be a powerful driving force for human cancer research and treatment in the future. However, AI also has several limitations, including a high dependence on data and a limited explanation. The "black box" behind traditional AI models prevents scientists from using algorithms for hypothesis validation and mining the logic behind the data. Moreover, in the drug development process, predicting the underlying logic behind a model is critical to designing the right drug molecules. In the future, interpretable AI models will be the new development direction, and the close combination of data and computation will be a feature of AI-assisted cancer drug development. We believe that AI will bring profound changes to anti-cancer drug designs.

Our study is also subject to certain limitations. For instance, we only focused on articles published in the last ten years. In addition, the search was limited to the database of PubMed. We will address these limitations in future studies.

Author Contributions: Conceptualization, L.W., K.L. and L.C.; methodology, L.W. and Y.S.; investigation, L.W., H.W. and X.Z.; resources, M.W. and J.H.; writing—original draft preparation, L.W. and L.C.; writing—review and editing, L.W., S.L. and L.Z.; supervision, K.L. and L.C.; project administration, L.C. All authors contributed to conceptualization, writing—review and editing. All authors have read and agreed to the published version of the manuscript.

Funding: This work was funded by the National Natural Science Foundation of China (81973149).

Institutional Review Board Statement: Not applicable.

Informed Consent Statement: Not applicable.

Data Availability Statement: No new data were created or analyzed in this study. Data sharing is not applicable to this article.

Conflicts of Interest: The authors declare no conflict of interest.

References

1. Moreau, P.; Garfall, A.L.; van de Donk, N.W.C.J.; Nahi, H.; San-Miguel, J.F.; Oriol, A.; Nooka, A.K.; Martin, T.; Rosinol, L.; Chari, A.; et al. Teclistamab in Relapsed or Refractory Multiple Myeloma. *N. Engl. J. Med.* **2022**, *387*, 495–505. [CrossRef] [PubMed]
2. Avery, R.K.; Alain, S.; Alexander, B.D.; Blumberg, E.A.; Chemaly, R.F.; Cordonnier, C.; Duarte, R.F.; Florescu, D.F.; Kamar, N.; Kumar, D.; et al. Maribavir for Refractory Cytomegalovirus Infections With or Without Resistance Post-Transplant: Results From a Phase 3 Randomized Clinical Trial. *Clin. Infect. Dis.* **2022**, *75*, 690–701. [CrossRef] [PubMed]
3. Cercek, A.; Lumish, M.; Sinopoli, J.; Weiss, J.; Shia, J.; Lamendola-Essel, M.; El Dika, I.H.; Segal, N.; Shcherba, M.; Sugarman, R.; et al. PD-1 Blockade in Mismatch Repair-Deficient, Locally Advanced Rectal Cancer. *N. Engl. J. Med.* **2022**, *386*, 2363–2376. [CrossRef] [PubMed]
4. Siegel, R.L.; Miller, K.D.; Fuchs, H.E.; Jemal, A. Cancer statistics, 2022. *CA Cancer J. Clin.* **2022**, *72*, 7–33. [CrossRef] [PubMed]
5. Li, N.; Spetz, M.R.; Li, D.; Ho, M. Advances in immunotherapeutic targets for childhood cancers: A focus on glypican-2 and B7-H3. *Pharmacol. Ther.* **2021**, *223*, 107892. [CrossRef]
6. Corti, C.; Cobanaj, M.; Marian, F.; Dee, E.C.; Lloyd, M.R.; Marcu, S.; Dombrovschi, A.; Biondetti, G.P.; Batalini, F.; Celi, L.A.; et al. Artificial intelligence for prediction of treatment outcomes in breast cancer: Systematic review of design, reporting standards, and bias. *Cancer Treat. Rev.* **2022**, *108*, 102410. [CrossRef]
7. Chiu, H.-Y.; Chao, H.-S.; Chen, Y.-M. Application of Artificial Intelligence in Lung Cancer. *Cancers* **2022**, *14*, 1370. [CrossRef]
8. Hamilton, W.; Walter, F.M.; Rubin, G.; Neal, R.D. Improving early diagnosis of symptomatic cancer. *Nat. Rev. Clin. Oncol.* **2016**, *13*, 740–749. [CrossRef]
9. Fountzilas, E.; Tsimberidou, A.M.; Vo, H.H.; Kurzrock, R. Clinical trial design in the era of precision medicine. *Genome. Med.* **2022**, *14*, 101. [CrossRef]
10. Bannigan, P.; Aldeghi, M.; Bao, Z.; Häse, F.; Aspuru-Guzik, A.; Allen, C. Machine learning directed drug formulation development. *Adv. Drug Deliv. Rev.* **2021**, *175*, 113806. [CrossRef]
11. Olgen, S. Overview on Anticancer Drug Design and Development. *Curr. Med. Chem.* **2018**, *25*, 1704–1719. [CrossRef]
12. Grandori, C.; Kemp, C.J. Personalized Cancer Models for Target Discovery and Precision Medicine. *Trends Cancer* **2018**, *4*, 634–642. [CrossRef]
13. Hoelder, S.; Clarke, P.A.; Workman, P. Discovery of small molecule cancer drugs: Successes, challenges and opportunities. *Mol. Oncol.* **2012**, *6*, 155–176. [CrossRef]
14. Schneider, P.; Walters, W.P.; Plowright, A.T.; Sieroka, N.; Listgarten, J.; Goodnow, R.A.; Fisher, J.; Jansen, J.M.; Duca, J.S.; Rush, T.S.; et al. Rethinking drug design in the artificial intelligence era. *Nat. Rev. Drug Discov.* **2020**, *19*, 353–364. [CrossRef]
15. Pandiyan, S.; Wang, L. A comprehensive review on recent approaches for cancer drug discovery associated with artificial intelligence. *Comput. Biol. Med.* **2022**, *150*, 106140. [CrossRef]
16. Wouters, O.J.; McKee, M.; Luyten, J. Estimated Research and Development Investment Needed to Bring a New Medicine to Market, 2009–2018. *JAMA* **2020**, *323*, 844–853. [CrossRef]
17. Dang, C.V.; Reddy, E.P.; Shokat, K.M.; Soucek, L. Drugging the 'undruggable' cancer targets. *Nat. Rev. Cancer* **2017**, *17*, 502–508. [CrossRef]
18. Bleker de Oliveira, M.; Koshkin, V.; Liu, G.; Krylov, S.N. Analytical Challenges in Development of Chemoresistance Predictors for Precision Oncology. *Anal. Chem.* **2020**, *92*, 12101–12110. [CrossRef]
19. Dentro, S.C.; Leshchiner, I.; Haase, K.; Tarabichi, M.; Wintersinger, J.; Deshwar, A.G.; Yu, K.; Rubanova, Y.; Macintyre, G.; Demeulemeester, J.; et al. Characterizing genetic intra-tumor heterogeneity across 2658 human cancer genomes. *Cell* **2021**, *184*. [CrossRef]
20. Suhail, Y.; Cain, M.P.; Vanaja, K.; Kurywchak, P.A.; Levchenko, A.; Kalluri, R.; Kshitiz. Systems Biology of Cancer Metastasis. *Cell Syst.* **2019**, *9*, 109–127. [CrossRef]
21. Fleming, N. How artificial intelligence is changing drug discovery. *Nature* **2018**, *557*, S55–S57. [CrossRef] [PubMed]
22. Boniolo, F.; Dorigatti, E.; Ohnmacht, A.J.; Saur, D.; Schubert, B.; Menden, M.P. Artificial intelligence in early drug discovery enabling precision medicine. *Expert. Opin. Drug Discov.* **2021**, *16*. [CrossRef] [PubMed]
23. Huwaimel, B.; Alobaida, A. Anti-Cancer Drug Solubility Development within a Green Solvent: Design of Novel and Robust Mathematical Models Based on Artificial Intelligence. *Molecules* **2022**, *27*, 5140. [CrossRef] [PubMed]
24. Yang, X.; Wang, Y.; Byrne, R.; Schneider, G.; Yang, S. Concepts of Artificial Intelligence for Computer-Assisted Drug Discovery. *Chem. Rev.* **2019**, *119*, 10520–10594. [CrossRef] [PubMed]
25. Kapetanovic, I.M. Computer-aided drug discovery and development (CADDD): In silico-chemico-biological approach. *Chem. Biol. Interact.* **2008**, *171*, 165–176. [CrossRef]
26. Hansen, S.G.; Wu, H.L.; Burwitz, B.J.; Hughes, C.M.; Hammond, K.B.; Ventura, A.B.; Reed, J.S.; Gilbride, R.M.; Ainslie, E.; Morrow, D.W.; et al. Broadly targeted $CD8^+$ T cell responses restricted by major histocompatibility complex E. *Science* **2016**, *351*, 714–720. [CrossRef]
27. Campos, K.R.; Coleman, P.J.; Alvarez, J.C.; Dreher, S.D.; Garbaccio, R.M.; Terrett, N.K.; Tillyer, R.D.; Truppo, M.D.; Parmee, E.R. The importance of synthetic chemistry in the pharmaceutical industry. *Science* **2019**, *363*. [CrossRef]
28. Woo, M. An AI boost for clinical trials. *Nature* **2019**, *573*, S100–S102. [CrossRef]
29. Zwicker, M. Understanding spatial environments from images. *Science* **2018**, *360*, 1188. [CrossRef]

30. Cao, L.; Yang, J.; Rong, Z.; Li, L.; Xia, B.; You, C.; Lou, G.; Jiang, L.; Du, C.; Meng, H.; et al. A novel attention-guided convolutional network for the detection of abnormal cervical cells in cervical cancer screening. *Med. Image Anal.* **2021**, *73*, 102197. [CrossRef]
31. Abriata, L.A.; Dal Peraro, M. State-of-the-art web services for de novo protein structure prediction. *Brief. Bioinform.* **2021**, *22*. [CrossRef]
32. Xuan, P.; Zhang, Y.; Cui, H.; Zhang, T.; Guo, M.; Nakaguchi, T. Integrating multi-scale neighbouring topologies and cross-modal similarities for drug-protein interaction prediction. *Brief. Bioinform.* **2021**, *22*. [CrossRef]
33. You, J.; Liu, B.; Ying, Z.; Pande, V.; Leskovec, J. Graph convolutional policy network for goal-directed molecular graph generation. *Adv. Neural Inform. Process. Syst.* **2018**, *31*. [CrossRef]
34. Ghimire, A.; Tayara, H.; Xuan, Z.; Chong, K.T. CSatDTA: Prediction of Drug-Target Binding Affinity Using Convolution Model with Self-Attention. *Int. J. Mol. Sci.* **2022**, *23*, 8453. [CrossRef]
35. Peng, J.; Wang, Y.; Guan, J.; Li, J.; Han, R.; Hao, J.; Wei, Z.; Shang, X. An end-to-end heterogeneous graph representation learning-based framework for drug-target interaction prediction. *Brief. Bioinform.* **2021**, *22*, bbaa430. [CrossRef]
36. Shao, K.; Zhang, Y.; Wen, Y.; Zhang, Z.; He, S.; Bo, X. DTI-HETA: Prediction of drug-target interactions based on GCN and GAT on heterogeneous graph. *Brief. Bioinform.* **2022**, *23*. [CrossRef]
37. Yang, Z.; Zhong, W.; Zhao, L.; Chen, C.Y.-C. ML-DTI: Mutual Learning Mechanism for Interpretable Drug-Target Interaction Prediction. *J. Phys. Chem. Lett.* **2021**, *12*, 4247–4261. [CrossRef]
38. Dara, S.; Dhamercherla, S.; Jadav, S.S.; Babu, C.M.; Ahsan, M.J. Machine Learning in Drug Discovery: A Review. *Artif. Intell. Rev.* **2022**, *55*, 1947–1999. [CrossRef]
39. Madhukar, N.S.; Khade, P.K.; Huang, L.; Gayvert, K.; Galletti, G.; Stogniew, M.; Allen, J.E.; Giannakakou, P.; Elemento, O. A Bayesian machine learning approach for drug target identification using diverse data types. *Nat. Commun.* **2019**, *10*, 5221. [CrossRef]
40. Olayan, R.S.; Ashoor, H.; Bajic, V.B. DDR: Efficient computational method to predict drug-target interactions using graph mining and machine learning approaches. *Bioinformatics* **2018**, *34*, 1164–1173. [CrossRef]
41. Raies, A.; Tulodziecka, E.; Stainer, J.; Middleton, L.; Dhindsa, R.S.; Hill, P.; Engkvist, O.; Harper, A.R.; Petrovski, S.; Vitsios, D. DrugnomeAI is an ensemble machine-learning framework for predicting druggability of candidate drug targets. *Commun. Biol.* **2022**, *5*, 1291. [CrossRef] [PubMed]
42. Huang, A.; Garraway, L.A.; Ashworth, A.; Weber, B. Synthetic lethality as an engine for cancer drug target discovery. *Nat. Rev. Drug Discov.* **2020**, *19*, 23–38. [CrossRef] [PubMed]
43. Wang, S.; Xu, F.; Li, Y.; Wang, J.; Zhang, K.; Liu, Y.; Wu, M.; Zheng, J. KG4SL: Knowledge graph neural network for synthetic lethality prediction in human cancers. *Bioinformatics* **2021**, *37*, i418–i425. [CrossRef] [PubMed]
44. Luo, Y.; Zhao, X.; Zhou, J.; Yang, J.; Zhang, Y.; Kuang, W.; Peng, J.; Chen, L.; Zeng, J. A network integration approach for drug-target interaction prediction and computational drug repositioning from heterogeneous information. *Nat. Commun.* **2017**, *8*, 573. [CrossRef]
45. Yamanishi, Y.; Araki, M.; Gutteridge, A.; Honda, W.; Kanehisa, M. Prediction of drug-target interaction networks from the integration of chemical and genomic spaces. *Bioinformatics* **2008**, *24*, i232–i240. [CrossRef]
46. Tang, J.; Szwajda, A.; Shakyawar, S.; Xu, T.; Hintsanen, P.; Wennerberg, K.; Aittokallio, T. Making sense of large-scale kinase inhibitor bioactivity data sets: A comparative and integrative analysis. *J. Chem. Inf. Model* **2014**, *54*, 735–743. [CrossRef]
47. Davis, M.I.; Hunt, J.P.; Herrgard, S.; Ciceri, P.; Wodicka, L.M.; Pallares, G.; Hocker, M.; Treiber, D.K.; Zarrinkar, P.P. Comprehensive analysis of kinase inhibitor selectivity. *Nat. Biotechnol.* **2011**, *29*, 1046–1051. [CrossRef]
48. Metz, J.T.; Johnson, E.F.; Soni, N.B.; Merta, P.J.; Kifle, L.; Hajduk, P.J. Navigating the kinome. *Nat. Chem. Biol.* **2011**, *7*, 200–202. [CrossRef]
49. Chen, H.; Engkvist, O. Has Drug Design Augmented by Artificial Intelligence Become a Reality? *Trends Pharmacol. Sci.* **2019**, *40*, 806–809. [CrossRef]
50. Lu, S.-H.; Wu, J.W.; Liu, H.-L.; Zhao, J.-H.; Liu, K.-T.; Chuang, C.-K.; Lin, H.-Y.; Tsai, W.-B.; Ho, Y. The discovery of potential acetylcholinesterase inhibitors: A combination of pharmacophore modeling, virtual screening, and molecular docking studies. *J. Biomed. Sci.* **2011**, *18*, 8. [CrossRef]
51. Paricharak, S.; Méndez-Lucio, O.; Chavan Ravindranath, A.; Bender, A.; Ijzerman, A.P.; van Westen, G.J.P. Data-driven approaches used for compound library design, hit triage and bioactivity modeling in high-throughput screening. *Brief. Bioinform.* **2018**, *19*, 277–285. [CrossRef]
52. Blay, V.; Tolani, B.; Ho, S.P.; Arkin, M.R. High-Throughput Screening: Today's biochemical and cell-based approaches. *Drug Discov. Today* **2020**, *25*, 1807–1821. [CrossRef]
53. Sheng, C.; Dong, G.; Miao, Z.; Zhang, W.; Wang, W. State-of-the-art strategies for targeting protein-protein interactions by small-molecule inhibitors. *Chem. Soc. Rev.* **2015**, *44*, 8238–8259. [CrossRef]
54. Lu, C.; Liu, S.; Shi, W.; Yu, J.; Zhou, Z.; Zhang, X.; Lu, X.; Cai, F.; Xia, N.; Wang, Y. Systemic evolutionary chemical space exploration for drug discovery. *J. Cheminform.* **2022**, *14*, 19. [CrossRef]
55. Yasuo, N.; Sekijima, M. Improved Method of Structure-Based Virtual Screening via Interaction-Energy-Based Learning. *J. Chem. Inf. Model.* **2019**, *59*, 1050–1061. [CrossRef]
56. Krasoulis, A.; Antonopoulos, N.; Pitsikalis, V.; Theodorakis, S. DENVIS: Scalable and High-Throughput Virtual Screening Using Graph Neural Networks with Atomic and Surface Protein Pocket Features. *J. Chem. Inf. Model.* **2022**, *62*, 4642–4659. [CrossRef]

57. Li, B.-H.; Ge, J.-Q.; Wang, Y.-L.; Wang, L.-J.; Zhang, Q.; Bian, C. Ligand-Based and Docking-Based Virtual Screening of MDM2 Inhibitors as Potent Anticancer Agents. *Comput. Math. Methods Med.* **2021**, *2021*, 3195957. [CrossRef]
58. Hutter, M.C. Differential Multimolecule Fingerprint for Similarity Search–Making Use of Active and Inactive Compound Sets in Virtual Screening. *J. Chem. Inf. Model.* **2022**, *62*, 2726–2736. [CrossRef]
59. Liu, R.; Li, X.; Lam, K.S. Combinatorial chemistry in drug discovery. *Curr. Opin. Chem. Biol.* **2017**, *38*, 117–126. [CrossRef]
60. Wang, Z.-Z.; Shi, X.-X.; Huang, G.-Y.; Hao, G.-F.; Yang, G.-F. Fragment-based drug design facilitates selective kinase inhibitor discovery. *Trends Pharmacol. Sci.* **2021**, *42*, 551–565. [CrossRef]
61. Erlanson, D.A. Introduction to fragment-based drug discovery. *Top. Curr. Chem.* **2012**, *317*. [CrossRef]
62. Bon, M.; Bilsland, A.; Bower, J.; McAulay, K. Fragment-based drug discovery-the importance of high-quality molecule libraries. *Mol. Oncol.* **2022**, *16*, 3761–3777. [CrossRef] [PubMed]
63. Eguida, M.; Schmitt-Valencia, C.; Hibert, M.; Villa, P.; Rognan, D. Target-Focused Library Design by Pocket-Applied Computer Vision and Fragment Deep Generative Linking. *J. Med. Chem.* **2022**, *65*, 13771–13783. [CrossRef] [PubMed]
64. Yang, Y.; Zheng, S.; Su, S.; Zhao, C.; Xu, J.; Chen, H. SyntaLinker: Automatic fragment linking with deep conditional transformer neural networks. *Chem. Sci.* **2020**, *11*, 8312–8322. [CrossRef]
65. Caburet, J.; Boucherle, B.; Bourdillon, S.; Simoncelli, G.; Verdirosa, F.; Docquier, J.-D.; Moreau, Y.; Krimm, I.; Crouzy, S.; Peuchmaur, M. A fragment-based drug discovery strategy applied to the identification of NDM-1 β-lactamase inhibitors. *Eur. J. Med. Chem.* **2022**, *240*, 114599. [CrossRef]
66. Yang, T.; Li, Z.; Chen, Y.; Feng, D.; Wang, G.; Fu, Z.; Ding, X.; Tan, X.; Zhao, J.; Luo, X.; et al. DrugSpaceX: A large screenable and synthetically tractable database extending drug space. *Nucleic. Acids Res.* **2021**, *49*, D1170–D1178. [CrossRef]
67. Vamathevan, J.; Clark, D.; Czodrowski, P.; Dunham, I.; Ferran, E.; Lee, G.; Li, B.; Madabhushi, A.; Shah, P.; Spitzer, M.; et al. Applications of machine learning in drug discovery and development. *Nat. Rev. Drug. Discov.* **2019**, *18*, 463–477. [CrossRef]
68. Paul, D.; Sanap, G.; Shenoy, S.; Kalyane, D.; Kalia, K.; Tekade, R.K. Artificial intelligence in drug discovery and development. *Drug. Discov. Today* **2021**, *26*, 80–93. [CrossRef]
69. Henson, A.B.; Gromski, P.S.; Cronin, L. Designing Algorithms To Aid Discovery by Chemical Robots. *ACS Cent. Sci.* **2018**, *4*, 793–804. [CrossRef]
70. Mouchlis, V.D.; Afantitis, A.; Serra, A.; Fratello, M.; Papadiamantis, A.G.; Aidinis, V.; Lynch, I.; Greco, D.; Melagraki, G. Advances in de Novo Drug Design: From Conventional to Machine Learning Methods. *Int. J. Mol. Sci.* **2021**, *22*, 1676. [CrossRef]
71. Tong, X.; Liu, X.; Tan, X.; Li, X.; Jiang, J.; Xiong, Z.; Xu, T.; Jiang, H.; Qiao, N.; Zheng, M. Generative Models for De Novo Drug Design. *J. Med. Chem.* **2021**, *64*, 14011–14027. [CrossRef]
72. Kingma, D.P.; Welling, M. Auto-encoding variational bayes. *arXiv* **2013**. [CrossRef]
73. Born, J.; Manica, M.; Oskooei, A.; Cadow, J.; Markert, G.; Rodríguez Martínez, M. PaccMann: De novo generation of hit-like anticancer molecules from transcriptomic data via reinforcement learning. *iScience* **2021**, *24*, 102269. [CrossRef]
74. Hong, S.H.; Ryu, S.; Lim, J.; Kim, W.Y. Molecular Generative Model Based on an Adversarially Regularized Autoencoder. *J. Chem. Inf. Model.* **2020**, *60*, 29–36. [CrossRef]
75. Samanta, B.; De, A.; Jana, G.; Gómez, V.; Chattaraj, P.K.; Ganguly, N.; Gomez-Rodriguez, M. Nevae: A deep generative model for molecular graphs. *J. Mach. Learn. Res.* **2020**, *21*, 1–33. [CrossRef]
76. Kriegeskorte, N.; Golan, T. Neural network models and deep learning. *Curr. Biol.* **2019**, *29*, R231–R236. [CrossRef]
77. Grisoni, F.; Moret, M.; Lingwood, R.; Schneider, G. Bidirectional Molecule Generation with Recurrent Neural Networks. *J. Chem. Inf. Model.* **2020**, *60*, 1175–1183. [CrossRef]
78. Mou, L.; Yan, R.; Li, G.; Zhang, L.; Jin, Z. Backward and forward language modeling for constrained sentence generation. *arXiv* **2015**. [CrossRef]
79. Berglund, M.; Raiko, T.; Honkala, M.; Kärkkäinen, L.; Vetek, A.; Karhunen, J.T. Bidirectional recurrent neural networks as generative models. *Adv. Neural Inform. Process. Syst.* **2015**, *28*. [CrossRef]
80. Krishnan, S.R.; Bung, N.; Bulusu, G.; Roy, A. Accelerating Drug Design against Novel Proteins Using Deep Learning. *J. Chem. Inf. Model.* **2021**, *61*, 621–630. [CrossRef]
81. Moret, M.; Friedrich, L.; Grisoni, F.; Merk, D.; Schneider, G. Generative molecular design in low data regimes. *Nature Mach. Intell.* **2020**, *2*, 171–180. [CrossRef]
82. Goodfellow, I.; Pouget-Abadie, J.; Mirza, M.; Xu, B.; Warde-Farley, D.; Ozair, S.; Courville, A.; Bengio, Y. Generative adversarial networks. *Commun. ACM* **2020**, *63*, 139–144. [CrossRef]
83. Maziarka, Ł.; Pocha, A.; Kaczmarczyk, J.; Rataj, K.; Danel, T.; Warchoł, M. Mol-CycleGAN: A generative model for molecular optimization. *J. Cheminform.* **2020**, *12*, 2. [CrossRef] [PubMed]
84. Abbasi, M.; Santos, B.P.; Pereira, T.C.; Sofia, R.; Monteiro, N.R.C.; Simões, C.J.V.; Brito, R.M.M.; Ribeiro, B.; Oliveira, J.L.; Arrais, J.P. Designing optimized drug candidates with Generative Adversarial Network. *J. Cheminform.* **2022**, *14*, 40. [CrossRef] [PubMed]
85. Arulkumaran, K.; Deisenroth, M.P.; Brundage, M.; Bharath, A.A. Deep reinforcement learning: A brief survey. *IEEE Signal Process. Magazine* **2017**, *34*, 26–38. [CrossRef]
86. Popova, M.; Isayev, O.; Tropsha, A. Deep reinforcement learning for de novo drug design. *Sci. Adv.* **2018**, *4*, eaap7885. [CrossRef]
87. Goel, M.; Raghunathan, S.; Laghuvarapu, S.; Priyakumar, U.D. MoleGuLAR: Molecule Generation Using Reinforcement Learning with Alternating Rewards. *J. Chem. Inf. Model.* **2021**, *61*, 5815–5826. [CrossRef]

88. Cheng, F.; Lu, W.; Liu, C.; Fang, J.; Hou, Y.; Handy, D.E.; Wang, R.; Zhao, Y.; Yang, Y.; Huang, J.; et al. A genome-wide positioning systems network algorithm for in silico drug repurposing. *Nat. Commun.* **2019**, *10*, 3476. [CrossRef]
89. Wang, Y.; Yang, Y.; Chen, S.; Wang, J. DeepDRK: A deep learning framework for drug repurposing through kernel-based multi-omics integration. *Brief. Bioinform.* **2021**, *22*, bbab048. [CrossRef]
90. Zhang, H.; Cui, H.; Zhang, T.; Cao, Y.; Xuan, P. Learning multi-scale heterogenous network topologies and various pairwise attributes for drug-disease association prediction. *Brief. Bioinform.* **2022**, *23*, bbac009. [CrossRef]
91. Jarada, T.N.; Rokne, J.G.; Alhajj, R. SNF-NN: Computational method to predict drug-disease interactions using similarity network fusion and neural networks. *BMC Bioinform.* **2021**, *22*, 28. [CrossRef]
92. Luo, H.; Wang, J.; Li, M.; Luo, J.; Peng, X.; Wu, F.-X.; Pan, Y. Drug repositioning based on comprehensive similarity measures and Bi-Random walk algorithm. *Bioinformatics* **2016**, *32*, 2664–2671. [CrossRef]
93. Sadeghi, S.; Lu, J.; Ngom, A. An Integrative Heterogeneous Graph Neural Network-Based Method for Multi-Labeled Drug Repurposing. *Front. Pharmacol.* **2022**, *13*, 908549. [CrossRef]
94. Doshi, S.; Chepuri, S.P. A computational approach to drug repurposing using graph neural networks. *Comput. Biol. Med.* **2022**, *150*, 105992. [CrossRef]
95. Hodgson, J. ADMET–turning chemicals into drugs. *Nat. Biotechnol.* **2001**, *19*, 722–726. [CrossRef]
96. Niu, J.; Straubinger, R.M.; Mager, D.E. Pharmacodynamic Drug-Drug Interactions. *Clin. Pharmacol. Ther.* **2019**, *105*, 1395–1406. [CrossRef]
97. Selvaraj, C.; Chandra, I.; Singh, S.K. Artificial intelligence and machine learning approaches for drug design: Challenges and opportunities for the pharmaceutical industries. *Mol. Divers* **2022**, *26*, 1893–1913. [CrossRef]
98. Li, T.; Tong, W.; Roberts, R.; Liu, Z.; Thakkar, S. DeepCarc: Deep Learning-Powered Carcinogenicity Prediction Using Model-Level Representation. *Front. Artif. Intell.* **2021**, *4*, 757780. [CrossRef]
99. Vatansever, S.; Schlessinger, A.; Wacker, D.; Kaniskan, H.Ü.; Jin, J.; Zhou, M.-M.; Zhang, B. Artificial intelligence and machine learning-aided drug discovery in central nervous system diseases: State-of-the-arts and future directions. *Med. Res. Rev.* **2021**, *41*, 1427–1473. [CrossRef]
100. Mulpuru, V.; Mishra, N. In Silico Prediction of Fraction Unbound in Human Plasma from Chemical Fingerprint Using Automated Machine Learning. *ACS Omega* **2021**, *6*, 6791–6797. [CrossRef]
101. Chawla, S.; Rockstroh, A.; Lehman, M.; Ratther, E.; Jain, A.; Anand, A.; Gupta, A.; Bhattacharya, N.; Poonia, S.; Rai, P.; et al. Gene expression based inference of cancer drug sensitivity. *Nat. Commun.* **2022**, *13*, 5680. [CrossRef] [PubMed]
102. Gogleva, A.; Polychronopoulos, D.; Pfeifer, M.; Poroshin, V.; Ughetto, M.; Martin, M.J.; Thorpe, H.; Bornot, A.; Smith, P.D.; Sidders, B.; et al. Knowledge graph-based recommendation framework identifies drivers of resistance in EGFR mutant non-small cell lung cancer. *Nat. Commun.* **2022**, *13*, 1667. [CrossRef] [PubMed]
103. Gerdes, H.; Casado, P.; Dokal, A.; Hijazi, M.; Akhtar, N.; Osuntola, R.; Rajeeve, V.; Fitzgibbon, J.; Travers, J.; Britton, D.; et al. Drug ranking using machine learning systematically predicts the efficacy of anti-cancer drugs. *Nat. Commun.* **2021**, *12*, 1850. [CrossRef] [PubMed]
104. Wang, Z.; Clark, N.R.; Ma'ayan, A. Drug-induced adverse events prediction with the LINCS L1000 data. *Bioinformatics* **2016**, *32*, 2338–2345. [CrossRef] [PubMed]
105. Zhao, H.; Zheng, K.; Li, Y.; Wang, J. A novel graph attention model for predicting frequencies of drug-side effects from multi-view data. *Brief. Bioinform.* **2021**, *22*, bbab239. [CrossRef]
106. Zhu, J.; Liu, Y.; Zhang, Y.; Chen, Z.; Wu, X. Multi-Attribute Discriminative Representation Learning for Prediction of Adverse Drug-Drug Interaction. *IEEE Trans. Pattern Anal. Mach. Intell.* **2022**, *44*, 10129–10144. [CrossRef]
107. Deng, Y.; Xu, X.; Qiu, Y.; Xia, J.; Zhang, W.; Liu, S. A multimodal deep learning framework for predicting drug-drug interaction events. *Bioinformatics* **2020**, *36*, 4316–4322. [CrossRef]
108. He, H.; Chen, G.; Yu-Chian Chen, C. 3DGT-DDI: 3D graph and text based neural network for drug-drug interaction prediction. *Brief Bioinform.* **2022**, *23*, bbac134. [CrossRef]
109. Wunnava, S.; Qin, X.; Kakar, T.; Sen, C.; Rundensteiner, E.A.; Kong, X. Adverse Drug Event Detection from Electronic Health Records Using Hierarchical Recurrent Neural Networks with Dual-Level Embedding. *Drug Saf.* **2019**, *42*, 113–122. [CrossRef]
110. Zhang, T.; Leng, J.; Liu, Y. Deep learning for drug-drug interaction extraction from the literature: A review. *Brief. Bioinform.* **2020**, *21*, 1609–1627. [CrossRef]
111. Basile, A.O.; Yahi, A.; Tatonetti, N.P. Artificial Intelligence for Drug Toxicity and Safety. *Trends Pharmacol. Sci.* **2019**, *40*, 624–635. [CrossRef]
112. Jing, Y.; Bian, Y.; Hu, Z.; Wang, L.; Xie, X.-Q. Deep Learning for Drug Design: An Artificial Intelligence Paradigm for Drug Discovery in the Big Data Era. *AAPS J.* **2018**, *20*, 58. [CrossRef]
113. Harari, Y.N. Reboot for the AI revolution. *Nature* **2017**, *550*, 324–327. [CrossRef]
114. Welling, D.B.; Collier, K.A.; Burns, S.S.; Oblinger, J.L.; Shu, E.; Miles-Markley, B.A.; Hofmeister, C.C.; Makary, M.S.; Slone, H.W.; Blakeley, J.O.; et al. Early phase clinical studies of AR-42, a histone deacetylase inhibitor, for neurofibromatosis type 2-associated vestibular schwannomas and meningiomas. *Laryngoscope Investig. Otolaryngol.* **2021**, *6*, 1008–1019. [CrossRef]
115. Casaletto, J.; Maglic, D.; Toure, B.B.; Taylor, A.; Schoenherr, H.; Hudson, B.; Bruderek, K.; Zhao, S.; O'Hearn, P.; Gerami-Moayed, N. RLY-4008, a novel precision therapy for FGFR2-driven cancers designed to potently and selectively inhibit FGFR2 and FGFR2 resistance mutations. *Cancer Res.* **2021**, *81*, 1455. [CrossRef]

116. Sun, J.; Patel, C.B.; Jang, T.; Merchant, M.; Chen, C.; Kazerounian, S.; Diers, A.R.; Kiebish, M.A.; Vishnudas, V.K.; Gesta, S.; et al. High levels of ubidecarenone (oxidized CoQ10) delivered using a drug-lipid conjugate nanodispersion (BPM31510) differentially affect redox status and growth in malignant glioma versus non-tumor cells. *Sci. Rep.* **2020**, *10*, 13899. [CrossRef]
117. Vladimer, G.; Alt, I.; Sehlke, R.; Lobley, A.; Baumgärtler, C.; Stulic, M.; Hackner, K.; Dzurillova, L.; Petru, E.; Hadjari, L. 23P Enriching for response: Patient selection criteria for A2AR inhibition by EXS-21546 through ex vivo modelling in primary patient material. *Immuno-Oncol. Technol.* **2022**, *16*, 100128. [CrossRef]
118. Park, S.J.; Chang, S.-J.; Suh, D.H.; Kong, T.W.; Song, H.; Kim, T.H.; Kim, J.-W.; Kim, H.S.; Lee, S.-J. A phase IA dose-escalation study of PHI-101, a new checkpoint kinase 2 inhibitor, for platinum-resistant recurrent ovarian cancer. *BMC Cancer* **2022**, *22*, 28. [CrossRef]

pharmaceuticals

Review

Computer-Aided Identification of Kinase-Targeted Small Molecules for Cancer: A Review on AKT Protein

Erika Primavera, Deborah Palazzotti, Maria Letizia Barreca and Andrea Astolfi *

Department of Pharmaceutical Sciences, "Department of Excellence 2018–2022", University of Perugia, 06123 Perugia, Italy; erika.primavera@studenti.unipg.it (E.P.); deborah.palazzotti@studenti.unipg.it (D.P.); maria.barreca@unipg.it (M.L.B.)
* Correspondence: andrea.astolfi@unipg.it

Abstract: AKT (also known as PKB) is a serine/threonine kinase that plays a pivotal regulatory role in the PI3K/AKT/mTOR signaling pathway. Dysregulation of AKT activity, especially its hyperactivation, is closely associated with the development of various human cancers and resistance to chemotherapy. Over the years, a wide array of AKT inhibitors has been discovered through experimental and computational approaches. In this regard, herein we present a comprehensive overview of AKT inhibitors identified using computer-assisted drug design methodologies (including docking-based and pharmacophore-based virtual screening, machine learning, and quantitative structure–activity relationships) and successfully validated small molecules endowed with anticancer activity. Thus, this review provides valuable insights to support scientists focused on AKT inhibition for cancer treatment and suggests untapped directions for future computer-aided drug discovery efforts.

Keywords: cancer; AKT; computer-aided drug discovery; virtual screening; docking; pharmacophore; kinase inhibitors; machine learning; QSAR

check for updates

Citation: Primavera, E.; Palazzotti, D.; Barreca, M.L.; Astolfi, A. Computer-Aided Identification of Kinase-Targeted Small Molecules for Cancer: A Review on AKT Protein. *Pharmaceuticals* **2023**, *16*, 993. https://doi.org/10.3390/ph16070993

Academic Editors: Marialuigia Fantacuzzi and Mariangela Agamennone

Received: 13 June 2023
Revised: 6 July 2023
Accepted: 8 July 2023
Published: 11 July 2023

1. Introduction

AKTs are a group of serine/threonine kinases, also known as protein kinase B or PKBs, which play an important role in the regulation of a wide range of cellular functions, such as cell growth and proliferation, glucose metabolism, genome stability, transcription and protein synthesis, and neovascularization [1,2]. Specifically, this enzyme is a key component of the PI3K/AKTs/mTOR signaling pathway, whose overactivation contributes to the development of many human cancers and resistance to chemotherapeutic drugs [3,4].

Three different isoforms are composing the AKT family. The first isoform to be discovered and characterized was AKT1 (also known as PKBα) [5], followed by AKT2 and AKT3 (also named PKBβ and PKBγ, respectively) [6,7]. A differential tissue-specific expression and cellular localization is observed for the three isoforms. AKT1 is expressed ubiquitously in the cytosol, as well as at the plasma membrane, whereas AKT2 and AKT3 can be especially found in muscle tissue and in neurons, respectively [8]. Interestingly, the three isoforms play non-overlapping or opposing functions in pathological conditions. As an example, AKT1 has been shown to suppress breast cancer cell migration and invasion in in vitro studies, while AKT2 promotes these processes, potentially facilitating cancer metastasis [9,10]. These opposing roles of AKT1 and AKT2 in cell migration and invasion have also been validated in mouse models in vivo [11].

Structurally, the three isoforms share a generally conserved sequence (overall identity higher than 75%; Figure 1) and common structural organizations characterized by the presence of an N-terminal pleckstrin homology (PH) domain, an interdomain linker, a kinase catalytic domain, and a C-terminal hydrophobic motif.

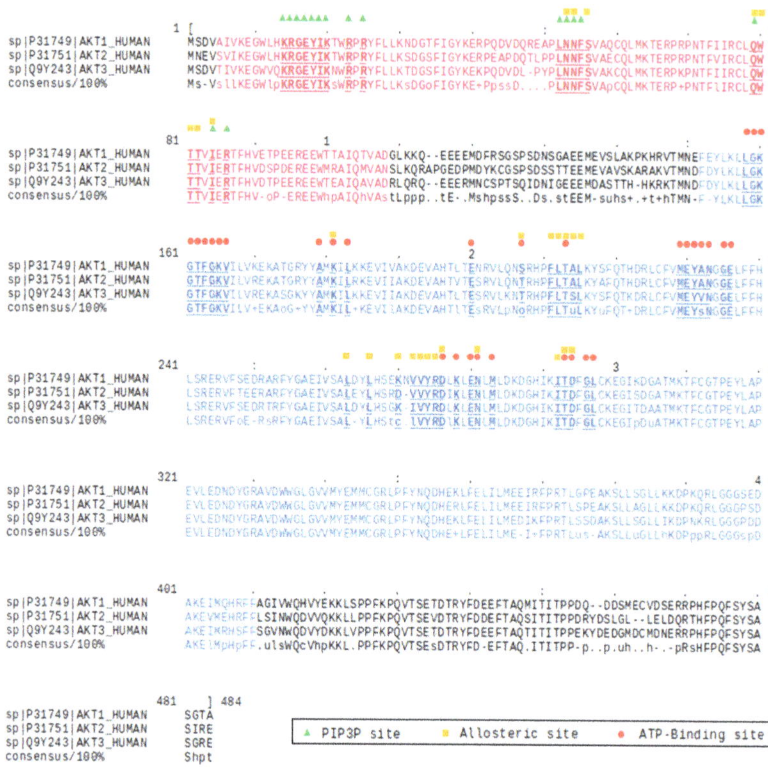

Figure 1. Sequence alignment of the three isoforms of AKT, i.e., AKT1, AKT2 and AKT3. The residues composing a ligand binding pocket are represented in bold and underlined. The specific pockets are differentiated by using different symbols. The PH domain and the kinase catalytic domain are color-coded in purple and blue, respectively. In the consensus sequence, conserved residues are indicated as capital letters, while non-conserved residues are represented as follows: c, charged; h, hydrophobic; l, aliphatic; o, alcohol; p, polar; s, small; t: turnlike; u, tiny; +, positive; -, negative. The alignment was performed by using the Cluster Omega server (https://www.ebi.ac.uk/Tools/msa/clustalo/ accessed on 1 June 2023).

As expected, in the three isoforms the most conserved domain is the catalytic domain (90% of sequence identity), while significative lower conservation is observed in the interdomain linker (25% of sequence identity) [12]. AKT activation occurs following the conversion in the plasma membrane of phosphatidylinositol (4,5)-bisphosphate (PI2P) into phosphatidylinositol (3,4,5)-trisphosphate (PI3P) by phosphatidylinositol 3-kinase (PI3K) (Figure 2A). The PH domain recognizes the charged head of the PI3P thanks to the presence on its structure of a proper site called the PI3P-binding site, allowing the translocation of AKT from the cytosol to the plasma membrane and subsequently activating the phosphorylation of Thr308 and Ser473. These double phosphorylations lead to increased accessibility of the ATP-binding site in the catalytic domain, thus resulting in an improvement of the kinase activity [13,14].

The PH domain can also negatively regulate the function of AKT1. Indeed, the interaction of the PH domain with the catalytic domain generates an autoinhibited AKT form that prevents accessibility of the ATP-binding site to the ATP molecules.

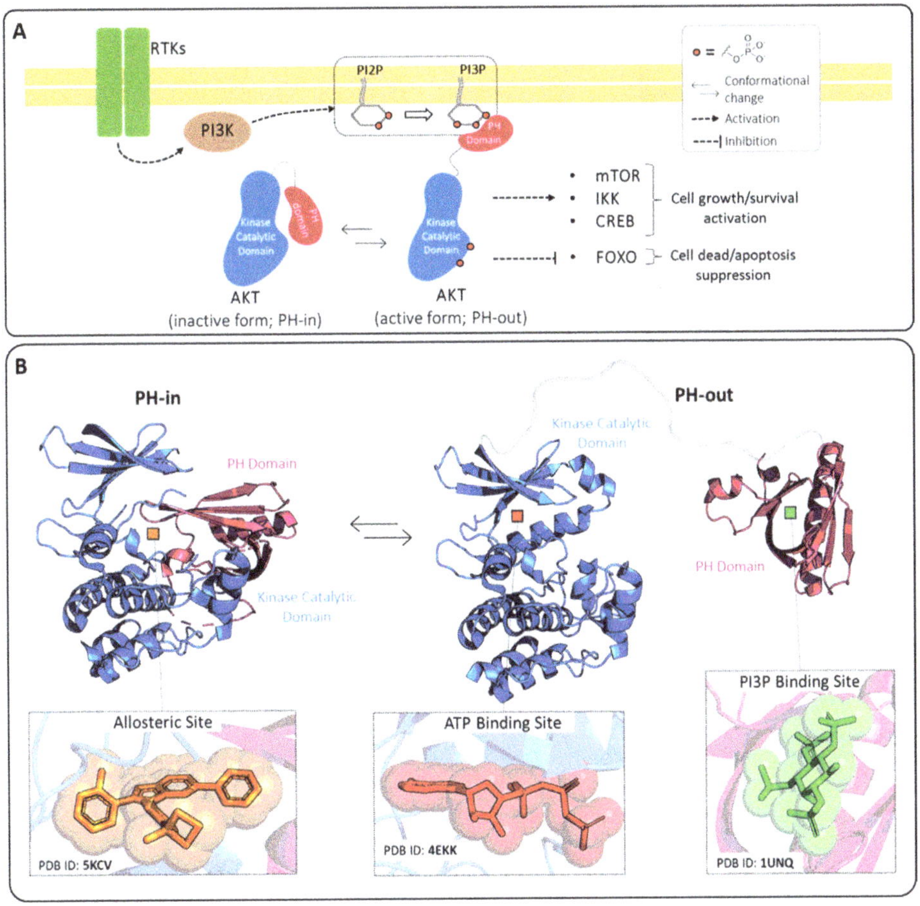

Figure 2. (**A**) Mechanisms of AKT regulation. The stimulation of receptor tyrosine kinases (RTKs) leads to activation of phosphatidylinositol 3-kinase (PI3K) and the subsequent conversion of the phosphatidylinositol (4,5)-bisphosphate (PI2P) into the phosphatidylinositol (3,4,5)-trisphosphate (PI3P). The inactive form of AKT (PH-in) is then recruited from the cytosol and translocated to the membrane thanks to the interaction between the AKT PH domain and PI3P. At this point, the double phosphorylation of AKT fully activates the protein (PH-out) that in turn modulates the downstream signaling proteins such as mammalian target of rapamycin (mTOR), IkappaB kinase (IKK), cAMP response element-binding protein (CREB) and Forkhead box O (FoxO). (**B**) AKT closed "PH-in" and opened "PH-out" conformations, with highlighted the three known ligand binding sites.

This inactive closed conformation is generally called "PH-in" which is distinguished from the opened active conformation "PH-out" (Figure 2). Several types of AKT inhibitors have been developed as a result of the presence of three well-known ligand-binding pockets in the AKT1-3 structures: the ATP-binding site, the allosteric site, and the PI3P-binding site. Detailed structural information on these pockets has been reported in the literature for isoform AKT1 and is summarized herein.

Regarding the ATP-binding site, no structures have been released of ATP bound to this pocket in the RCSB protein data bank (PDB) [15], but important clues can be gathered from the complex between AKT1 and adenylyl-imidodiphosphate (AMP-PNP; PDB ID: 4EKK [16]). Specifically, the ATP-binding pocket is composed of different structural elements that surround the central cavity, in which is accommodated the ATP substrate [16].

The hinge region (residues 227–230) connects the C-terminal and N-terminal domains and stabilizes the ATP adenine ring by two hydrogen bonds. Additionally, the glycine-rich loop (residues 157–162), the DFG motif (residues 292–294), and the αC-helix (residues 191–104) supply some amino acids which contribute to the constitution of the catalytic site. Compounds able to interact with this pocket prevent the binding of the ATP molecules, acting as competitive inhibitors.

The PI3P-binding site is mainly composed of positively charged residues, in accordance with the specificity of the site for the PI3P head (i.e., inositol-1,3,4,5-tetraphosphate; 4IP). One of the crystal structures of the AKT1 PH-domain in complex with 4IP is available under the PDB ID 1UNQ [17]. The phosphate groups of 4IP established a complex network of hydrogen bonds and salt bridges with Lys14, Glu17, Tyr18, Ile19, Arg23, Arg25, Asn53, and Arg86. Also in this case, small molecules able to interfere with this site can prevent the binding event between AKT1 and PI3P, acting as competitive inhibitors and blocking the protein translocation.

Finally, a third druggable site is formed in the surface involved in the protein-protein interaction between the kinase catalytic and PH domains. This pocket is only accessible in the "PH-in" conformation and can be targeted by AKT allosteric inhibitors. The binding of proper small molecules to this site allows for the (i) stabilization of the inactive conformation of AKT in which the accessibility of the ATP-binding site is reduced and (ii) stabilization of the PH-in conformation in which the PI3P cavity is shielded from the solvent and unavailable for PI3P recognition.

Following our interest in the kinase inhibitors' field [18,19], we recently investigated the AKT1, reporting the identification of a novel ATP-site-directed AKT1 inhibitor with anticancer activity [20]. Even though, to date, no FDA-approved drugs against this protein are on the market, many efforts have been made in the discovery of molecules capable of targeting AKT by binding to one of its three pockets [4,13,21], with particular emphasis on small molecules for cancer treatment [3,4,14]. In this context, we herein provide a comprehensive literature survey of published papers reporting the use of in silico approaches to identify AKT-targeted small molecules with experimentally validated anticancer activity.

The collected works were classified based on the targeted site (i.e., ATP-binding site, allosteric site, or PI3P-binding site) and the main applied methodology, namely docking-based, pharmacophore-based, machine learning (ML), or quantitative structure-activity relationship (QSAR) as reported in Table 1. For instance, most of the examined manuscripts employed molecular docking experiments either alone or in combination. In particular, docking simulations were performed with three different aims: (a) to carry out virtual screening of compound libraries; (b) to refine the results obtained in pharmacophore-based, ML, or QSAR studies, and (c) to predict the binding mode of specific compounds. In light of this consideration, we classified the docking-based approach only to those works in which this computational technique was applied as the main methodology. Furthermore, three-dimensional (3D) pharmacophore models were also developed by applying two well-known strategies: (i) ligand-based pharmacophore modeling, which uses small ligands with different binding affinities to build predictive models in the absence of drug-target structural data, and (ii) structure-based pharmacophore modeling, which uses structure-based data and the bioactive conformations of known ligands to build models. Finally, for the best-characterized compound(s) reported in each work, we report the 2D structures and the available biological activities in terms of half-maximal inhibitory concentration (IC_{50}), half-maximal effective concentration (EC_{50}), half-maximal growth inhibition concentration (GI_{50}), dissociation constant (Kd), inhibition constant (Ki) and percentage of inhibition (%inh). The reported activities can be referred to as assays performed on the isolated enzyme (i.e., AKT1, AKT2 or AKT3) or cellular system. In the latter case, the cellular line employed in the biological test is also indicated.

Table 1. Overview of collected AKT-targeted in silico studies according to the ligand binding site and the main computational methodology.

Site	Computational Methodology	Refs
ATP-binding site	docking-based	[20,22–27]
	pharmacophore-based	[28–31]
	ML	[32]
	QSAR	[33]
Allosteric site	docking-based	[34,35]
	pharmacophore-based	[36]
PI3P-binding site	pharmacophore-based	[37,38]

2. ATP-Binding Site

2.1. Docking-Based Approaches

From a historical point of view, the first published works were oriented towards targeting the AKT2 isoform.

In 2007, Donald et al. [22] started their work looking for a fragment that could act as a hinge-binder at the AKT2 ATP-binding site. Virtual screening of a fragment library composed of about 300,000 low molecular weight compounds (≤250 Da) allowed the selection of the 7-azaindole molecule **1** (Figure 3) as a possible hinge binder. Co-crystallization experiments performed with the PKA-AKT2 chimeras validated the docking predictions, and in its experimental binding mode, compound **1** established a double hydrogen bond interaction with the hinge residues (Glu121 and Ala123 in the PKA-AKT2 chimera, PDB ID: 2UVX [39]). Notably, the PKA-AKT2 chimera was previously validated as a valuable AKT2 surrogate for ATP-binder discovery [40]. Using **1** as the starting point, the authors coupled structure-based design and protein-ligand crystallography to rapidly identify novel potent AKT2 inhibitors (Figure 3). Specifically, to improve the synthetic accessibility of the scaffold and maintain the ability to form hydrogen bonds in the hinge region, the 7-azaindole core was replaced with a purine system. The fragment was suitably decorated and focused analogs were designed. Based on the crystallographic data provided by AKT2 bound to the known isoquinoline-5-sulfonamide inhibitor **2** (PDB ID: 2JDO [41]), the authors designed a series of derivatives (**3–5**, Figure 3) to mimic the interactions described as crucial between compound **2** and AKT2. The introduction of the basic amine in **3** enabled favorable polar contacts with the acidic residue Glu127 and the backbone carbonyl of Glu170. Furthermore, the addition of a terminal hydrophobic group (i.e., a benzyl moiety in compound **4**) resulted in a greater than 15-fold increase in AKT2 enzyme inhibition (IC_{50} 0.40 μM) compared to analog **3** (IC_{50} 6.9 μM). The final activity boost was obtained with the halogenation of the terminal aromatic ring (compound **5** with IC_{50} of 0.009 μM). The authors explained that the superior activity of **5** with respect to **2** was due to (i) the two hydrogen bonds interacting with the hinge region and (ii) the presence of a more rigid molecular structure bearing the same key pharmacophoric elements. Indeed, given the high number of rotatable bonds, the binding of compound **2** was associated with a higher entropic penalty. Unfortunately, despite the increased inhibitory potency, compound **5** showed low anticancer activity when tested in the PC3 cell lines. The authors hypothesized that the purine derivatives might have had a cellular permeability issue due to the low ratio of lipophilicity (CLogP) to topological polar surface area (TPSA). To overcome this limitation, they designed analog **6** which was predicted to have a more pronounced lipophilic character (ClogP: 4.81). The newly designed compound maintained a nanomolar IC_{50} and an optimal ligand efficiency (0.38 kcal mol^{-1} per non-H atom) against AKT2. Additionally, derivative **6** induced growth inhibition on different cancer cell lines in the low micromolar range (4.5, 5.7, and 8.7 μM in PC3, HCT116, and U87MG cell lines, respectively) and decreased the levels of pGSK3, pS6, and pFKHR proteins, that are downstream targets of the AKT2 pathway.

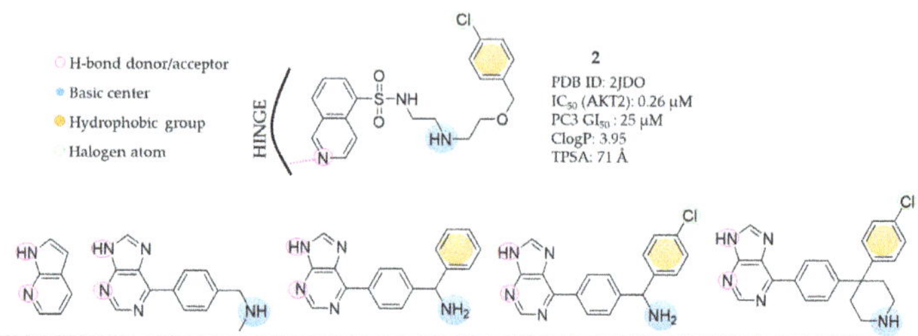

	1	3	4	5	6
PDB ID	2UVX	2UVY	2UVZ	\	2UV0
IC_{50} (AKT2)	> 100 μM	6.9 μM	0.40 μM	0.009 μM	0.020 μM
PC3 GI_{50}	NC	>50 μM	32 μM	28 μM	4.5 μM
ClogP	1.18	1.25	2.18	2.89	4.81
TPSA	24 Å	61 Å	75 Å	75 Å	61 Å

Figure 3. Targeting ATP-binding site. Rational identification of derivative **6** starting from fragment **1**. The AKT2-ligand interactions emerged from co-crystallization experiments are highlighted.

In 2009, Medina-Franco et al. [23] described the use of a consensus docking strategy in the structure-based VS, focusing on the selection of novel AKT2 inhibitors. Generally speaking, the consensus approach entails the use of more than one docking software/scoring function in the same VS pipeline to improve the overall docking performance [12,13]. Specifically, two docking software (i.e., FRED [42] and GOLD [43]) were used to explore AKT2 by using the two protein conformations co-crystallized with the known ATP-binding site inhibitors **AT7867** (PDB ID: 2UW9 [44]) and **A443654** (PDB ID: 2JDR [41]; Figure 4A). These structure-based strategies enabled the virtual screening of 105,937 lead-like compounds and the selection of nineteen putative AKT2 inhibitors that matched at least one of the following two criteria: top-ranked position based on the computed docking scores and/or the ability to make hydrogen bonds with the two key hinge residues Glu230 and Ala232. Notably, the co-crystallized inhibitors **AT7867** and **A443654** performed this double interaction (Figure 4A). Compound **7** (Figure 4B) was selected as the top-ranked molecule in the docking experiments against the 2UW9 protein model and was validated as a real hit in biological assays (IC_{50} = 1.1 μM). Interestingly, no interaction with the hinge residues was observed between this small molecule and the kinase domain. Conversely, the benzoxazolinone system performed two hydrogen bonds with the side chain atoms of Thr213 and Thr292, the 4*H*-1,4-benzoxazin-3-one moiety and Asp293 established a third hydrogen bond, and a π-π interaction emerged between the phenyl ring of **7** and Phe443. In the biological assays, compound **7** showed a pan-AKT inhibition activity with IC_{50} values of 2.6 μM, 1.1 μM, and 4.0 μM for AKT1, AKT2, and AKT3, respectively. Furthermore, this molecule exhibited interesting inhibition activities on the growth of MDA-MB-468 (EC_{50}: 3.8 μM) and MDA-MB-453 (EC_{50}: 10 μM) cancer cell lines.

It is well known that competitive kinase inhibitors mimic the interaction of the ATP adenine group with the hinge region, generally through a heteroaromatic system that can ensure an effective binding to the ATP-binding site [45–47].

Starting from this knowledge, in 2015 Chuang et al. [24] filtered the SPECS commercial library to keep only compounds that (i) included in their structure heteroaromatic ring systems able to form hydrogen bonds with the protein and (ii) were commercially available. The retained subset (35,367 compounds) was docked against the AKT1 ATP-binding site (PDB ID: 3MVH [48]) using the DOCK program [49], and the results were filtered considering energy score values, visual inspection, and chemical diversity. Finally, forty-eight compounds were selected as virtual hits for biological validation. The in silico procedure led to the identification of compound **8** (Figure 5), which was able to inhibit about 75% of AKT1 activity at a concentration of 100 μM.

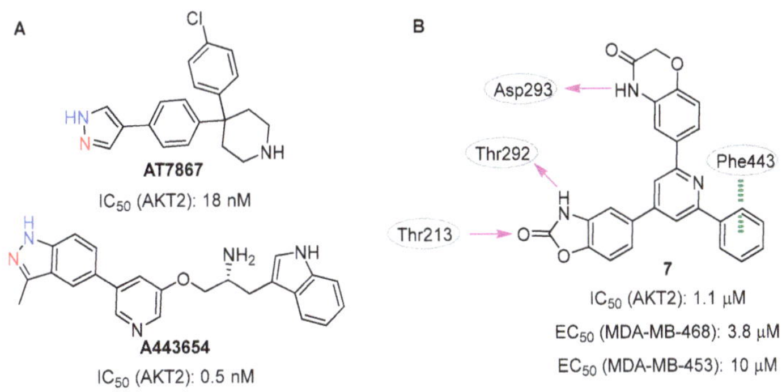

Figure 4. Targeting ATP-binding site. Co-crystallized AKT2 inhibitors **AT7867** and **A443654** (**A**) and validated hit **7** (**B**). The atoms involved in the hydrogen bond interactions with the AKT2 hinge residues Glu230 and Ala232 are highlighted in blue and red, respectively. Predicted polar intermolecular interactions are illustrated as follows: magenta arrow, hydrogen bond; green dotted line, π-π interaction.

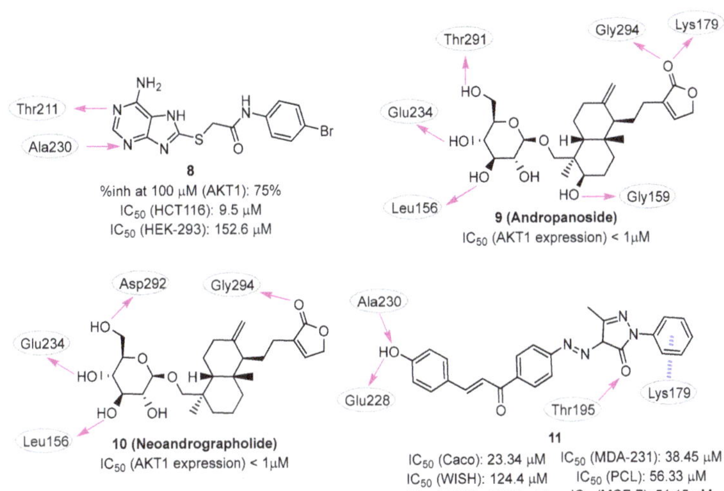

Figure 5. Targeting ATP-binding site. Validated hits **8–11** with the predicted polar intermolecular interactions illustrated as follows: magenta arrow, hydrogen bond; purple dotted lines: π-cation interaction.

The compound was further characterized by cytotoxicity evaluation on HCT-116 human colon cancer cells and HEK-293 normal cells, in which the inhibitor showed IC_{50} values of 9.5 and 152.6 μM, respectively, resulting in a selectivity index (IC_{50}-HEK-293/IC_{50}-HCT-116) of 16.1. In the proposed binding mode, **8** established two hydrogen bonds with the AKT1 hinge residues, Thr211 and Ala230. This small molecule also showed multiple hydrophobic interactions with surrounding residues, including Leu156, Val164, Met227, Tyr229, Phe237, Met281, Phe438, and Phe442.

In 2022, three computer-aided drug discovery (CADD) approaches described the identification of new competitive AKT1 small molecule inhibitors endowed with anticancer activities [20,25,26].

In the first work, Zhong et al. [26] queried the ZINC collection to rationally identify novel AKT1 inhibitors from compounds of natural origin. The original database (ZINC15 natural collection) consisted of 17,931 commercially available molecules. LibDock module from Discovery Studio [50] was used to perform docking experiments (PDB ID: 4EKL [16]) and the twenty top-ranked hits were chosen for further in silico studies, e.g., absorption, distribution, metabolism, excretion (ADME), and toxicity predictions. Particular attention was paid to the aqueous solubility, the human intestinal absorption, and the cytochrome P450 2D6 inhibition. Additionally, the stability of the predicted ligand binding mode was assessed using molecular dynamics (MD) simulations. The two final candidates, **andropanoside (9)** and **neoandrographonide (10)** (Figure 5) were subjected to biological characterization by using the human osteosarcoma MG63 cell line and showed an effective ability in inhibiting the cell growth, although no explicit data values (e.g., IC_{50} or EC_{50}) were provided. Additionally, both compounds inhibited the AKT1 expression in MG63 cells at a concentration in the submicromolar range. However, the authors did not provide direct evidence of AKT1 inhibition and/or binding for the two molecules.

In the second work performed by Noser et. al., docking studies were applied for the identification of a PI3K/AKT1 dual inhibitor [25]. The advantage of developing dual inhibitors of AKT1 and PI3K is that both proteins are involved in the same signaling pathway and play critical roles in the proliferation and survival of cancer cells. Therefore, targeting these two proteins simultaneously can lead to a more potent and efficient anticancer effect compared to targeting either protein alone. The authors used the PyRx platform [51] to virtually screen a small set of twenty-one in-house compounds for the ability to interact with both AKT1 and PI3K. The results highlighted that compound **11** (Figure 5) had favorable binding energy against both PI3K and AKT1 targets and promising in silico drug-like properties. In the predicted binding mode, the compound interacted with the AKT1 hinge region by hydrogen bonding Ala230, Glu228, and Thr195, while a π-cation interaction was predicted with the basic side chain of Lys179. Unfortunately, no information about the ability of **11** in inhibiting the isolated enzymes was provided, but the molecule was able to reduce the phosphorylation levels of both AKT1 and PI3K in a dose-dependent manner in the Caco colon cancer cell line. This small molecule also inhibited tumor growth in A549, MDA-231, Caco, PCL, and MCF-7 cancer cell lines with IC_{50} values of 40.91, 38.45, 23.34, 56.33, and 50.15 μM, respectively, and was characterized by low cytotoxic effects on WISH normal cells (IC_{50} = 124.4 μM).

Finally, we have recently reported the identification of novel competitive AKT1 inhibitors as possible agents against acute myeloid leukemia (AML) [20]. The serendipitous discovery of compound **12** (Figure 6) as an AKT1 inhibitor prompted us to carry out modeling studies to aid the selection of **12**-like compounds for biological testing. As the first step, **12** was used as a query molecule in the BioSolveIT Feature Trees (FTrees) [52] software, a fast 2D-similarity screening tool used to mine a library of in-house compounds. The resulting compounds were then docked with Glide [53,54] using the AKT1 conformation co-crystallized with the clinical trial inhibitor **capivasertib** (PDB ID: 4GV1 [55]). This protocol led to the identification of the 5,6,7,8-tetrahydrobenzo[4,5]thieno[2,3-*d*]pyrimidin-4(3*H*)-one derivative **13** (Figure 6, **T126** in the original paper) as the most interesting compound. Based on the predicted binding pose generated by the docking protocol and subsequent MD simulations, **13** interacted with the hinge region by using the para-hydroxyl group of the catechol moiety, while the second hydroxyl group was in contact with Thr291. Interestingly, the endocyclic amide established well-conserved, water-mediated interactions with residues Thr291 and Asp292. The compound inhibited the AKT1 activity with IC_{50} and K_i values of 1.99 and 0.41 μM, respectively. Accordingly, a clear effect on the growth inhibition and induction of apoptosis in AML cells at low micromolar concentrations was observed. Indeed, **13** showed IC_{50} values 4.2, 4.3, 2.4, 9.2, and 6.9 μM for OCI-AML3, IMS-M2, OCI-AML2, MOLM-13, and SKM-1 cell lines, respectively. Additionally, given the presence of a catechol moiety that is counted as one of the pan-assay interference compounds' (PAINS) motifs, we performed specific experiments to rule out the possibility that the compound interfered with the biochemical and cellular assays by the mechanism of intrinsic fluorescence effect, metal chelation, and chemical aggregation. Because of these encouraging data, the biologically validated hit **13** is now the subject of additional biological investigations and hit-to-lead efforts.

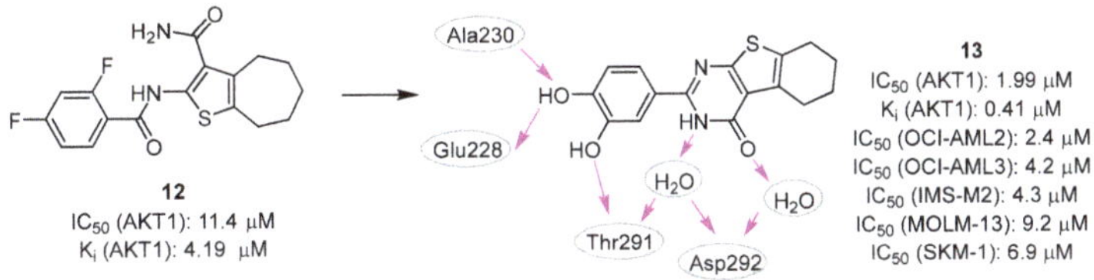

Figure 6. Targeting ATP-binding site. Validated AKT1 inhibitors **12** and **13** with the predicted polar intermolecular interactions for the latter compound illustrated as follows: magenta arrow, hydrogen bond.

Still, in 2022 Zhu et al. [27] reported a new potent and selective AKT1 proteolysis targeting chimera (PROTAC). PROTAC is a heterobifunctional molecule made up of two heads and a linker that can be used to degrade undesirable proteins. Indeed, while one PROTAC head binds to the target protein intended for degradation, the second head interacts with the E3 ubiquitin ligase, leading to ubiquitination and proteasome-mediated degradation of the former protein [56]. In this context, structure-based approaches are especially useful in the rational design of the PROTAC linker, where they may support the choice of the preferred chain length and/or composition. In particular, computational protocols, such as the "protein-protein docking & double clustering" method [57] can be applied in the prediction of PROTAC-mediated ternary complex model (i.e., PROTAC + E3 ligase + target protein). Using this strategy, the authors aimed to create a PROTAC molecule by linking the known AKT1 inhibitor **14** and the E3 ubiquitin ligase binder **pomalidomide** (Figure 7). Among the designed and tested compounds, derivative **15** (Figure 7) rapidly and completely removed AKT1 protein, also exhibiting efficacious antiproliferative effects on haematological cancer cells. From a chemistry point of view, this PROTAC molecule was characterized by a phenyl ring in the linker that was able to establish a favorable π-π interaction with the Phe236 of AKT1. The authors suggested that **15** and its analogues could be valuable for studying AKT1 biological functions and developing drugs to treat AKT-associated human cancers. Further evaluation of **15**'s safety benefits and therapeutic potential in mantle cell lymphoma (MCL) treatment is ongoing.

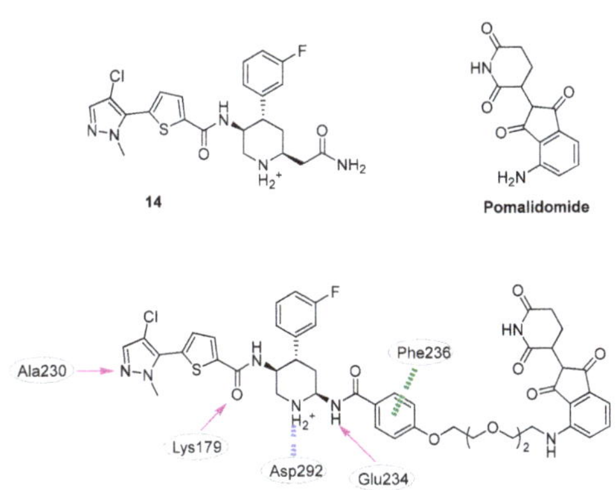

Figure 7. Targeting ATP-binding site. Compound **14** and **pomalidomide** were linked to generate the PROTAC molecule **15**. The predicted polar intermolecular interactions are illustrated as follows: magenta arrow, hydrogen bond; green dotted line, π-π interaction; purple dotted lines: π-cation interaction.

2.2. Pharmacophore-Based Approaches

In 2011, Dong et al. [28] reported the first successful example of AKT1-targeting molecules for cancer that was identified by ligand-based pharmacophore modeling. The authors collected, from the literature, seventy-four competitive AKT1 inhibitors with IC_{50s} spanning over six orders of magnitude (from 0.00016 to 32.25 μM). This library was then split into twenty-four and fifty compounds constituting the training- and the test set, respectively. The former set was used to generate a HypoGen [58] model composed of four pharmacophoric features, including one hydrophobic group, one hydrogen bond acceptor, and two hydrogen bond donors (Figure 8; compound **16** is reported as a representative of the training set). The model showed an area under the ROC curve (AUC) of 0.957, and a correlation coefficient of 0.803 between the experimental and predicted AKT1 IC_{50s} of the compounds in the test set. Encouraged by these results, the authors used the developed pharmacophore to screen 1024 and about 60,000 compounds from an *in-house* and the Maybridge databases, respectively. Among them, eighty candidates showed a good fit with the pharmacophoric features, and their selection was refined by using molecular docking experiments. To select both the proper docking algorithm and protein conformation, the authors performed self-docking studies on six different AKT1 protein structures (i.e., PDB IDs: 3CQU [59], 3CQW [59], 3MV5 [48], 3MVH [48], 3OCB [60], and 3OW4 [61]) employing five molecular docking software (i.e., Libdock [50], LigandFit [62], FlexX [63], CDOCKER [64], and flexiDOCK [65]). The best result was obtained by combining the 3OCB conformation with the flexiDOCK software (average RMSD in self-docking studies of 0.11 Å). Nine compounds were finally selected after docking studies and subsequently subjected to biochemical assays (HTScan PKB/Akt1 Kinase Assay Kit) and cytotoxicity studies on PC3, OVCAR-8, and HL-60 cell lines. The best inhibitory activities against AKT1 were shown by the flavonoid compound **17** with an IC_{50} value on the isolated AKT1 of 5.4 μM (Figure 8). The cellular assays also disclosed promising antiproliferative activity concerning all tested cell lines (PC3M, OVCAR-8 human ovarian carcinoma cells, and HL60 human leukemia cell line), with an IC_{50} ranging from 2.5 to 23.8 μM. Additionally, the apoptotic ability of **17** toward the OVCAR-8 cell line was tested as well. After 72 h of exposure, induction of apoptosis was observed in 40.19% of the treated cells. As proposed by the authors, computational studies suggested two main interactions between the flavonoid compound and AKT1: (i) the catechol moiety formed two hydrogen bonds with the residues Ala230 and Glu228, and (ii) the 5-hydroxyl group of the carboxylic oxygen of the chromanone ring interacted with the Lys179 residue. In an effort to enhance the inhibition activity of this chemical family, **17** was submitted to chemical optimization [31]. Based on docking and MD simulations, only derivative **18** was selected for further studies, as this compound was predicted to establish an additional hydrogen bond involving the 7-hydroxyl group and a π-π stacking interaction between one phenyl functionality and the Phe161 residue (Figure 8). Unfortunately, derivative **18** did not show any improvement compared to the parent compound **17** when submitted for biological experiments.

In a sequent study published in 2013 [30], the same research group applied the in silico workflow developed to identify compound **17** to investigate analogues of the previous published compound **5** (Figure 3). Specifically, a series of diphenyl methylamine derivatives were rationally designed in a stepwise manner by first decorating the primary amine with substituents of a different nature, and then replacing the purine system with a pyrazole ring. The generated molecules were submitted to pharmacophore screening and molecular docking studies. The best analog produced by this work was compound **19** (Figure 8), endowed with an IC_{50} of 0.038 μM on isolated AKT1 and low micromolar anti-proliferative activity on OVCAR-8, HL60, and HCT-116 cancer cell lines (8.1, 5.3, and 8.9 μM, respectively). Additionally, kinase selectivity studies underlined the excellent selectivity of this derivative against Aurora A, Drak, IKKb, GSK3b, SYK and JAK2 kinases. Finally, the information gathered by the produced biological results was used to refine the original pharmacophore model, shown in Figure 8 for the AKT1 inhibitor.

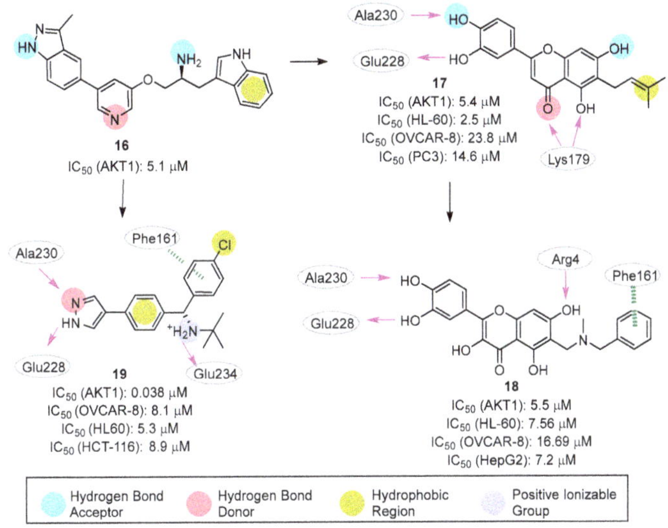

Figure 8. Targeting ATP-binding site. Known AKT1 inhibitor **16** and validated hits **17–19**. The predicted polar intermolecular interactions are illustrated as follows: magenta arrow, hydrogen bond; green dotted lines, π-π interaction. The original (**16** and **17**) [28] and the refined (**19**) [30] pharmacophore models are also illustrated.

Unlike the previous ligand-based approach, in 2020 Fratev et al. [29] proposed a structure-based pharmacophore model which was developed by combining the application of the e-Pharmacophore tool [66] from Schrödinger (https://www.schrodinger.com/), with fragment-based virtual screening. Specifically, the 3D model was generated based on docking results obtained for several hundred fragments oriented into the ATP-binding site of AKT1 (PDB ID: 3QKK [67]). The position of the well-scored fragments was used to define a specific pharmacophore feature (i.e., hydrogen bond acceptor, hydrogen bond donor, hydrophobic, negative ionizable, positive ionizable, or aromatic ring) according to the chemical nature of the fragment. The best-obtained model included four aromatic rings, three hydrogen-bond acceptors, and two hydrogen bond donors, and was integrated into a virtual screening protocol to query 3.5 million "lead-like" ZINC compounds. The top 1% of retrieved compounds (i.e., 35,000 ligands) were then docked and rescored by using Glide-SP. Finally, nine candidates emerged as virtual hits and were subjected to biochemical and cellular assays. Despite the efforts, the most interesting compound (**20**; Figure 9) was endowed with only a low ability to inhibit AKT1 activity and cancer cell growth. No information was provided about the fitting of derivative **20** on the developed pharmacophoric model.

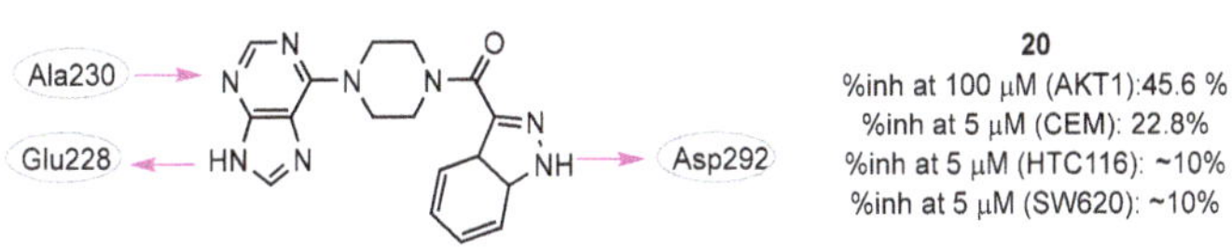

Figure 9. Targeting ATP-binding site. Compound **20** with the predicted polar intermolecular interactions illustrated as follows: magenta arrow, hydrogen bond.

2.3. Machine Learning Approaches

In 2021, Wang et al. applied a deep learning strategy called SyntaLinker to design novel AKT1 inhibitors [32]. SyntaLinker connects molecular fragments using syntactic pattern recognition and deep conditional transformer neural networks to design compound libraries by learning from the vast knowledge contained in chemogenomic repositories, such as ChEMBL [68]. This process enables a "scaffold hopping" approach for the identification of novel molecules with a potentially high binding affinity against the selected target. The authors started by decomposing the structure of the known AKT1 inhibitor **capivasertib** into three parts: a central piperidine scaffold, the pyrrolopyrimidine ring, and the chlorophenyl substituted propanol as terminal fragments (TFs) (Figure 10). They then applied the SyntaLinker strategy and generated new scaffolds able to link the two TFs as well as to satisfy some structural filters (e.g., maximal bond distance ranging from three to seven atoms and only one aromatic ring). The retrieved compounds were submitted to docking experiments leading to the selection of twenty-four molecules with ligand binding that have similar poses to **capivasertib**. Among these molecules, the authors selected the aminobenzamide derivative **21** (Figure 10) for synthesis and kinase activity validation, obtaining a moderate IC_{50} value of 7.2 μM. Subsequent chemical optimization strategies were carried out by (i) replacing the hydroxyethyl group with a basic methylamine moiety and (ii) removing the amine between the benzene ring and the pyrrolopyrimidine to shorten the length of the molecule. These modifications (compound **22**; Figure 11) significantly improved the inhibitory activities of AKT1, AKT2, and AKT3 (i.e., IC_{50}s of 0.088, 0.7, and 0.092 μM, respectively). The predicted pose of **22** in the ATP binding site of AKT1 showed that this inhibitor performed the conserved hydrogen bond interactions with the hinge residues Glu228 and Ala230, and an additional hydrogen bond with Asn279. Furthermore, hydrophobic interactions with Val164, Phe161, and Leu181, were present. Of note, **22** exhibited selective anticancer activity against U937 (IC_{50} value of 0.39 μM) cancer cells, while displaying less potent activities against other cancer cell lines (activities against HEPG-2, HEK293 and HCT116 with IC_{50} values of >10 μM, 1.96 μM, and 2.25 μM, respectively).

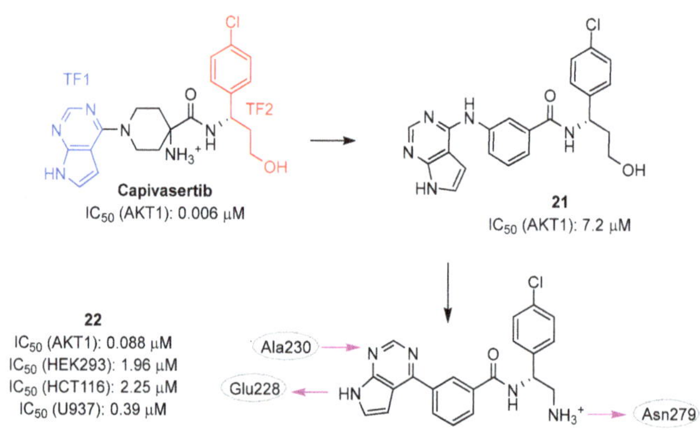

Figure 10. Targeting ATP-binding site. Known AKT1 inhibitor **capivasertib** and validated hits **21** and **22**. The predicted polar intermolecular interactions for compound **22** are illustrated as follows: magenta arrow, hydrogen bond. TF, terminal fragment.

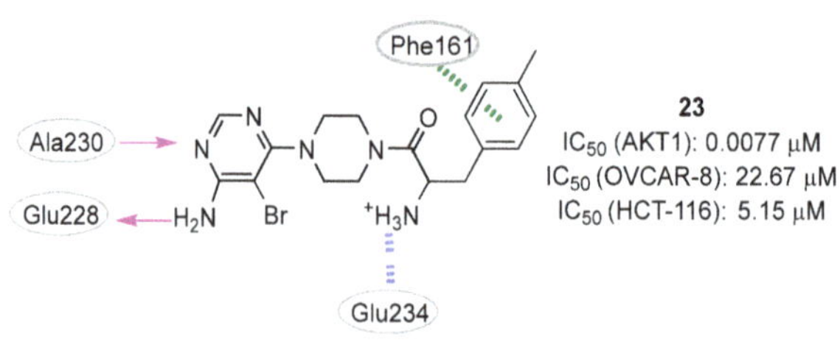

Figure 11. Targeting ATP-binding site. Validated hit **25** with the predicted polar intermolecular interactions illustrated as follows: magenta arrow, hydrogen bond; green dotted line, π-π interaction; purple dotted lines, π-cation interaction.

2.4. QSAR Modelling

In 2015, Zhan et al. [33] described the development of a molecular docking-based QSAR model for AKT1 inhibitors. The model was trained with forty-seven molecules collected from the literature data by applying the support vector regression (SVR) method. The collected set was subjected to docking simulations using the LigandFit program [62] and several scoring functions were employed to calculate the ligand binding scores. Additionally, key interaction profiles referred to the distances between the docked compounds and specific amino acid residues in the AKT1 were derived for each compound pose. Examples of these key amino acid residues included Glu228, Ala230, Glu234, Glu292, and Phe163. The combination of docking scores, key interaction profiles, and molecular descriptors (e.g., clogP and tPSA) allowed the generation of a QSAR model for bioactivity estimation with a very interesting predictability (R^2 = 0.934) that encompassed the predictability of the single LigandFit scoring function (R^2 < 0.417). The generated model was then used to select the most promising AKT1 inhibitors within a virtual library of 4-aminopyrimidine derivatives. Compound **25** (Figure 11) emerged as the most promising virtual hit (predicted pIC_{50}: 7.9), and its experimental activity on the isolated enzyme (IC_{50}: 7.7 nM; pIC_{50}: 8.1) confirmed the in silico prediction. Additionally, this small molecule showed promising antiproliferative effects against the HCT116 and OVCAR-8 cancer cell lines (IC_{50} of 5.15 and 22.67 μM, respectively). According to the docking studies reported by the authors, the molecular interactions between **25** and AKT1 involved two hydrogen bonds between the compound 4-aminopyrimidine group and Glu228 and Ala230 residues. Moreover, a hydrophobic interaction was hypothesized between the 4-chlorophenyl ring and the amino acid Phe161. Finally, an ionic interaction was predicted between the primary amino group of **25** and the Glu292 residue.

3. Allosteric Site

3.1. Docking-Based Approaches

Two papers published in 2022 described the application of docking-based virtual screening to discover potential allosteric AKT1 inhibitors. The SiBioLead web-tool (https://sibiolead.com/) was used in both works to identify a single lead molecule able to act as a dual inhibitor.

The first study, published by Al Shahrani et al. [35], was aimed at finding new molecules that could overcome resistance to **vemurafenib** in the treatment of melanoma. **Vemurafenib** is an approved kinase inhibitor that specifically inhibits the activity of the mutated form of the BRAF protein, known as BRAF-V600E [69]. As resistance to this drug is often due to reactivation of the MAPK and PI3K/AKT signaling pathways, the researchers planned to identify a dual inhibitor that could target both BRAF-V600E and AKT1 proteins. KINAcore and KINA-set libraries from ChemBridge were selected as the compound source and merged to create a collection of 23,365 compounds sharing 3D pharmacophore fingerprints with well-validated and published kinase inhibitors. The virtual screening of the two libraries against the BRAF-V600E (PDB ID: 1UWH [70]) led to the selection of the top 15 compounds for which more stringent docking calculations were produced, among which the evaluation of the potential affinity for the AKT1 target (PDB ID: 6HHG [71]). Compound **244** (Figure 12) was found to potentially bind both BRAF-V600E and AKT1 with high binding energy and affinity.

Specifically, docking experiments predicted a conserved π–π stacking interaction with Trp80, and three hydrogen bonds with Ser205, Asn54 and Gln79. Biological validation showed that this small molecule dose-dependently inhibited BRAF-V600E and AKT1 with IC_{50} values of 635 and 154.3 nM, respectively, and also controlled the proliferation of normal (A375-N cell line) and **vemurafenib**-resistant (A375-R cell line) melanoma cells. Indeed, while the effect of **vemurafenib** was reduced by about 3-fold in resistant cells (GI_{50} of 13.73 and 34.60 μM for A375-N and A375-R, respectively), no activity decrease was observed in the case of treatment with **24** (GI_{50} of 222.3 and 230.5 nM for A375-N and A375-R, respectively). Additionally, the compound was able to induce cell cycle arrest and apoptosis and to reduce the number of cells with high levels of pERK and pAKT, a condition generally associated with insurgency of resistance mechanisms.

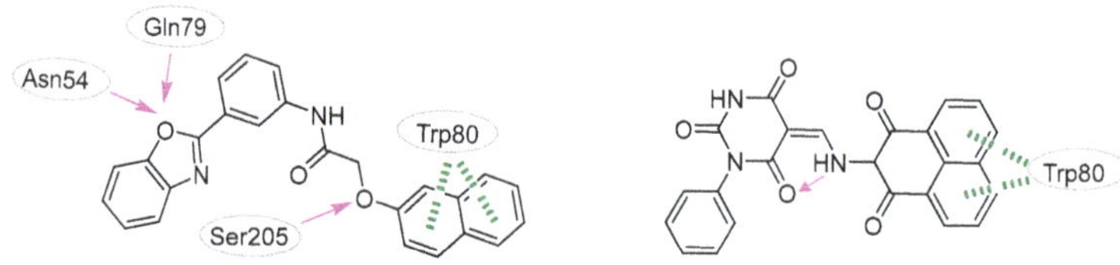

24
IC_{50} (AKT1): 0.154 μM
GI_{50} (A375-N): 0.222 μM
GI_{50} (A375-R): 0.230 μM

25
IC_{50} (AKT1): 0.189 μM
GI_{50} (THP-1): 0.077 μM
GI_{50} (A375-R): 0.049 μM

Figure 12. Targeting ATP-binding site. Validated AKT1 inhibitors **24** and **25** with the predicted polar intermolecular interactions for the latter compound illustrated as follows: magenta arrow, hydrogen bond; green dotted line, π-π interaction.

In the second study, Abohassan et al. [34] reported their efforts in identifying a single lead molecule able to inhibit both PI3K and AKT1 for the treatment of AML. Because the majority of potent kinase inhibitors have molecular weights between 350 and 750 Da, the authors built a chemical library for virtual screening composed of about 800,000 ChemBridge compounds respecting this range. To accelerate the screening process, diversity-based high-throughput virtual screening (D-HTVS) was applied. In the first step, structurally diverse compounds selected from the original library were docked against the target protein, with each compound taken as a representative of a given scaffold. In the second step, each scaffold of the top 10 compounds was used to collect a focused library for further docking simulations. First, the D-HTVS workflow was applied to explore the PI3Kγ protein. Indeed, PI3K is expressed in several isoforms, and previous studies indicated that the molecules targeting both PI3Kγ and PI3Kδ subunits showed maximum efficacy in various cancers where PI3K is implicated [72]. Then, the obtained virtual hits were subjected to a funnel-like process in which docking experiments were used to evaluate the affinity of the queried library again PI3Kα, PI3Kβ, PI3Kδ, and AKT1, each time discarding the molecules that did not meet the required criteria. Specifically, the authors selected only compounds with high affinity against PI3Kγ, PI3Kδ, and AKT1 and low affinity against PI3Kα and PI3Kβ. Four molecules matched these criteria, with compound **27** (Figure 12) chosen as the most promising hit based on the binding energies and docking poses calculated against all the protein targets considered in this study. Regarding the predicted interactions with AKT1, the benzo[*de*]isoquinoine system of **25** formed a π–π stacking interaction with Trp80 as well as additional hydrophobic contacts with Leu264, Ile84, Val270, and Arg270 residues. No intermolecular hydrogen bonds were predicted, while one intramolecular hydrogen bond was hypothesized between the nitrogen atom of the aminic linker and one of the oxygens of the barbituric ring. The in vitro kinase assays corroborated with computational predictions, showing that **25** inhibited PI3Kγ, PI3Kδ, and AKT1 kinases in the nanomolar range. Of note, this small molecule also showed potent anticancer activity when tested against THP-1 and HL-60 AML cell lines, validating the hypothesis that dual inhibitors targeting PI3K and AKT can efficiently inhibit AML cell proliferation.

3.2. Pharmacophore-Based Approaches

In 2017, Lakshmi et al. [36] gathered thirty-six known AKT1 pyridopyrimidine biphenyl derivative inhibitors (exemplified by compound **26**, Figure 13) to create a 3D pharmacophore model with Phase [73] composed of three hydrogen-bond acceptors, two positive groups and two aromatic rings. This model was then used to screen a library of about 5000 natural compounds and the 708 retained molecules were docked into the allosteric site of AKT1 (PDB ID: 3O96 [74]). Specifically, the authors employed a funnel-like pipeline, including Glide [54] HTVS, SP, XP, and Prime [75] MMGBSA steps. This process resulted in the retrieval of forty-five compounds, which were visually inspected leading to the final selection of twenty-three molecules able to potentially interact with the residues Ser205 and Trp80. Further MD computational analysis, based on the ligand hydrogen bond occupancy with key kinase residues, persistent polar contacts, and favorable binding free energy, suggested quercetin-7-O-*d*-glucopyranoside (**27**, Figure 13) as the most promising virtual hit. Biological validation of **27** revealed that this small molecule induced dose-dependent inhibition of breast cancer cells (MDA MB-231) and down-regulated the expression of p-AKT1 (Ser473). The authors further confirmed the ability of **29** to bind protein kinase AKT1 by measuring a Kd of 0.246 μM.

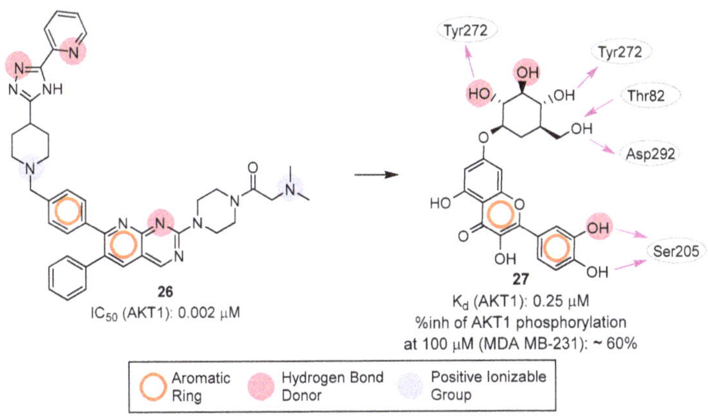

Figure 13. Targeting allosteric site. Known AKT1 inhibitor **26** and validated hit **27** and **24**. The predicted polar intermolecular interactions for the later compound are illustrated as follows: magenta arrow, hydrogen bond. The generated pharmacophore model is also highlighted.

4. PI3P-Binding Site

Pharmacophore-Based Approaches

In 2008, Mahadevan et al. described an in silico strategy for the development of PI3P site binders starting from the crystal structure of the AKT1 PH domain (PDB ID: 1H10 [76]) in complex with the head of PI3P (i.e., 4IP) [37]. The intermolecular interactions performed by the ligand with the protein allowed the generation of a pharmacophore query generated with UNITY (Tripos, L.P.) [77], although the authors did not explicitly describe the selected pharmacophore features. The generated pharmacophore query was used to screen the National Cancer Institute (NCI) database, and the retrieved compounds were submitted to docking simulations. Based on the predicted binding mode, molecule **28** (Figure 14) emerged as the most promising AKT1 potential binder as it was able to mimic the natural substrate. Indeed, the ligand sulfonamido and the diazopyrazolyl moieties formed hydrogen bond interactions with Arg86 and Arg23, respectively, which are two residues already known to establish strong contacts with the phosphate groups of PI3P. Additionally, other hydrogen bonds were hypothesized between the backbone of Ile19 and Asn53 with the sulfonamide function of molecule **31**. By using surface plasmon resonance (SPR) experiments, the authors validated the molecule as a true binder of the AKT1 PH domain with a K_d value of 0.37 μM. Additionally, this PH-domain binder inhibited the interaction of PI3P to the PH domain of AKT1 with an IC_{50} of 0.08 μM. In cellular assays, compound **31** inhibited both the AKT1 phosphorylation with IC_{50} values of 4 and 13 μM in NIH3T3 and HT-29 cell lines, respectively, and the growth of HT-29 cells with an IC_{50} of 24 μM. Unfortunately, no appreciable effects were observed in in vivo experiments, probably due to the rapid metabolism/elimination of this compound.

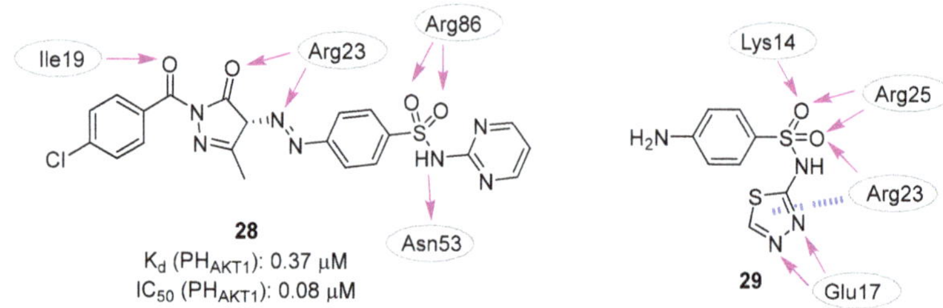

K_d (PH_{AKT1}): 0.37 μM
IC_{50} (PH_{AKT1}): 0.08 μM
IC_{50} (AKT1 phosphorilation in NIH3T3 cells): 4 μM
IC_{50} (AKT1 phosphorilation in HT-29 cells): 13 μM
IC_{50} (HT-29): 24 μM

Figure 14. Targeting PI3P-binding site. Validated hits **28** and **29** with the predicted polar intermolecular interactions illustrated as follows: magenta arrow, hydrogen bond; blue dotted line, cation-π interaction.

In 2009, the same research group (Moses et al. in 2009) [38] applied the pharmacophore-based approach described above to query a library of about 300,000 compounds derived from different compound collections (NCI Chemical and Natural Products Library, the Maybridge Available Chemicals Directory, and the Lead-Quest Chemical Library). The best 20 compounds of each collection were kept, clustered based on the 2D structure, and reduced to 4 virtual hits submitted for biological evaluation. All the selected compounds showed a binding affinity for the PH domain of AKT1 in the low micromolar range, with K_d values ranging from 0.39 to 6.27 μM. Even though the four compounds belonged to different chemical families, they shared the presence of at least one sulphone containing a linker that connected two aromatic rings. With the aim to expand the structure-activity relationship of each chemical family, commercially available analogues of each one of the four validated hits were selected for further testing, but only the N-(1,3,4-thiadiazol-2-yl) benzenesulfonamide derivatives emerged as promising scaffolds in the development of PI3P-site binders. In particular, compound **29** (Figure 14) showed a K_d value of 0.45 μM, but no evident effects of inhibiting the recognition of PI3P (K_i > 50.0 μM) and preventing tumoral cell survival were observed (Table 2). The authors hypothesized that the inactivity of **29** in cellular assays could be imputed to the low predicted LogP and Caco-2 permeability of the compound, and thus they rationally designed some analogues to overcome this issue.

Table 2. Summary of calculated and experimental parameters for compounds **29**–**31**.

Compound	LogP [a]	Caco-2 [a]	K_d (μM)	K_i (μM)	pAKT1 Inhibition (IC_{50}, μM)	Cellular Growth Inhibition (IC_{50}, μM)
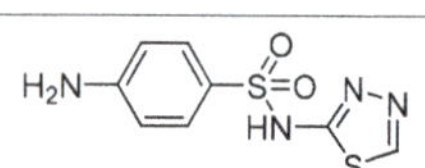 29	0.13	0.3	0.45 ± 0.1	>50.0	20 [b]/25 [c]	NI [b]/NI [c]
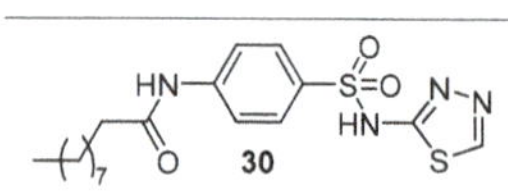 30	4.93	10.1	19.6 ± 4.9	21.8 ± 1.8	10 [b]/15 [c]	127 [b]/90 [c]
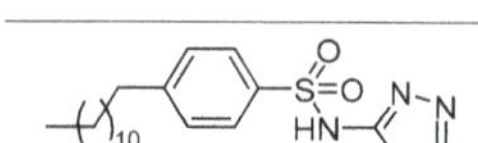 31	7.54	0.1	40.8 ± 2.5	2.4 ± 0.6	6 [b]/10 [c]	65 [b]/30 [c]

[a] Predicted value; [b] Panc-1 (human pancreatic cancer cells); [c] MiaPaCa-2 (human pancreatic cancer cells).

Docking studies of the hit binder performed with GOLD [43] suggested a complex network of polar interactions established with Lys14, Glu17, Arg23, and Arg25 (Figure 15). Interestingly, the phenyl moiety protruded the para-amino substituent towards the solvent and this position was explored by introducing modifications aimed to increase the calculated LogP and Caco-2 permeability. The best results were obtained for derivative **30** (Table 2) by adding a hydrophobic moiety, obtaining an improved K_i and cellular activity. To rule out the hypothesis that analogue **30** could be embedded into the cellular membrane and, subsequently to the cleavage of the amide bond, released into the cytoplasm, the non-cleavable compound **31** (Table 2) was synthesized.

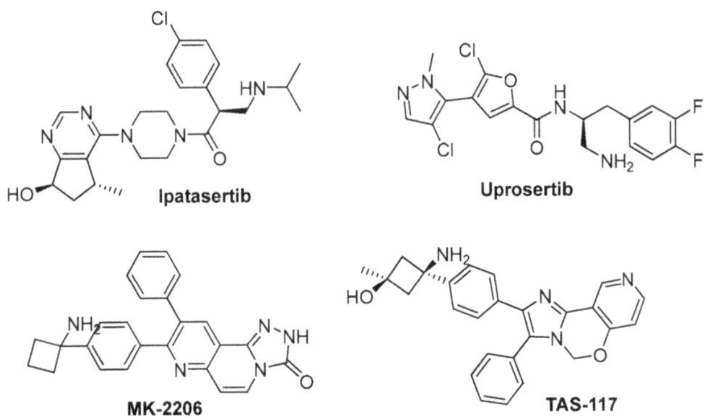

Figure 15. Chemical structure of the AKT clinical candidate inhibitors **Ipatasertib**, **Uprosertib**, **MK-2206**, and **TAS-117**.

Despite the predicted high LogP and low Caco-2 permeability properties, this derivative exhibited the most effective reduction of cell phospho-Ser473-AKT and cellular growth. Additionally, by using a fluorescent analog of **30**, the authors demonstrated that in the cellular environment, the compound was mainly located in the cytosol and/or lipid vesicles, potentially trapping AKT in the former matrix. Finally, this PI3P-site binder was also tested for antitumor activity against BxPC-3 pancreatic cancer xenografts at a dose of 125 mg/kg i.p., twice a day for 5 days, and demonstrated significant antitumor activity, with complete cessation of tumor growth and even regression during treatment.

5. Conclusions

In the era of precision oncology, AKT represents an attractive therapeutic target for the discovery of pathway-based targeted therapies selectively hitting cancer cells characterized by aberrant activation of PI3K/AKTs/mTOR signaling pathway.

Against this backdrop, we have conducted a literature survey focused on computer-aided approaches applied to the rational identification of AKT-targeting small molecules endowed with anticancer activity. Several considerations can be drawn from the analysis of the collected works and are listed below.

First, molecular docking and pharmacophore modeling emerged as the most commonly employed methods, whereas other approaches such as the development of QSAR models were less prevalent. Additionally, although ML algorithms, such as support vector machines (SVM) and deep learning models, have shown promise in predicting the activity and the properties of potential inhibitors [78,79], in the context of AKT for cancer, these approaches have been relatively underexplored. In this regard, accessing and analyzing the vast amount of data available in databases, such as PubChem and ChEMBL, becomes crucial. For instance, our research group recently released a KNIME workflow that can effectively aggregate information on the biological activities of compounds from different public/proprietary depositories. As a case study, we reported the application of such a tool for the generation of a curated dataset of over 358,000 compounds associated with experimental activity against AKT1 [80]. Access to such a wealth of data can support the construction of ML models and, more broadly, the use of ligand-based approaches to aid the development of novel AKT1 inhibitors.

Second, most of the reviewed studies focused on hit identification rather than hit-to-lead optimization. Specifically, the newly described hit compounds showed IC_{50} values ranging from 0.54 to 7.2 μM. Although these activities fall significantly short of the low nanomolar range exhibited by AKT1 inhibitors in clinical trials (Table 3; Figures 10 and 15), it is important to acknowledge that the latter are the results of extensive optimization efforts. Nevertheless, computer-aided drug design methods can strongly support lead identification as shown by Wang et al. [32] in the successful optimization of compound **22** (IC_{50}: 7.2 μM) to obtain derivative **23** (IC_{50}: 88 nM).

Table 3. Active clinical trial studies on AKT inhibitors for cancer therapy.

Compound	Figure	IC_{50}	Ligand Binding Site	Clinical Phase	NCT Number
Capivasertib (AZD5363)	Figure 10	6 nM (AKT1)	ATP-binding site	Phase I	NCT01226316, NCT04556773
				Phase I/II	NCT01992952, NCT02208375, NCT03742102
				Phase II	NCT02117167, NCT02299999, NCT02465060, NCT02664935, NCT04439123
				Phase III	NCT03997123, NCT04305496
Ipatasertib (GDC-0068)	Figure 15	5 nM (AKT1)	ATP-binding site	Phase I	NCT03959891
				Phase I/II	NCT03280563, NCT03424005, NCT03853707
				Phase II	NCT02465060, NCT03395899, NCT03498521, NCT04464174, NCT04591431, NCT04632992, NCT05498896
				Phase III	NCT03072238, NCT04060862
Uprosertib (GSK2141795)	Figure 15	9.6 nM (AKT1)	ATP-binding site	Phase I/II	NCT01902173
MK-2206	Figure 15	8 nM (AKT1)	Allosteric site	Phase I	NCT01480154
				Phase II	NCT01251861, NCT01306045
TAS-117	Figure 15	0.55 nM (AKT1)	Allosteric site	Phase II	NCT04770246

Third, we assessed whether the described studies had evaluated, in silico and/or in vitro, the propensity of the newly identified hit compounds to form aggregates or exhibit promiscuous activity as PAINS. In this context, in the last years, several computational tools have been developed to mitigate the selection of false positives resulting from the interference of the small molecule with the biological assay rather than the specific ligand interaction with the target protein. For instance, the aggregator advisor [81] and SwissADME [82] servers represent two valuable freely accessible web interfaces that allow the recognition of potential aggregators or PAINS, respectively. Regrettably, the majority of the analyzed studies failed to address this critical aspect. Therefore, we attempted to scrutinize the compounds featured in this review with the two previously mentioned servers, noting that several molecules were flagged as potential aggregators (cLogP $\geq$ 5 and Tanimoto similarity with known aggregators $\geq$ 0.5: compounds **11** and **24**) or PAINS (presence in the chemical structure of (i) azo group: compounds **11** and **28**; (ii) catechol moiety: compounds **13**, **17**, **18**, and **27**; (iii) barbiturate ring: compound **25**). Against this backdrop, we would like to remind researchers that it should be standard practice to first carefully predict these compound's characteristics and then conduct specific orthogonal tests to rule out artifacts that may arise during biological evaluations.

Fourth, of particular interest, is the absence of publications reporting on the computer-assisted design of covalent inhibitors. It is worth noting that two specific cysteine residues, well conserved among the three AKT isoforms (corresponding to Cys296 and Cys310 in AKT1), are located in the proximity of the allosteric site and have been already exploited in the identification of covalent inhibitors [8,71,83–85]. These small molecules possess both the selectivity resulting from targeting the allosteric pocket and the prolonged target residence time ensured by the covalent modification of the two non-catalytic cysteines in the AKT activation loop. As a result, this class of inhibitors is more effective at inhibiting AKT activity than non-covalent binders [86,87].

Fifth, it has been demonstrated that inhibiting both AKT1 and AKT2 is essential for obtaining a greater sensitization of tumor cells to apoptotic stimuli and reduceing AKT phosphorylation in vivo [88,89]. Despite this evidence, most of the papers reviewed here have tested their molecules only on one of the two isoforms, thus not considering the validated aspect of the dual AKT1/2 inhibition.

Sixth, our examination reveals that there are considerably more publications focused on the rational identification of competitive inhibitors as compared to allosteric inhibitors. While the abundance of research on competitive inhibitors is not surprising, given their historical prominence, it is important to recognize the potential advantages offered by allosteric modulation.

Finally, protein degraders have emerged as a promising class of therapeutics in drug discovery [56,90–92]. Even though some attempts have been made to discover AKT-targeted degraders [93–96], there is still ample room for further research and development that can be accelerated by the support of CADD approaches.

In conclusion, while several computational strategies have supported multidisciplinary efforts aimed at the rational identification of AKT inhibitors as anticancer agents, there remain intriguing avenues for future in silico research, particularly in the design of covalent inhibitors or small molecule degraders. These strategies offer promising possibilities for overcoming drug resistance and expanding the scope of personalized medicine in cancer treatment. In support of these objectives, the RCSB PDB is a valuable source of 3D data for structure-based CADD investigations. To provide additional support to the scientific community interested in the AKT target, we report here a comprehensive table (Table 4) summarizing the available crystal structures of AKT, either unbound or bound to a ligand, organized according to the isoform and the binding pocket.

Table 4. List of available PDB structures for AKT1, AKT2 and AKT3.

Pocket	Domain	PDB ID	Release Date	Resolution (Å)	Exp. IC_{50} (nM)	Notes
			AKT1			
no ligands	PH	1UNP	2004	1.65		
	PH	1UNR	2004	1.25		
	PH	2UZR	2007	1.94		
	Kinase	6BUU	2018	2.4		
	Kinase	6NPZ	2019	2.12		
	Full-length	7APJ	2021	2.05		Complexed with antibody
	PH	7MYX	2022	1.39		
ATP-binding site	Kinase	3CQU	2008	2.2	151	
	Kinase	3CQW	2008	2	318	
	Kinase	3MV5	2010	2.47	180	
	Kinase	3MVH	2010	2.01	0.5	

Table 4. *Cont.*

Pocket	Domain	PDB ID	Release Date	Resolution (Å)	Exp. IC_{50} (nM)	Notes
	Kinase	3OCB	2010	2.7	5	
	Kinase	3OW4	2010	2.6	22	
	Kinase	3QKK	2011	2.3	300	
	Kinase	3QKL	2011	1.9	9	
	Kinase	3QKM	2011	2.2	38	
	Kinase	4EKK	2012	2.8	*NR* [a]	AMP-PNP
	Kinase	4EKL	2012	2	36.9	Clinical candidate for cancer (**Ipatasertib**)
	Kinase	4GV1	2013	1.49	4	Clinical candidate for cancer (**Capivasertib**)
	Kinase	6CCY	2018	2.18	3	
Allosteric site	Full-length	3O96	2010	2.7	58	
	Full-length	4EJN	2012	2.19	5	
	Full-length	5KCV	2016	2.7	8	Clinical candidate Proteus syndrome (**Miransertib**)
	Full-length	6HHF	2019	2.9	0.5	Covalent binder
	Full-length	6HHG	2019	2.3	9.1	Covalent binder
	Full-length	6HHH	2019	2.7	126	Covalent binder
	Full-length	6HHI	2019	2.7	3.6	Covalent binder
	Full-length	6HHJ	2019	2.3	3	Covalent binder
	Full-length	6S9W	2019	2.3	39	Covalent binder
	Full-length	6S9X	2019	2.6	381	Covalent binder
	Full-length	7NH4	2021	2.3	44	Covalent binder
	Full-length	7NH5	2021	1.9	112	Covalent binder
PI3P-binding site	PH	1H10	2003	1.4	*NR* [a]	4IP
	PH	1UNQ	2004	0.98	*NR* [a]	4IP
	PH	2UVM	2007	1.94	Ki = 80 nM	
	PH	2UZS	2007	2.46	*NR* [a]	4IP, E17K mutation
			AKT2			
no ligands	Kinase	1GZK	2003	2.3		
	Kinase	1GZN	2003	2.5		
	Kinase	1GZO	2003	2.75		
	Kinase	1MRV	2003	2.8		
	Kinase	1MRY	2003	2.8		
	PH	1P6S	2004	NMR		
ATP-binding site	Kinase	1O6K	2002	1.7	*NR* [a]	ANP
	Kinase	1O6L	2002	1.6	*NR* [a]	ANP
	Kinase	2JDO	2007	1.8	230	
	Kinase	2JDR	2007	2.3	0.5	
	Kinase	2UW9	2007	2.1	18	
	Kinase	2X39	2010	1.93	6	
	Kinase	2XH5	2010	2.72	27	
	Kinase	3D0E	2008	2	Ki: 4 nM	
	Kinase	3E87	2008	2.3	*NR* [a]	
	Kinase	3E88	2008	2.5	0.6	
	Kinase	3E8D	2008	2.7	2	
			AKT3			
no ligands	PH	2X18	2010	1.46		

[a] NR: not reported.

Author Contributions: Conceptualization, M.L.B. and A.A.; writing—original draft preparation, E.P., D.P. and A.A.; writing—review and editing, M.L.B. and A.A. All authors have read and agreed to the published version of the manuscript.

Funding: This research received no external funding.

Acknowledgments: A.A. and M.L.B. would like to acknowledge the Fondazione Umberto Veronesi for the "Post-doctoral fellowship 2021" assigned to A.A. for the project entitled "Tuning the Precision Oncology on the PI3K/AKT/mTOR pathway". Currently, A.A. is a temporary researcher (RTD-A) supported by PON "Ricerca e innovazione" 2014–2020, Azione IV.6 (tematiche green) cod. 23-G15435-1. E.P would like to acknowledge her PhD scholarship at the University of Perugia funded by PNRR-DM 352/2022 and by Sibylla Biotech S.p.A.

Conflicts of Interest: The authors declare no conflict of interest.

References

1. Duggal, S.; Jailkhani, N.; Midha, M.K.; Agrawal, N.; Rao, K.V.S.; Kumar, A. Defining the Akt1 interactome and its role in regulating the cell cycle. *Sci. Rep.* **2018**, *8*, 1303. [CrossRef]
2. Manning, B.D.; Toker, A. AKT/PKB Signaling: Navigating the Network. *Cell* **2017**, *169*, 381–405. [CrossRef]
3. He, Y.; Sun, M.M.; Zhang, G.G.; Yang, J.; Chen, K.S.; Xu, W.W.; Li, B. Targeting PI3K/Akt signal transduction for cancer therapy. *Signal Transduct. Target. Ther.* **2021**, *6*, 425. [CrossRef]
4. Martorana, F.; Motta, G.; Pavone, G.; Motta, L.; Stella, S.; Vitale, S.R.; Manzella, L.; Vigneri, P. AKT Inhibitors: New Weapons in the Fight Against Breast Cancer? *Front. Pharmacol.* **2021**, *12*, 662232. [CrossRef] [PubMed]
5. Bellacosa, A.; Testa, J.R.; Staal, S.P.; Tsichlis, P.N. A retroviral oncogene, akt, encoding a serine-threonine kinase containing an SH2-like region. *Science* **1991**, *254*, 274–277. [CrossRef] [PubMed]
6. Cheng, J.Q.; Godwin, A.K.; Bellacosa, A.; Taguchi, T.; Franke, T.F.; Hamilton, T.C.; Tsichlis, P.N.; Testa, J.R. AKT2, a putative oncogene encoding a member of a subfamily of protein-serine/threonine kinases, is amplified in human ovarian carcinomas. *Proc. Natl. Acad. Sci. USA* **1992**, *89*, 9267–9271. [CrossRef]
7. Konishi, H.; Kuroda, S.; Tanaka, M.; Matsuzaki, H.; Ono, Y.; Kameyama, K.; Haga, T.; Kikkawa, U. Molecular cloning and characterization of a new member of the RAC protein kinase family: Association of the pleckstrin homology domain of three types of RAC protein kinase with protein kinase C subspecies and beta gamma subunits of G proteins. *Biochem. Biophys. Res. Commun.* **1995**, *216*, 526–534. [CrossRef]
8. Quambusch, L.; Depta, L.; Landel, I.; Lubeck, M.; Kirschner, T.; Nabert, J.; Uhlenbrock, N.; Weisner, J.; Kostka, M.; Levy, L.M.; et al. Cellular model system to dissect the isoform-selectivity of Akt inhibitors. *Nat. Commun.* **2021**, *12*, 5297. [CrossRef]
9. Chin, Y.R.; Toker, A. The actin-bundling protein palladin is an Akt1-specific substrate that regulates breast cancer cell migration. *Mol. Cell* **2010**, *38*, 333–344. [CrossRef]
10. Irie, H.Y.; Pearline, R.V.; Grueneberg, D.; Hsia, M.; Ravichandran, P.; Kothari, N.; Natesan, S.; Brugge, J.S. Distinct roles of Akt1 and Akt2 in regulating cell migration and epithelial-mesenchymal transition. *J. Cell Biol.* **2005**, *171*, 1023–1034. [CrossRef] [PubMed]
11. Wang, J.; Zhao, W.; Guo, H.; Fang, Y.; Stockman, S.E.; Bai, S.; Ng, P.K.; Li, Y.; Yu, Q.; Lu, Y.; et al. AKT isoform-specific expression and activation across cancer lineages. *BMC Cancer* **2018**, *18*, 742. [CrossRef] [PubMed]
12. Mattmann, M.E.; Stoops, S.L.; Lindsley, C.W. Inhibition of Akt with small molecules and biologics: Historical perspective and current status of the patent landscape. *Expert. Opin. Ther. Pat.* **2011**, *21*, 1309–1338. [CrossRef]
13. Garcia-Echeverria, C.; Sellers, W.R. Drug discovery approaches targeting the PI3K/Akt pathway in cancer. *Oncogene* **2008**, *27*, 5511–5526. [CrossRef] [PubMed]
14. Cheng, J.Q.; Lindsley, C.W.; Cheng, G.Z.; Yang, H.; Nicosia, S.V. The Akt/PKB pathway: Molecular target for cancer drug discovery. *Oncogene* **2005**, *24*, 7482–7492. [CrossRef]
15. Berman, H.M.; Battistuz, T.; Bhat, T.N.; Bluhm, W.F.; Bourne, P.E.; Burkhardt, K.; Feng, Z.; Gilliland, G.L.; Iype, L.; Jain, S.; et al. The Protein Data Bank. *Acta Crystallogr. D Biol. Crystallogr.* **2002**, *58*, 899–907. [CrossRef] [PubMed]
16. Lin, K.; Lin, J.; Wu, W.I.; Ballard, J.; Lee, B.B.; Gloor, S.L.; Vigers, G.P.; Morales, T.H.; Friedman, L.S.; Skelton, N.; et al. An ATP-site on-off switch that restricts phosphatase accessibility of Akt. *Sci. Signal* **2012**, *5*, ra37. [CrossRef]
17. Milburn, C.C.; Deak, M.; Kelly, S.M.; Price, N.C.; Alessi, D.R.; Van Aalten, D.M. Binding of phosphatidylinositol 3,4,5-trisphosphate to the pleckstrin homology domain of protein kinase B induces a conformational change. *Biochem. J.* **2003**, *375*, 531–538. [CrossRef]
18. Astolfi, A.; Iraci, N.; Manfroni, G.; Barreca, M.L.; Cecchetti, V. A Comprehensive Structural Overview of p38alpha MAPK in Complex with Type I Inhibitors. *ChemMedChem* **2015**, *10*, 957–969. [CrossRef]
19. Astolfi, A.; Kudolo, M.; Brea, J.; Manni, G.; Manfroni, G.; Palazzotti, D.; Sabatini, S.; Cecchetti, F.; Felicetti, T.; Cannalire, R.; et al. Discovery of potent p38alpha MAPK inhibitors through a funnel like workflow combining in silico screening and in vitro validation. *Eur. J. Med. Chem.* **2019**, *182*, 111624. [CrossRef]
20. Astolfi, A.; Milano, F.; Palazzotti, D.; Brea, J.; Pismataro, M.C.; Morlando, M.; Tabarrini, O.; Loza, M.I.; Massari, S.; Martelli, M.P.; et al. From Serendipity to Rational Identification of the 5,6,7,8-Tetrahydrobenzo[4,5]thieno[2,3-d]pyrimidin-4(3H)-one Core as a New Chemotype of AKT1 Inhibitors for Acute Myeloid Leukemia. *Pharmaceutics* **2022**, *14*, 2295. [CrossRef]
21. Uko, N.E.; Guner, O.F.; Matesic, D.F.; Bowen, J.P. Akt Pathway Inhibitors. *Curr. Top. Med. Chem.* **2020**, *20*, 883–900. [CrossRef]
22. Dotolo, S.; Cervellera, C.; Russo, M.; Russo, G.L.; Facchiano, A. Virtual Screening of Natural Compounds as Potential PI(3)K-AKT1 Signaling Pathway Inhibitors and Experimental Validation. *Molecules* **2021**, *26*, 492. [CrossRef] [PubMed]

23. Medina-Franco, J.L.; Giulianotti, M.A.; Yu, Y.; Shen, L.; Yao, L.; Singh, N. Discovery of a novel protein kinase B inhibitor by structure-based virtual screening. *Bioorg. Med. Chem. Lett.* **2009**, *19*, 4634–4638. [CrossRef] [PubMed]
24. Chuang, C.H.; Cheng, T.C.; Leu, Y.L.; Chuang, K.H.; Tzou, S.C.; Chen, C.S. Discovery of Akt kinase inhibitors through structure-based virtual screening and their evaluation as potential anticancer agents. *Int. J. Mol. Sci.* **2015**, *16*, 3202–3212. [CrossRef] [PubMed]
25. Noser, A.A.; Shehadi, I.A.; Abdelmonsef, A.H.; Salem, M.M. Newly Synthesized Pyrazolinone Chalcones as Anticancer Agents via Inhibiting the PI3K/Akt/ERK1/2 Signaling Pathway. *ACS Omega* **2022**, *7*, 25265–25277. [CrossRef]
26. Zhong, S.; Zhang, Z.; Guo, Z.; Yang, W.; Dou, G.; Lv, X.; Wang, X.; Ge, J.; Wu, B.; Pan, X.; et al. Identification of novel natural inhibitors targeting AKT Serine/Threonine Kinase 1 (AKT1) by computational study. *Bioengineered* **2022**, *13*, 12003–12020. [CrossRef]
27. Zhu, C.L.; Luo, X.; Tian, T.; Rao, Z.; Wang, H.; Zhou, Z.; Mi, T.; Chen, D.; Xu, Y.; Wu, Y.; et al. Structure-based rational design enables efficient discovery of a new selective and potent AKT PROTAC degrader. *Eur. J. Med. Chem.* **2022**, *238*, 114459. [CrossRef]
28. Dong, X.; Zhou, X.; Jing, H.; Chen, J.; Liu, T.; Yang, B.; He, Q.; Hu, Y. Pharmacophore identification, virtual screening and biological evaluation of prenylated flavonoids derivatives as PKB/Akt1 inhibitors. *Eur. J. Med. Chem.* **2011**, *46*, 5949–5958. [CrossRef]
29. Fratev, F.; Gutierrez, D.A.; Aguilera, R.J.; Tyagi, A.; Damodaran, C.; Sirimulla, S. Discovery of new AKT1 inhibitors by combination of in silico structure based virtual screening approaches and biological evaluations. *J. Biomol. Struct. Dyn.* **2021**, *39*, 368–377. [CrossRef]
30. Liu, T.; Zhan, W.; Wang, Y.; Zhang, L.; Yang, B.; Dong, X.; Hu, Y. Structure-based design, synthesis and biological evaluation of diphenylmethylamine derivatives as novel Akt1 inhibitors. *Eur. J. Med. Chem.* **2014**, *73*, 167–176. [CrossRef]
31. Zhan, W.; Lin, S.; Chen, J.; Dong, X.; Chu, J.; Du, W. Design, synthesis, biological evaluation, and molecular docking of novel benzopyran and phenylpyrazole derivatives as Akt inhibitors. *Chem. Biol. Drug Des.* **2015**, *85*, 770–779. [CrossRef] [PubMed]
32. Wang, Z.; Ran, T.; Xu, F.; Wen, C.; Song, S.; Zhou, Y.; Chen, H.; Lu, X. Deep learning-driven scaffold hopping in the discovery of Akt kinase inhibitors. *Chem. Commun.* **2021**, *57*, 10588–10591. [CrossRef]
33. Zhan, W.; Li, D.; Che, J.; Zhang, L.; Yang, B.; Hu, Y.; Liu, T.; Dong, X. Integrating docking scores, interaction profiles and molecular descriptors to improve the accuracy of molecular docking: Toward the discovery of novel Akt1 inhibitors. *Eur. J. Med. Chem.* **2014**, *75*, 11–20. [CrossRef] [PubMed]
34. Abohassan, M.; Alshahrani, M.; Alshahrani, M.Y.; Rajagopalan, P. Insilco and Invitro approaches identify novel dual PI3K/AKT pathway inhibitors to control acute myeloid leukemia cell proliferations. *Med. Oncol.* **2022**, *39*, 249. [CrossRef] [PubMed]
35. Al Shahrani, M.; Rajagopalan, P.; Abohassan, M.; Alshahrani, M.; Alraey, Y. CB-RAF600E-1 exerts efficacy in vemurafenib-resistant and non-resistant-melanoma cells via dual inhibition of RAS/RAF/MEK/ERK and PI3K/Akt signaling pathways. *Saudi J. Biol. Sci.* **2022**, *29*, 103285. [CrossRef]
36. Pragna Lakshmi, T.; Kumar, A.; Vijaykumar, V.; Natarajan, S.; Krishna, R. Identification of natural allosteric inhibitor for Akt1 protein through computational approaches and in vitro evaluation. *Int. J. Biol. Macromol.* **2017**, *96*, 200–213. [CrossRef]
37. Mahadevan, D.; Powis, G.; Mash, E.A.; George, B.; Gokhale, V.M.; Zhang, S.; Shakalya, K.; Du-Cuny, L.; Berggren, M.; Ali, M.A.; et al. Discovery of a novel class of AKT pleckstrin homology domain inhibitors. *Mol. Cancer Ther.* **2008**, *7*, 2621–2632. [CrossRef]
38. Moses, S.A.; Ali, M.A.; Zuohe, S.; Du-Cuny, L.; Zhou, L.L.; Lemos, R.; Ihle, N.; Skillman, A.G.; Zhang, S.; Mash, E.A.; et al. In vitro and in vivo activity of novel small-molecule inhibitors targeting the pleckstrin homology domain of protein kinase B/AKT. *Cancer Res.* **2009**, *69*, 5073–5081. [CrossRef]
39. Donald, A.; McHardy, T.; Rowlands, M.G.; Hunter, L.J.; Davies, T.G.; Berdini, V.; Boyle, R.G.; Aherne, G.W.; Garrett, M.D.; Collins, I. Rapid evolution of 6-phenylpurine inhibitors of protein kinase B through structure-based design. *J. Med. Chem.* **2007**, *50*, 2289–2292. [CrossRef]
40. Gassel, M.; Breitenlechner, C.B.; Ruger, P.; Jucknischke, U.; Schneider, T.; Huber, R.; Bossemeyer, D.; Engh, R.A. Mutants of protein kinase a that mimic the ATP-binding site of protein kinase B (AKT). *J. Mol. Biol.* **2003**, *329*, 1021–1034. [CrossRef]
41. Davies, T.G.; Verdonk, M.L.; Graham, B.; Saalau-Bethell, S.; Hamlett, C.C.; McHardy, T.; Collins, I.; Garrett, M.D.; Workman, P.; Woodhead, S.J.; et al. A structural comparison of inhibitor binding to PKB, PKA and PKA-PKB chimera. *J. Mol. Biol.* **2007**, *367*, 882–894. [CrossRef] [PubMed]
42. McGann, M. FRED pose prediction and virtual screening accuracy. *J. Chem. Inf. Model.* **2011**, *51*, 578–596. [CrossRef] [PubMed]
43. Jones, G.; Willett, P.; Glen, R.C.; Leach, A.R.; Taylor, R. Development and validation of a genetic algorithm for flexible docking. *J. Mol. Biol.* **1997**, *267*, 727–748. [CrossRef]
44. Saxty, G.; Woodhead, S.J.; Berdini, V.; Davies, T.G.; Verdonk, M.L.; Wyatt, P.G.; Boyle, R.G.; Barford, D.; Downham, R.; Garrett, M.D.; et al. Identification of inhibitors of protein kinase B using fragment-based lead discovery. *J. Med. Chem.* **2007**, *50*, 2293–2296. [CrossRef] [PubMed]
45. Akhtar, N.; Jabeen, I. Pharmacoinformatic Approaches to Design Novel Inhibitors of Protein Kinase B Pathways in Cancer. *Curr. Cancer Drug Targets* **2018**, *18*, 830–846. [CrossRef] [PubMed]
46. Sharma, V.; Gupta, M. Designing of kinase hinge binders: A medicinal chemistry perspective. *Chem. Biol. Drug Des.* **2022**, *100*, 968–980. [CrossRef]
47. Xing, L.; Klug-Mcleod, J.; Rai, B.; Lunney, E.A. Kinase hinge binding scaffolds and their hydrogen bond patterns. *Bioorg. Med. Chem.* **2015**, *23*, 6520–6527. [CrossRef]

48. Freeman-Cook, K.D.; Autry, C.; Borzillo, G.; Gordon, D.; Barbacci-Tobin, E.; Bernardo, V.; Briere, D.; Clark, T.; Corbett, M.; Jakubczak, J.; et al. Design of selective, ATP-competitive inhibitors of Akt. *J. Med. Chem.* **2010**, *53*, 4615–4622. [CrossRef]
49. Meng, E.C.; Shoichet, B.K.; Kuntz, I.D. Automated Docking with Grid-Based Energy Evaluation. *J. Comput. Chem.* **1992**, *13*, 505–524. [CrossRef]
50. Rao, S.N.; Head, M.S.; Kulkarni, A.; LaLonde, J.M. Validation studies of the site-directed docking program LibDock. *J. Chem. Inf. Model.* **2007**, *47*, 2159–2171. [CrossRef]
51. Dallakyan, S.; Olson, A.J. Small-molecule library screening by docking with PyRx. *Methods Mol. Biol.* **2015**, *1263*, 243–250. [CrossRef]
52. Rarey, M.; Dixon, J.S. Feature trees: A new molecular similarity measure based on tree matching. *J. Comput. Aided Mol. Des.* **1998**, *12*, 471–490. [CrossRef]
53. Friesner, R.A.; Murphy, R.B.; Repasky, M.P.; Frye, L.L.; Greenwood, J.R.; Halgren, T.A.; Sanschagrin, P.C.; Mainz, D.T. Extra precision glide: Docking and scoring incorporating a model of hydrophobic enclosure for protein-ligand complexes. *J. Med. Chem.* **2006**, *49*, 6177–6196. [CrossRef]
54. Friesner, R.A.; Banks, J.L.; Murphy, R.B.; Halgren, T.A.; Klicic, J.J.; Mainz, D.T.; Repasky, M.P.; Knoll, E.H.; Shelley, M.; Perry, J.K.; et al. Glide: A new approach for rapid, accurate docking and scoring. 1. Method and assessment of docking accuracy. *J. Med. Chem.* **2004**, *47*, 1739–1749. [CrossRef]
55. Addie, M.; Ballard, P.; Buttar, D.; Crafter, C.; Currie, G.; Davies, B.R.; Debreczeni, J.; Dry, H.; Dudley, P.; Greenwood, R.; et al. Discovery of 4-amino-N-[(1S)-1-(4-chlorophenyl)-3-hydroxypropyl]-1-(7H-pyrrolo[2,3-d]pyrimidin-4-yl)piperidine-4-carboxamide (AZD5363), an orally bioavailable, potent inhibitor of Akt kinases. *J. Med. Chem.* **2013**, *56*, 2059–2073. [CrossRef]
56. Bekes, M.; Langley, D.R.; Crews, C.M. PROTAC targeted protein degraders: The past is prologue. *Nat. Rev. Drug Discov.* **2022**, *21*, 181–200. [CrossRef]
57. Drummond, M.L.; Henry, A.; Li, H.; Williams, C.I. Improved Accuracy for Modeling PROTAC-Mediated Ternary Complex Formation and Targeted Protein Degradation via New In Silico Methodologies. *J. Chem. Inf. Model.* **2020**, *60*, 5234–5254. [CrossRef]
58. Li, H.; Sutter, J.; Hoffmann, R. HypoGen: An automated system for generating 3D predictive pharmacophore models. In *Pharmacophore Perception, Development, and Use in Drug Design*; Li, H., Sutter, J., Hoffmann, R., Eds.; Tsigelny, I. San Diego Supercomputer Center, University of California: San Diego, CA, USA, 2000; p. 171.
59. Lippa, B.; Pan, G.; Corbett, M.; Li, C.; Kauffman, G.S.; Pandit, J.; Robinson, S.; Wei, L.; Kozina, E.; Marr, E.S.; et al. Synthesis and structure based optimization of novel Akt inhibitors. *Bioorg. Med. Chem. Lett.* **2008**, *18*, 3359–3363. [CrossRef]
60. Baell, J.B.; Holloway, G.A. New substructure filters for removal of pan assay interference compounds (PAINS) from screening libraries and for their exclusion in bioassays. *J. Med. Chem.* **2010**, *53*, 2719–2740. [CrossRef]
61. Bencsik, J.R.; Xiao, D.; Blake, J.F.; Kallan, N.C.; Mitchell, I.S.; Spencer, K.L.; Xu, R.; Gloor, S.L.; Martinson, M.; Risom, T.; et al. Discovery of dihydrothieno- and dihydrofuropyrimidines as potent pan Akt inhibitors. *Bioorg. Med. Chem. Lett.* **2010**, *20*, 7037–7041. [CrossRef]
62. Venkatachalam, C.M.; Jiang, X.; Oldfield, T.; Waldman, M. LigandFit: A novel method for the shape-directed rapid docking of ligands to protein active sites. *J. Mol. Graph. Model.* **2003**, *21*, 289–307. [CrossRef]
63. Rarey, M.; Kramer, B.; Lengauer, T.; Klebe, G. A fast flexible docking method using an incremental construction algorithm. *J. Mol. Biol.* **1996**, *261*, 470–489. [CrossRef] [PubMed]
64. Vieth, M.; Hirst, J.D.; Dominy, B.N.; Daigler, H.; Brooks, C.L. Assessing search strategies for flexible docking. *J. Comput. Chem.* **1998**, *19*, 1623–1631. [CrossRef]
65. flexiDOCK, Sybyl 6.9 (2002), Molecular Modelling System, Tripos Associates. Available online: https://www.certara.com/ (accessed on 10 July 2023).
66. Salam, N.K.; Nuti, R.; Sherman, W. Novel method for generating structure-based pharmacophores using energetic analysis. *J. Chem. Inf. Model.* **2009**, *49*, 2356–2368. [CrossRef]
67. Kallan, N.C.; Spencer, K.L.; Blake, J.F.; Xu, R.; Heizer, J.; Bencsik, J.R.; Mitchell, I.S.; Gloor, S.L.; Martinson, M.; Risom, T.; et al. Discovery and SAR of spirochromane Akt inhibitors. *Bioorg. Med. Chem. Lett.* **2011**, *21*, 2410–2414. [CrossRef] [PubMed]
68. Yang, Y.; Zheng, S.; Su, S.; Zhao, C.; Xu, J.; Chen, H. SyntaLinker: Automatic fragment linking with deep conditional transformer neural networks. *Chem. Sci.* **2020**, *11*, 8312–8322. [CrossRef]
69. Bollag, G.; Tsai, J.; Zhang, J.; Zhang, C.; Ibrahim, P.; Nolop, K.; Hirth, P. Vemurafenib: The first drug approved for BRAF-mutant cancer. *Nat. Rev. Drug Discov.* **2012**, *11*, 873–886. [CrossRef]
70. Wan, P.T.; Garnett, M.J.; Roe, S.M.; Lee, S.; Niculescu-Duvaz, D.; Good, V.M.; Jones, C.M.; Marshall, C.J.; Springer, C.J.; Barford, D.; et al. Mechanism of activation of the RAF-ERK signaling pathway by oncogenic mutations of B-RAF. *Cell* **2004**, *116*, 855–867. [CrossRef]
71. Uhlenbrock, N.; Smith, S.; Weisner, J.; Landel, I.; Lindemann, M.; Le, T.A.; Hardick, J.; Gontla, R.; Scheinpflug, R.; Czodrowski, P.; et al. Structural and chemical insights into the covalent-allosteric inhibition of the protein kinase Akt. *Chem. Sci.* **2019**, *10*, 3573–3585. [CrossRef]
72. Yang, J.; Nie, J.; Ma, X.; Wei, Y.; Peng, Y.; Wei, X. Targeting PI3K in cancer: Mechanisms and advances in clinical trials. *Mol. Cancer* **2019**, *18*, 26. [CrossRef]
73. Dixon, S.L.; Smondyrev, A.M.; Rao, S.N. PHASE: A novel approach to pharmacophore modeling and 3D database searching. *Chem. Biol. Drug Des.* **2006**, *67*, 370–372. [CrossRef]

74. Wu, W.I.; Voegtli, W.C.; Sturgis, H.L.; Dizon, F.P.; Vigers, G.P.; Brandhuber, B.J. Crystal structure of human AKT1 with an allosteric inhibitor reveals a new mode of kinase inhibition. *PLoS ONE* **2010**, *5*, e12913. [CrossRef]
75. Jacobson, M.P.; Pincus, D.L.; Rapp, C.S.; Day, T.J.; Honig, B.; Shaw, D.E.; Friesner, R.A. A hierarchical approach to all-atom protein loop prediction. *Proteins* **2004**, *55*, 351–367. [CrossRef]
76. Thomas, C.C.; Deak, M.; Alessi, D.R.; van Aalten, D.M. High-resolution structure of the pleckstrin homology domain of protein kinase b/akt bound to phosphatidylinositol (3,4,5)-trisphosphate. *Curr. Biol.* **2002**, *12*, 1256–1262. [CrossRef]
77. *Certara*; UNITY, SYBYL-X.: St. Louis, MO, USA, 2013.
78. Moingeon, P.; Kuenemann, M.; Guedj, M. Artificial intelligence-enhanced drug design and development: Toward a computational precision medicine. *Drug Discov. Today* **2022**, *27*, 215–222. [CrossRef]
79. Urbina, F.; Lentzos, F.; Invernizzi, C.; Ekins, S. Dual Use of Artificial Intelligence-powered Drug Discovery. *Nat. Mach. Intell.* **2022**, *4*, 189–191. [CrossRef]
80. Palazzotti, D.; Fiorelli, M.; Sabatini, S.; Massari, S.; Barreca, M.L.; Astolfi, A. Q-raKtion: A Semiautomated KNIME Workflow for Bioactivity Data Points Curation. *J. Chem. Inf. Model.* **2022**, *62*, 6309–6315. [CrossRef]
81. Irwin, J.J.; Duan, D.; Torosyan, H.; Doak, A.K.; Ziebart, K.T.; Sterling, T.; Tumanian, G.; Shoichet, B.K. An Aggregation Advisor for Ligand Discovery. *J. Med. Chem.* **2015**, *58*, 7076–7087. [CrossRef]
82. Daina, A.; Michielin, O.; Zoete, V. SwissADME: A free web tool to evaluate pharmacokinetics, drug-likeness and medicinal chemistry friendliness of small molecules. *Sci. Rep.* **2017**, *7*, 42717. [CrossRef]
83. Quambusch, L.; Landel, I.; Depta, L.; Weisner, J.; Uhlenbrock, N.; Muller, M.P.; Glanemann, F.; Althoff, K.; Siveke, J.T.; Rauh, D. Covalent-Allosteric Inhibitors to Achieve Akt Isoform-Selectivity. *Angew. Chem. Int. Ed. Engl.* **2019**, *58*, 18823–18829. [CrossRef]
84. van der Westhuizen, L.; Weisner, J.; Taher, A.; Landel, I.; Quambusch, L.; Lindemann, M.; Uhlenbrock, N.; Muller, M.P.; Green, I.R.; Pelly, S.C.; et al. Covalent Allosteric Inhibitors of Akt Generated Using a Click Fragment Approach. *ChemMedChem* **2022**, *17*, e202100776. [CrossRef] [PubMed]
85. Weisner, J.; Landel, I.; Reintjes, C.; Uhlenbrock, N.; Trajkovic-Arsic, M.; Dienstbier, N.; Hardick, J.; Ladigan, S.; Lindemann, M.; Smith, S.; et al. Preclinical Efficacy of Covalent-Allosteric AKT Inhibitor Borussertib in Combination with Trametinib in KRAS-Mutant Pancreatic and Colorectal Cancer. *Cancer Res.* **2019**, *79*, 2367–2378. [CrossRef]
86. Landel, I.; Quambusch, L.; Depta, L.; Rauh, D. Spotlight on AKT: Current Therapeutic Challenges. *ACS Med. Chem. Lett.* **2020**, *11*, 225–227. [CrossRef] [PubMed]
87. Lazaro, G.; Kostaras, E.; Vivanco, I. Inhibitors in AKTion: ATP-competitive vs allosteric. *Biochem. Soc. Trans.* **2020**, *48*, 933–943. [CrossRef] [PubMed]
88. Lindsley, C.W.; Zhao, Z.; Leister, W.H.; Robinson, R.G.; Barnett, S.F.; Defeo-Jones, D.; Jones, R.E.; Hartman, G.D.; Huff, J.R.; Huber, H.E.; et al. Allosteric Akt (PKB) inhibitors: Discovery and SAR of isozyme selective inhibitors. *Bioorg Med. Chem. Lett.* **2005**, *15*, 761–764. [CrossRef]
89. DeFeo-Jones, D.; Barnett, S.F.; Fu, S.; Hancock, P.J.; Haskell, K.M.; Leander, K.R.; McAvoy, E.; Robinson, R.G.; Duggan, M.E.; Lindsley, C.W.; et al. Tumor cell sensitization to apoptotic stimuli by selective inhibition of specific Akt/PKB family members. *Mol. Cancer Ther.* **2005**, *4*, 271–279. [CrossRef]
90. Chamberlain, P.P.; Hamann, L.G. Development of targeted protein degradation therapeutics. *Nat. Chem. Biol.* **2019**, *15*, 937–944. [CrossRef]
91. Mayor-Ruiz, C.; Bauer, S.; Brand, M.; Kozicka, Z.; Siklos, M.; Imrichova, H.; Kaltheuner, I.H.; Hahn, E.; Seiler, K.; Koren, A.; et al. Rational discovery of molecular glue degraders via scalable chemical profiling. *Nat. Chem. Biol.* **2020**, *16*, 1199–1207. [CrossRef]
92. Spagnolli, G.; Massignan, T.; Astolfi, A.; Biggi, S.; Rigoli, M.; Brunelli, P.; Libergoli, M.; Ianeselli, A.; Orioli, S.; Boldrini, A.; et al. Pharmacological inactivation of the prion protein by targeting a folding intermediate. *Commun. Biol.* **2021**, *4*, 62. [CrossRef]
93. Henning, R.K.; Varghese, J.O.; Das, S.; Nag, A.; Tang, G.; Tang, K.; Sutherland, A.M.; Heath, J.R. Degradation of Akt using protein-catalyzed capture agents. *J. Pept. Sci.* **2016**, *22*, 196–200. [CrossRef]
94. Yu, X.; Xu, J.; Shen, Y.; Cahuzac, K.M.; Park, K.S.; Dale, B.; Liu, J.; Parsons, R.E.; Jin, J. Discovery of Potent, Selective, and In Vivo Efficacious AKT Kinase Protein Degraders via Structure-Activity Relationship Studies. *J. Med. Chem.* **2022**, *65*, 3644–3666. [CrossRef] [PubMed]
95. Yu, X.; Xu, J.; Xie, L.; Wang, L.; Shen, Y.; Cahuzac, K.M.; Chen, X.; Liu, J.; Parsons, R.E.; Jin, J. Design, Synthesis, and Evaluation of Potent, Selective, and Bioavailable AKT Kinase Degraders. *J. Med. Chem.* **2021**, *64*, 18054–18081. [CrossRef]
96. Yu, X.; Xu, J.; Cahuzac, K.M.; Xie, L.; Shen, Y.; Chen, X.; Liu, J.; Parsons, R.E.; Jin, J. Novel Allosteric Inhibitor-Derived AKT Proteolysis Targeting Chimeras (PROTACs) Enable Potent and Selective AKT Degradation in KRAS/BRAF Mutant Cells. *J. Med. Chem.* **2022**, *65*, 14237–14260. [CrossRef]

pharmaceuticals

Review

A Comprehensive Computational Insight into the PD-L1 Binding to PD-1 and Small Molecules

Marialuigia Fantacuzzi *, Roberto Paciotti and Mariangela Agamennone *

Department of Pharmacy, University "G. d'Annunzio" of Chieti-Pescara, Via Dei Vestini, 31, 66100 Chieti, Italy; r.paciotti@unich.it
* Correspondence: marialuigia.fantacuzzi@unich.it (M.F.); mariangela.agamennone@unich.it (M.A.)

Abstract: Immunotherapy has marked a revolution in cancer therapy. The most extensively studied target in this field is represented by the protein–protein interaction between PD-1 and its ligand, PD-L1. The promising results obtained with the clinical use of monoclonal antibodies (mAbs) directed against both PD-1 and PD-L1 have prompted the search for small-molecule binders capable of disrupting the protein–protein contact and overcoming the limitations presented by mAbs. The disclosure of the first X-ray complexes of PD-L1 with BMS ligands showed the protein in dimeric form, with the ligand in a symmetrical hydrophobic tunnel. These findings paved the way for the discovery of new ligands. To this end, and to understand the binding mechanism of small molecules to PD-L1 along with the dimerization process, many structure-based computational studies have been applied. In the present review, we examined the most relevant articles presenting computational analyses aimed at elucidating the binding mechanism of PD-L1 with PD-1 and small molecule ligands. Additionally, virtual screening studies that identified validated PD-L1 ligands were included. The relevance of the reported studies highlights the increasingly prominent role that these techniques can play in chemical biology and drug discovery.

Keywords: cancer immunotherapy; PD-1; PD-L1; computational studies; docking; molecular dynamics; virtual screening

Citation: Fantacuzzi, M.; Paciotti, R.; Agamennone, M. A Comprehensive Computational Insight into the PD-L1 Binding to PD-1 and Small Molecules. *Pharmaceuticals* **2024**, *17*, 316. https://doi.org/10.3390/ph17030316

Academic Editor: Osvaldo Andrade Santos-Filho

Received: 25 January 2024
Revised: 14 February 2024
Accepted: 26 February 2024
Published: 28 February 2024

1. Introduction

The connection between cancer and the immune system was suggested for the first time in 1863 when Virchow noticed the infiltration of leukocytes in cancer tissue [1]. Some years later, Coley administered a mix of bacteria, the "Coley's toxin", to treat inoperable tumors, obtaining, with a variable clinical response, a substantial reduction of tumor dimensions [2]. In 1971, Burnet and Thomas hypothesized that the immune system is able to control cancer development, recognizing and eliminating tumor cells [3]. Nowadays, the role of the immune system in cancer progression control is widely recognized. In particular, the interplay between the immune system and cancer cells has been better defined with the so defined immunoediting process: elimination, equilibrium, and escape [4].

1.1. Immune Checkpoints, PD-1, and Its Binders

The immune response, in particular T-cell activity, is regulated by a complex network of events and involves several actors [5], but a key role is played by immune checkpoints (ICs) that can have a co-inhibitory or co-stimulatory action. In physiological conditions, ICs are responsible for immune tolerance, avoiding autoimmune reactions and tissue damage due to prolonged inflammation [6]. Cancer tissue exploits this mechanism to prevent the immune system from eliminating cancer cells by silencing T-cells. Ipilimumab was approved by the FDA in 2011 as the first example of mAbs used in cancer therapy targeting an immune checkpoint (CTLA-4) and represents a revolution in cancer treatment. The disclosure of the first immunotherapeutic drug paved the way for the research of other tools targeting different ICs, also to overcome the severe immune-related side effects presented by ipilimumab [7]. Since then, the number of clinical trials in the immuno–oncology field has increased almost exponentially [8]. In this context, the most explored targets are PD-1 and its ligands PD-L1.

The programmed death protein PD-1 was identified for the first time by Ishida and coworkers in 1992 [9], even if its effective role was clarified later [10]. PD-1 belongs to the CD28 family and is encoded by the Pdcd1 gene on chromosome 2 (2q37) [11]. It is a glycoprotein expressed mainly on the surface of T- and B-cells, but also on myeloid cells, thymocytes, natural killer (NK) cells, dendritic cells, and monocytes, and its expression is promoted by T-cell activation.

PD-1 binds two endogenous ligands, PD-L1 and PD-L2, identified in 1999 and 2001, respectively, and encoded on the same chromosome 9p24.2 [12,13].

PD-L1 (B7-H1, CD274) is constitutively expressed on antigen-presenting cells (APC) but can be widely located on hematopoietic cells (B-cells, T-cells, monocytes, and dendritic cells), and peripheral non-hematopoietic tissues such as the heart, kidney, lung, placenta, and liver. PD-L1 expression is induced by several pro-inflammatory cytokines (e.g., INF-g, TNF-a, VEGF, and others). PD-1 is not the unique binder of PD-L1 that can interact also with CD-80 [14].

PD-L2 (B7-DC, CD273), the second identified PD-1 binder, has a similar profile to PD-L1 in terms of expression and function, even though it shares a limited identity with PD-L1 (almost 40%). Moreover, PD-L1 and PD-L2 present a sequence identity of 20% with B7-1 and B7-2 that bind CD28 and CTLA-4, respectively.

PD-L2 binds PD-1 differently and with stronger affinity than PD-L1 [15,16] thanks to the PD-1 adaptability and flexibility [17]. However, it is a less-explored target with respect to PD-L1 because of its limited expression [15]. Both ligands, along with PD-1, can be released in soluble form and represent negative prognostic markers in several tumors [18].

1.2. The PD-1/PD-L1 Pathway

In physiological conditions, PD-1 expression on the T-cell surface is promoted by T-cell activation. Its expression is accompanied by the release of interferon that fosters the expression of PD-L1 on APC or surrounding tissues. The PD-1 binding to endogenous ligands, PD-L1 and PD-L2, at the immunological synapse strongly inhibits TCR signal transduction and CD28/CD80 co-stimulation (Figure 1). In particular, PD-1/PD-L1(2) contact causes the phosphorylation of the Immunoreceptor Tyrosine-Based Inhibitory Motif (ITIM) and the Immunoreceptor Inhibitory Tyrosine-Based Switch Motif (ITSM) located at the intracellular PD-1 tail. Src homology 2 domain-containing protein tyrosine phosphatase 1 and 2 (SHP-1 and SHP-2) are then recruited by ITIM and ITSM and block the TCR signal transduction [19]. Another effect of the PD-1 binding to its ligands is the PTEN-mediated blocking of T-cell proliferation [20].

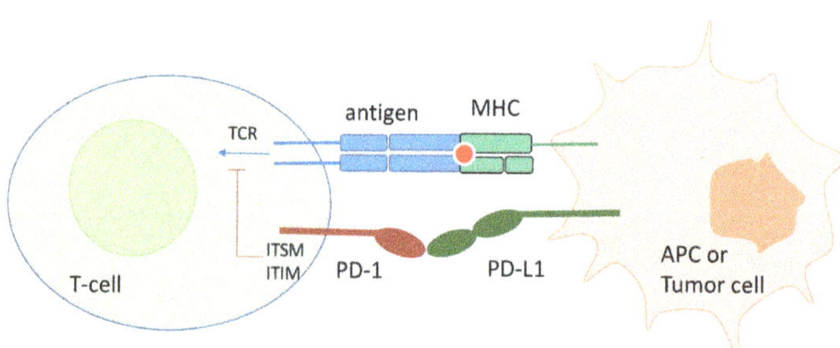

Figure 1. Schematic representation of the PD-1/PD-l1 contact in physiological conditions, and in tumor environment. Monoclonal antibodies directed toward PD-1 and PD-L1 can block the protein–protein contact and reactivate T-cell function.

The PD-1/PD-L1 interaction results in the reduction of the synthesis of cytokines, the blocking of the activation, proliferation, and acquisition of the effector capacities of the T-cells. In addition, activated PD-1 leads to a suppression of the consumption of oxygen. The oxidation of fatty acids, and no longer aerobic glycolysis, is used as the dominant energy source. Consequently, T-cells with activated PD-1 increase the production of reactive oxygen species helping to create an oxidative environment [21].

As already mentioned, cancer tissue can exploit this physiological mechanism to create an immunosuppressive environment favorable to tumor growth and progression by overexpressing PD-L1 in the escape phase. There are two main mechanisms of PD-L1 up-regulation: the innate immune resistance produced by oncogene suppression, and the adaptive immune resistance that exploits the INF-γ to induce the PD-L1 expression [22]. In both cases, the final result is the immune system silencing [23].

This evidence prompted the search for agents able to interfere with the PD-1/PD-L1 contact to be exploited in cancer treatment. The emergence of mAbs interacting with ICs allowed treating chemo-resistant tumors with surprising efficacy.

1.3. Current Drugs

As already mentioned, Ipilimumab was the first example of an immune checkpoint inhibitor (ICI) in clinical use and represented a breakthrough in cancer treatment, despite its severe side effects. Nowadays, the largely most exploited drugs in the immuno-oncology field are human or humanized mAbs targeting PD-1 or PD-L1. A panel of currently approved mAbs by the FDA is reported in Table 1. Apart from those that are available, the investigation of these compounds is still ongoing, with the number of enrolling clinical studies constantly increasing since 2014 [24].

Table 1. The FDA approved anti-PD1 or PD-L1 mAbs reported in the Drug Bank (https://go.drugbank.com/, accession date 22 January 2024).

DrugBank ID	Name	Year of Approval	Target	Commercializing Company
DB09035	Nivolumab	2014	PD-1	BMS
DB09037	Pembrolizumab	2014	PD-1	Merk
DB14707	Cemiplimab	2019	PD-1	Sanofi
DB15627	Dostarlimab	2021	PD-1	GSK
DB15766	Retifanlimab	2023	PD-1	Incyte Biosciences
DB11595	Atezolizumab	2016	PD-L1	Genentech
DB11945	Avelumab	2017	PD-L1	Merk
DB11714	Durvalumab	2017	PD-L1	Astra Zeneca

MAbs, in fact, represented a revolution in cancer treatment, and a tool providing a long-lasting resolution for drug-resistant and metastatic tumors. Nevertheless, ICIs have a series of limitations. Targeting PD-1 and PD-L1 can produce immune-related adverse events that can hamper the patient's treatment. Moreover, depending on the tumor type, just a limited percentage of patients (10–60%) is respondent to the therapy [25]. In this respect, combination therapies could help to reduce side effects and increase the number of patients with beneficial effects [26].

Apart from these aspects, mAbs present several limitations in terms of high production costs, side effects, missing oral bioavailability that forces intravenous administration, prolonged tissue retention, and low membrane permeability. To overcome these shortcomings, research has switched to the development of small molecule inhibitors.

A recent report analyzing clinical trials in the immuno–oncology field highlights a slight but substantial inversion of the previous trends where the number of Phase II trials involving the axis PD-1/PD-L1 is diminishing. The authors highlight the increased interest in other targets involved in cancer immunotherapy, the orientation toward different technologies, and the increased number of combination therapies under study to overcome ICI limits [8,27].

1.4. Small Molecule Binders of PD-L1

The identification of small molecules interfering with the PD-1/PD-L1 protein–protein interaction has lagged behind the discovery of the role played by these two important immune checkpoints. The reason for the difficult identification of effective PD-1/PD-L1 small molecule inhibitors can be attributed to the characteristics of the interacting surfaces of the partner proteins that are very flexible and flat, and so hardly druggable.

The first small molecules binding PD-L1 were disclosed in 2015 in two patents from the Bristol Mayer Squibb [28,29], and represent the paradigm compounds as PD-L1 inhibitors until today. Since then, the number of published PD-L1 ligands has increased constantly.

The most important contribution to the discovery of small molecule binders of PD-L1 field has been provided by Holak and coworkers, who investigated the BMS ligands binding mode on PD-L1. The Polish research group resolved the X-ray complexes of the PD-L1 extracellular domain with several BMS compounds. The crystallographic data revealed a ligand/protein ratio of 1:2, with PD-L1 binding the small molecule in dimeric form [30].

The availability of PD-L1 X-ray complexes with small molecules paved the way for the discovery of many other small molecules able to produce the same dimerization process.

Thus, the design strategy shifted from the search for a protein–protein interaction (PD-1/PD-L1) inhibitor to the identification of protein–protein (PD-L1 dimer) structure stabilizers, a new and fascinating field of research [31].

All ligands producing PD-L1 dimerization share a common structural feature consisting in a biphenyl/biaryl portion linked to another aromatic system. Skalniak et al. demonstrated, by NMR studies, that starting from BMS-1166, the minimum fragment-conserving activity is the biphenyl group (Figure 2) [32].

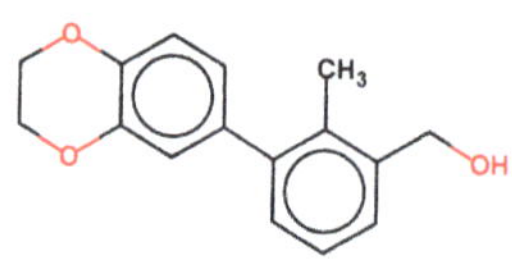

Figure 2. BMS-1166-derived fragment maintaining activity toward PD-L1 in ^{1}H-^{15}N HMQC NMR assay [32].

Many reviews published so far have focused on PD-L1 ligands [33–39]. A very recent review revised the application of computational methods to the discovery of PD-1/PD-L1 inhibitors [40].

The present work explores the computational approaches used to improve the understanding of the binding processes involving PD-L1. In particular, we focused on computational studies aimed at mapping the binding mechanism of PD-L1 with its natural binder PD-1 and small molecules. In addition, the application of in silico methods to the identification of new compounds through virtual screening campaigns was explored. Articles reporting computational studies, in particular docking, as ancillary to other investigations were not considered.

Computational approaches are now an integral part of drug discovery and chemical biology studies. Recent technological and scientific advances have promoted the role of computational methods as leading tools in the prediction of protein structure and function [41] and in the overall drug discovery process, also because of the application of machine learning approaches along with physics-based methods that can exploit increasingly powerful computing systems [42].

2. Structural Depiction of PD-L1 and Its Binders

Most of the computational methods presented are based on structure-based approaches. The availability of X-ray data of PD-L1 bound to its endogenous ligand and to small molecules, indeed, fed a plethora of studies exploiting docking, molecular dynamics (MD) and other target-based in silico analyses to understand PD-L1 interaction mechanism with its binders. Here, we review the structural features of the studied proteins, focusing on the available experimental data.

PD-L1 is a 290 amino acids (aa) protein belonging to the type I transmembrane protein family. It is composed of extracellular Ig-V- and Ig-C-like domains, a transmembrane portion, and a short intracellular tail of 30-aa (Figure 3B). The distal Ig-type V-like domain is responsible for the interaction with PD-1 and small molecules.

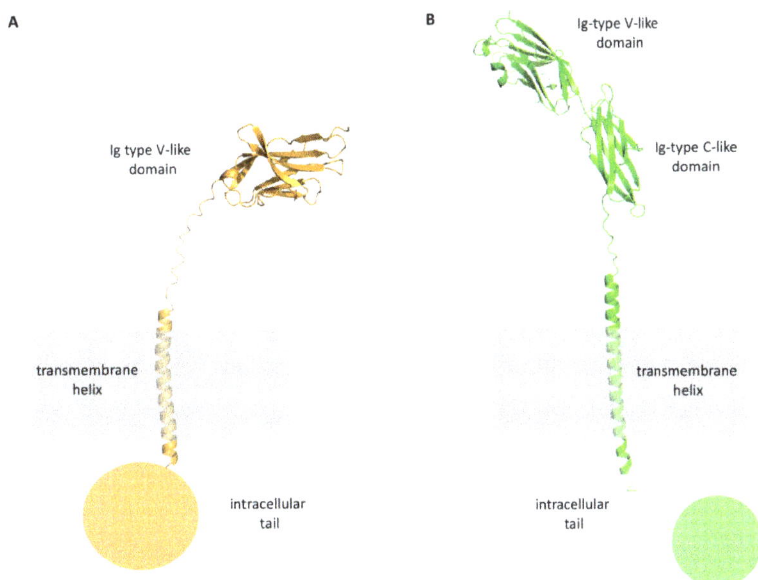

Figure 3. Structural representation of PD-1 (panel (**A**), orange cartoon), and PD-L1 (panel (**B**), green cartoon).

PD-1 is a 55 kDa type I transmembrane protein of the immune globulin superfamily, composed of an N-terminal Immunoglobulin Variable (Ig-V) extracellular domain, a transmembrane domain, and a cytoplasmic tail (Figure 3A). It is composed of 288 aa and shares the 21–33% sequence identity with CTLA-4, ICOS, and CD28.

The extracellular domain is responsible for the interactions with its ligands, while the intracellular tail has two phosphorylation sites, ITIM and ITSM, essential for its activity.

2.1. PD-L1 Structure in the Apo Form and in Complex with PD-1

The first crystallographic data of the hPD-1/mPD-L1 complex (PDB ID 3BIK) were published in 2008 [43], while the fully human complex (PDB ID 4ZQK) was published in 2015 [44]. A list of the PDB data of apo PD-L1 or PD-1/PD-L1 is provided in Table 2.

The contact surfaces between the Ig-V-like domains of hPD-1 and hPD-L1 are arranged orthogonally as a 1:1 complex. The proteins undergo a conformational change during complex formation. This is more evident for PD-1, where the loop CC′ adopts a closed conformation. The interaction area is large (1870 Å^2) and flat, and hydrophobic and polar interactions take place between PD-1 and PD-L1. (Figure 4).

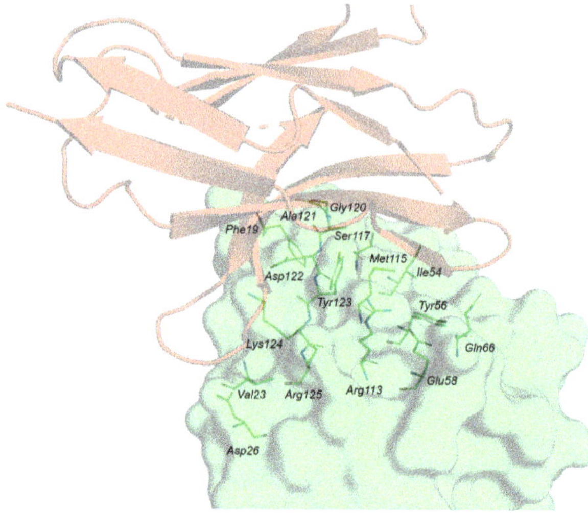

Figure 4. PD-1/PD-L1 complex. PD-1 is represented by the orange cartoon, and PD-L1 as the green surface with the contact residues in green sticks.

An insightful analysis of PD-1/PD-L1 structures, complexed with several binders, has been recently published by Boisgerault and Bertrand [45].

Table 2. Available X-ray structures of apo PD-L1 and PD-1/PD-L1 complex.

PDB ID	Protein	Resolution (Å)	Release	Reference
3BIS	Apo PD-L1	2.64	2008	[43]
4Z18	Apo PD-L1	1.95	2015	Fedorov, A.A., To be published
5C3T	PD-L1 binding domain	1.80	2015	[44]
5JDR	Apo PD-L1	2.70	2017	[46]
6NP9	Apo mutant PD-L1 (V76T)	1.27	2019	[47]
3FN3	Dimeric structure of PD-L1	2.70	2009	[48]
6L8R	Membrane-bound cytoplasmatic domain PD-L1	NMR	2020	[49]
7DCV	Transmembrane domain PD-L1	NMR	2022	[50]
3BIK	Complex mPD-1/hPD-L1	2.65	2008	[43]
3SBW	Complex mPD-1/hPD-L1	2.28	2011	Lazar-Molnar, To be published
4ZQK	Complex hPD-1/hPD-L1	2.45	2015	[44]
5IUS	Complex with high affinity mutated PD-1	2.89	2016	[51]

2.2. PD-L1 in Complex with Small Molecule Binders

In 2016, Holak and coworkers deposited the first X-ray data of complexes between PD-L1 and the small molecules BMS-202 and BMS-8 disclosed in the BMS patents (PDB IDs 5J89 and 5J8O, respectively) [30]. The stoichiometry of the complex is 2:1, with the ligand bound at the dimer interface (Figure 5A). The central part of the PD-1 interacting surface represents the small molecule binding site, and the displacement of Tyr56, Met115, and Tyr123 creates a cylindrical hydrophobic cavity that can accommodate the biphenyl group of BMS molecules. In Figure 5B, the superposition of the structures of PD-L1 bound to PD-1 and BMS-202 is shown, while in Figure 5C, the interacting residues of each PD-L1 protein and the contact surface with the ligands are shown.

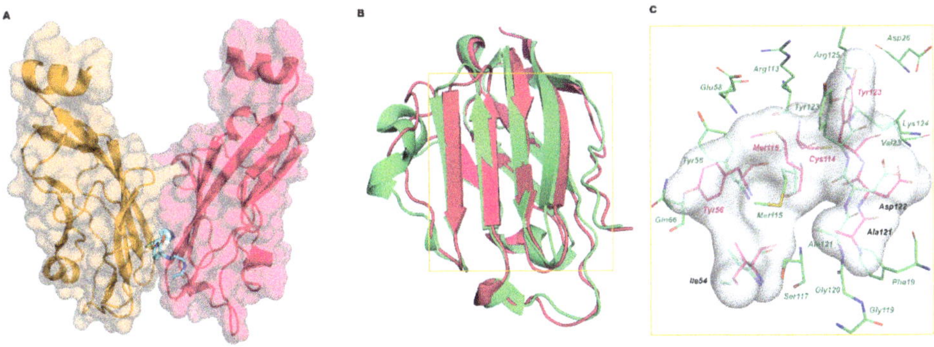

Figure 5. (**A**) Complex of BMS-202 (cyan stick) and PD-L1 dimer (chain A hot pink, chain B orange mixed carton/surface representation); (**B**) Superimposition of the crystal structure of PD-L1 bound to PD-1 (green carton, PDB ID 4ZQK) with PD-L1 bound to BMS-202 (hot-pink carton, PDB ID 5J89); (**C**) Zoom on the interaction residues of PD-L1/PD-1 (green) and of PD-L1/BMS-202 (hot pink). The interaction surface of PD-L1/BMS-202 is represented in grey.

A complete list of available experimental structures of PD-L1 in the complex, with small molecules reported in Table 3 and X-ray ligand structures represented in Figure 6.

Table 3. PDB data of crystal of the complex small molecule/PD-L1.

PDB ID	Resolution (Å)	Ligand	IC_{50} Value (nM)	Reference
5J89	2.20	6GX (BMS-202)	18	[30]
5J8O	2.30	6GZ (BMS-8)	146	[30]
5N2D	2.35	8J8 (BMS-37)	6–100	[52]
5N2F	1.70	8HW (BMS-200)	80.00	[52]
5NIU	2.01	8YZ (BMS-1001)	2.25	[32]
6NM7	2.43	22L	n.d.	[47]
6NM8	2.79	KSD	53.00	[47]
6NOJ	2.33	KW7	Kd = 1.9 mM	[47]
6NOS	2.70	KWA	Kd = 1.9 mM	[47]
6R3K	2.20	JQT (BMS-1166)	1.85	[53]
6RPG	2.70	KDW	3.00	[54]
6VQN	2.49	R81	0.4	[55]
7BEA	2.45	TK2	16.80	[56]
7DY7	2.42	HOU	27.80	[57]
7NLD	2.30	UGZ	2.07	[53]
7VUN	2.00	8H7	8.90	[58]
8OR1	3.50	VYC	2.4	[59]
8K5N	2.20	I7M	1.8	[60]
8R6Q	2.17	WEW	<0.5	[61]

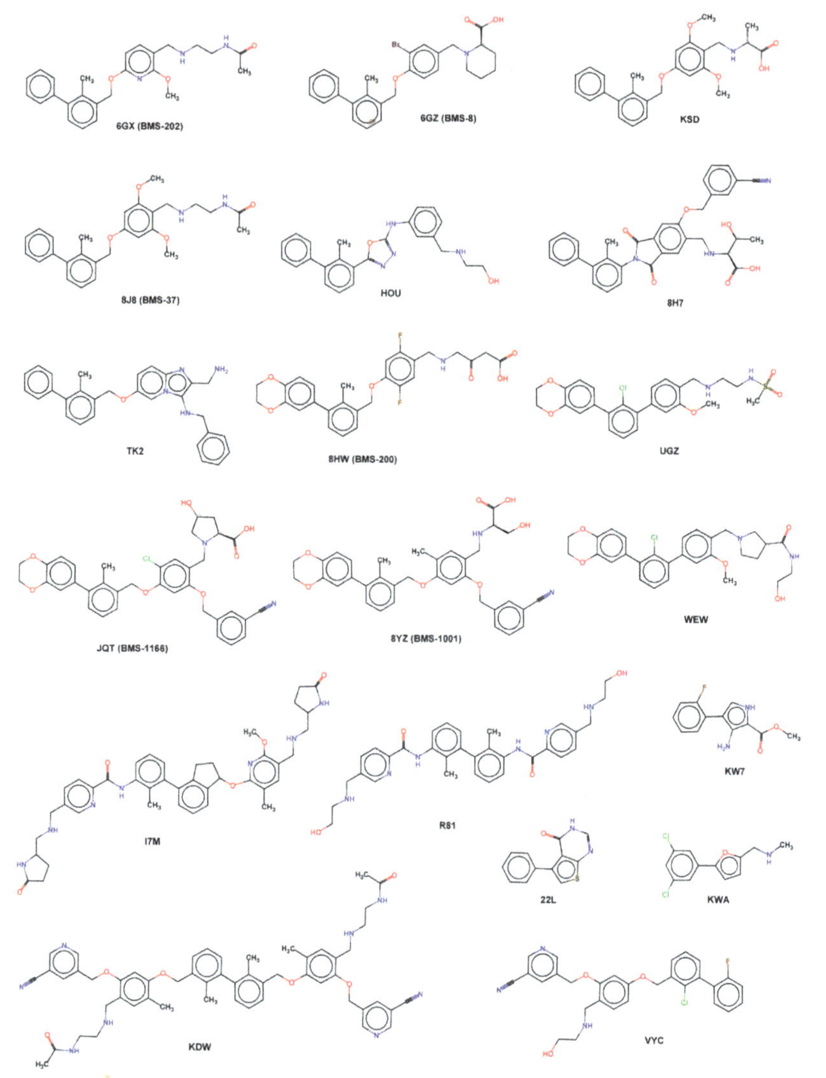

Figure 6. 2D structure of X-ray ligands.

3. Computational Studies Contributing to the Binding Mechanism Comprehension

Molecular dynamics (MD) is a powerful computational approach that allows for the exploration of the conformational space of large biological systems.

More specifically, in a classical MD protocol, it is possible to calculate the force applied to each atom in the system and then use Newton's laws of motion to predict the spatial position of each atom as a function of time [62]. The resulting trajectory describes the conformational change of the 3D structure during the simulation time. The length of a trajectory is, therefore, a critical point of the simulation.

Forces in an MD simulation are calculated using a molecular mechanics force field (FF), which contains structural parameters (for example, the length of each covalent bond), as well as a set of equations to define electrostatic interactions between atoms and other types of interatomic non-covalent interactions [62]. The type of FF adopted for MD simulation can significantly influence the computational results and, therefore, should be selected carefully.

MD is widely applied in structural biology and ligand–receptor interaction studies because it can provide important insight into protein flexibility and inter-domain interactions that can be difficult to study by experimental approaches [63]. MD simulations are often performed starting from X-ray and NMR data to refine the experimental structure and sample its configuration space in order to evaluate the energy changes induced by several stimuli such as mutations, pH, binding with small molecules, etc.

This important computational method was, therefore, extensively applied also to study PD-1, PD-L1, and their complex. Indeed, classical and accelerated MD calculations using the AMBERff14SB force field (FF) were performed [64] to investigate the conformational space of the 19–127 domain of the apo-PD-L1, which is the region involved in the binding with other proteins. The extended MD trajectories (1 μs) and principal component analysis (PCA) provided detailed information on structural displacements in apo-PD-L1, mainly associated with the movement of a specific region (C″D loop), suggesting that the PD-L1 binding process occurs basically by a conformational selection mechanism.

In the next paragraph, we reported some of the most relevant MD studies, often coupled with other computational methods, aimed to characterize the binding mechanism of hPD-L1 to PD-1, an essential step for the design of small molecules able to inhibit this immune checkpoint.

3.1. PD-1 Binding

To design effective small molecules able to disrupt the PD-1/PD-L1 pathway, it is crucial to know the 3D structure of the corresponding protein–protein complex. As already mentioned, the X-ray structure of the human PD-1/PD-L1 adduct (PDB ID 4ZQK) was reported for the first time by Holak and coworkers along with the apo-PD-1 binding domain from hPD-L1 (PDB ID 5C3T) [44].

Analysis of the crystal structure of the PD-1/PD-L1 complex (Figure 7A) revealed that the two subunits establish both hydrophobic and polar interactions, with a central hydrophobic core formed by non-polar residues of both PD-1 (Val64, Ile126, Leu128, Ala132, Ile134) and PD-L1 (Ile54, Tyr56, Met115, Ala121, Tyr123) units. The protein–protein interface is also characterized by a buried π–π stacking interaction between the aromatic moieties of Tyr68 (PD-1) and Tyr123 (PD-L1). The interaction region exposed to the solvent is instead characterized by hydrogen bonds and ionic interactions such as Ala132-Gln66 (PD-L1), Ile134-Tyr56 (PD-L1) and Glu136-Arg113 (PD-L1). Other important contacts include Thr76-Tyr123 (PD-L1), Gln75-Arg125 (PD-L1), Thr76-Lys124 (PD-L1), Lys78-Phe19 (PD-L1).

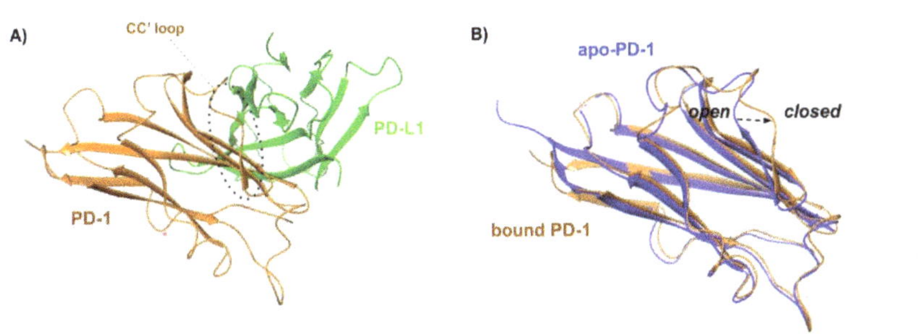

Figure 7. (**A**) Cartoon representation of the X-ray structure of the complex between human PD-1 (orange cartoon) and PD-L1 (green cartoon) resolved by Holak and coworkers (PDB ID 4ZQK). The CC′ loop of PD-1 in its closed conformation is surrounded by a dashed line. (**B**) Superposition of X-ray structures of the bound conformation of PD-1 (orange profile, PDB ID 4ZQK) and apo-PD-1 (blue profile, PDB ID 5C3T). The rearrangement from the open to the closed conformation, which occurs moving from the apo to the bound state, is marked by a black arrow.

Three main hot regions have been identified on hPD-L1. The first region is a hydrophobic pocket composed of the side chains of Tyr56, Glu58, Arg113, Met115, and Tyr123, which can fit Ile134 of PD-1. The second hot region, located near the previous hydrophobic pocket, is determined by Met115, Ala121, and Tyr123, and accommodates Ile126 of PD-1. The third hot region is formed by the charged side chains of Asp122-Arg125 and Asp26, representing a polar groove that can fit Tyr68, Gln75, and Thr76 of PD-1 (Table 4).

Table 4. The three main hot regions and the corresponding hot spots on the PD-L1 surface detected by Holak and co-workers [44] analyzing the X-ray structure of the hPD-1/PD-L1 complex (PDB ID: 4ZQK).

Hot Region	Hot Spots	Note
1	Tyr56, Glu58, Arg113, Tyr123, Met115	Deepest cleft with predominantly hydrophobic character
2	Met115, Ala121, and Tyr123	Located near the previous hot region
3	Asp122, Tyr123, Lys124, Arg125	Extended groove with multiple H-bond donor and acceptor groups

The X-ray data revealed additional structural details. Specifically, the formation of the PD-1/PD-L1 complex requires significant structural flexibility of the hPD-1 unit. Upon superimposing the apo-hPD-1 with the hPD-1 structure extracted from the previously described PD-1/PDL1 (PDB ID: 4ZQK), a high conformational similarity was observed, except for the Met70-Asp77 loop (CC′ loop), which shows an open conformation in the apo-PD-1 structure (Figure 7B). Conversely, the Met70-Asp77 loop undergoes a closed conformation around hPD-L1 in the complex formation (Figure 7A,B), promoting the formation of multiple hydrogen bonds between the two partner proteins. This indicates that the rearrangement of the CC′ loop stabilizes the initial transient interaction between hPD-1 and hPD-L1.

The role of the CC′ loop in the PD-1/PD-L1 complex formation was further investigated by performing MD simulations of apo-PD-1 and the PD-1/PD-L1 complex using the CHARMM36 FF by Liu et al. [65]. The MD results and cluster analysis suggest that the open and closed conformations coexist within a dynamic ensemble in apo-PD-1. The energy barrier is represented by the H-bond pattern that must be broken to allow for conformational repositioning. The PD-1 receptor, in its open conformation, binds to PD-L1 to form the initial encounter complex. This complex undergoes structural rearrangements, resulting in the final closed complex, as observed in the PD-1/PD-L1 crystal structure (PDB ID: 4ZQK). These findings suggest a complex binding process between PD-1 and PD-L1, involving both conformational selection and induced-fit mechanisms.

The PD-1/PD-L1 adduct was also investigated by Kenn et al. [66] performing MD simulations (three replicas of 600 ns) using the Amber99sb-ildn FF coupling with an unsupervised clustering method. They identified specific regions of the PD-1/PD-L1 complex forming stable clusters over time, which are hence named "semi-rigid domains".

Further insight into interactions between PD-L1 and PD-1 (and different monoclonal antibodies) was provided by Shi et al. [67], adopting an elaborated multi-layered computational approach based on MD simulations (Amber FF14SB FF, simulation time 200 ns), per-residue free energy decomposition, virtual alanine scanning mutagenesis and residue-residue contact analysis. In agreement with other studies previously mentioned, the virtual alanine scanning mutagenesis suggested that Tyr56, Gln66, Met115, Asp122, Tyr123, and Arg125 are the most important residues on the PD-L1 surface for PPI (hot spots). The residue–residue contact analysis further shows that PD-1 interacts with PD-L1 mainly by F and G strands.

The potential hot spots of PD-1/PD-L1 were characterized by using another alanine scanning approach based on single-trajectory MD calculations (MD-based computational alanine scanning) coupled with MM/GBSA/IE method [68]. The results reveal eight hot spot residues for both PD-1 (Gln75, Ile134, Ile126, Glu84, Lys78, Tyr68, Leu128, and Asn66) and PD-L1 ($_L$Tyr123, $_L$Tyr56, $_L$Arg125, $_L$Met115, $_L$Arg113, $_L$Gln66, $_L$Ile54 and $_L$Lys124). Among them, $_L$Tyr123 (PD-L1) was demonstrated to be one of the most significant residues in the PD-1/PD-L1 interaction since it establishes favorable contacts with Ile134, Tyr68, and Glu136 of the PD-1 receptor.

The apo-PD-1 and its complex with PD-L1 were also studied by Du et al. using the MD protocol (OPLS/AA FF, production run 50 ns) combined with computational mutagenesis studies providing crucial information for designing engineered PD-1 mutants to modulate the PD-1/PD-L1 pathway [69]. The MD simulation revealed that not all of the key residues identified in the crystal structure analysis contribute to the protein–protein interaction (PPI) throughout the entire trajectory, indicating their limited involvement in the binding process. Additionally, the solvent-accessible surface area (SASA) calculations showed that the binding surface expands from 220 Å in the X-ray structure to 440 Å after the MD simulations.

The decomposition analysis of the total binding energy computed with MM/PBSA revealed that Arg104, Lys131, and Lys135 are the most important residues on the hPD-1 surface for PPI. Several hPD-1 mutants, including Met70Ile, Ser87Trp, Ala132Leu, and Lys135Met, showed improved hPD-L1 binding ability compared to wild-type hPD-1. These mutants provide important details for modulating the interaction between hPD-1 and hPD-L1.

Another PD-1 mutant with ultra-high affinity for PD-L1 has been obtained and characterized, named high-affinity consensus (HAC) PD-1, showing superior therapeutic efficacy in mice compared with antibodies. The resolution of its X-Ray structure in the complex with PD-L1 (PDB ID 5IUS) showed that HAC PD-1 binds PD-L1, establishing polar interactions [51]. MD simulations (Amber ff14SB FF, 20 ns) revealed that the wild-type PD-1 is affected by a greater conformational variability compared with HAC PD-1 and that the mutations Tyr68His, Met70Glu, Lys78Thr allow the formation of favorable contacts with PD-L1, stabilizing the HAC PD-1/PD-L1 complex.

As previously described, tumor cells overexpress PD-L1 on their surface to elude the immune system. Notably, the tumor microenvironment is generally characterized by acidic pH which affects the protonation states of the residue side chains with consequent effects on PD-1/PD-L1 interactions. Interestingly, Pascolutti et al. found that the HAC PD-1 exhibits pH-dependent affinity, with strong binding at low pH conditions [51]. Indeed, the Tyr68His mutation allows for the formation of salt bridges with the Asp122 of PD-L1 due to protonation of His68 occurring at low pH, greatly improving the stability of the protein–protein adduct.

The effect of acidic pH on the mechanism and kinetics of the HAC PD-1/PD-L1 formation was also investigated by Klyukin et al. [70]. They used the infrequent metadynamics technique (Amber03 FF, production run 100 ns) considering two pH levels, 7.4 and 5.5, corresponding to physiological conditions and acidic tumor microenvironments, respectively [70]. In agreement with Pascolutti et al., their results showed that the PPIs of HAC PD-1/PD-L1 are significantly affected by pH changes, and also that small variations can induce a relevant increase in binding strength. In particular, His68 (PD-1) undergoing protonation at pH 5.5, greatly stabilizes the complex interacting with PD-L1 Asp122.

All the computational outcomes so far described have shed light on the binding mechanism of the PD-1/PD-L1 formation and confirmed the role of hot spot residues identified by Holak. As discussed in the next paragraphs, this crucial information is used for structure-based drug design studies aimed at identifying small molecule inhibitors of the PD-1/PD-L1 pattern.

Table 5 summarizes the main computational methods discussed in the present paragraph.

Table 5. Principal computational methods applied for the investigation of the hPD-1/PD-L1 binding mechanism.

Methods	Force Field	Thermodynamic Ensemble	Ref.
Classical and accelerated MD	AMBER FF14SB	NVT/NPT	[64]
Classical MD	AMBER03	NVT/NPT	[70]
Classical MD	CHARMM36	NPT	[65]
Classical MD	AMBER99SB-ILDN	NVT/NPT	[66]
Classical MD, MM/GBSA	AMBER FF14SB	NVT/NPT	[67]
Classical MD, MM/PBSA	OPLS/AA	NVT/NPT	[69]
Classical MD, infrequent MTD	AMBER03	NVT/NPT	[71]
Classical MD, MM/GBSA/IE	AMBER FF14SB	NVT/NPT	[68]

MD: molecular dynamics; MTD: metadynamics.

3.2. Small Molecule Binding

As previously stated, the discovery of the initial X-ray complexes of BMS ligands with PD-L1 revealed the dimerization effect caused by small molecules and prompted researchers to use structure-based computational methods to explain the binding and dimerization processes of biphenyl-based inhibitors.

In 2019, Almahmoud and Zhong conducted a study on the binding mode of 29 biphenyl derivatives extracted from the BMS patents through molecular docking studies. The docking calculations aimed to identify the most relevant residues interacting with ligands to design optimized binders. To ensure the consistency of the docking results, calculations were performed on two PD-L1 X-ray structures (PDB ID 5N2F and 5NIU). The computational analysis was conducted using the software packages MOE (Molecular Operating Environment (MOE), Chemical Computing Group ULC, 910-1010 Sherbrooke St. W., Montreal, QC H3A 2R7, Canada) and the Schrödinger Suite (Schrödinger Suite, Schrödinger, LLC, New York, NY, USA). The final results indicate the asymmetric binding of small molecule ligands to the two PD-L1 monomers, which is consistent with the findings of Zak et al. [30]. The most relevant residues are Tyr56 of both chains, which is retrieved in 100% of predicted complexes, followed by $_{B}$Asp122, $_{B}$Lys124, and $_{B}$Arg125 [71].

Sasmal et al. obtained a similar result when investigating the binding site of PD-L1 dimers in the complex with small molecules using more tools. At first, they exploited DoGSiteScorer and Pankweb to quantify the dimensions and properties of the large hydrophobic tunnel formed between the two PD-L1 monomers. Thus, BIOVIA Discovery Studio (BIOVIA, Dassault Systèmes, San Diego: Dassault Systèmes) was used to prepare and dock a comprehensive series of biphenyl derivatives from the literature and patents into the PD-L1 crystal structure (PDB ID 5N2F) to assess their binding mode in the PD-L1 tunnel. The main conclusions are that hydrophobicity is essential for PD-L1 inhibition, the biphenyl system is required to interact with $_{A}$Tyr56, while another aromatic portion is essential for contact with $_{B}$Tyr56 and $_{B}$Asp122. The third aromatic ring must be derivatized by flexible polar chains interacting with the PD-L1 groove [72].

The Alanine-Scanning-Interaction-Entropy (AS-IE) approach was applied to quantify the contribution of single residues to the global binding ΔG of BMS derivatives into the PD-L1 dimer. AS-IE is a computational method developed by Liu et al. [73] that determines the entropic contribution to binding free energy from fluctuations in individual residue–ligand interaction energies in a single MD trajectory. The relative values of ligand binding to the wild type compared to the mutated protein provide the calculated residue-specific binding free energies for each residue. A total of 35 BMS derivatives were divided into five groups based on their inhibition potency and subjected to MD simulation and subsequent AS-IE analysis using AMBER16 with the ff14SB and GAFF force fields.

The residues that contributed the most to the global binding ΔG were $_{A}$Tyr123 and $_{B}$Tyr56, followed by $_{B}$Met155, $_{A}$Met115, $_{B}$Gln66, and $_{A}$Asp122. The analysis suggests that modifications to the third aromatic ring (C) are important for designing new ligands, while substitutions on the A ring of the biphenyl system do not result in better inhibitors [74].

Partial Least Squares Discriminant Analysis and flexible docking studies were carried out by Kuang at al. to elicit features distinguishing between active and inactive inhibitors. The authors collected 2558 PD-L1 inhibitors from the literature and submitted them to the generation of a classification model adding 7674 non-inhibitors randomly selected from PubChem. The classification model, which presented good sensitivity and specificity, reports the most relevant contribution of intramolecular H-bonds, amphotericity, radius of gyration, non-bonded electrostatic energy, octanol—water partition coefficient, and fractional van der Waals surface area of H-bond donors in the discrimination between active and inactive compounds [75].

Shi and coworkers conducted a comprehensive computational study of PD-L1 to gain insight into the dimerization process and to identify the most relevant part of the BMS derivatives that bind to the target protein. They generated an R-group QSAR model that suggested the most relevant substituent position and the residues that mainly contribute to the ligand potency. The results indicate that substituents in the para position of both external aromatic rings have more influence on the ligand potency [76].

In 2019, Mejias and Guirola applied a co-solvent MD simulation to represent a pharmacophore model. The NAMD software with the CHARMM27 force field was used to perform 100 ns MD simulations with three different solvent mixtures (isopropanol/water, acetamide/water, and isopropylamine/acetate) on the PD-L1 monomeric structure (PDB ID 5C3T). The Volmap tool was used to define the occupancy of each probe. Finally, after an energy-based filtering procedure, a final pharmacophore of ten sites was defined, taking into account the properties of probes that interact more strongly with the protein. The obtained pharmacophore was superimposed with known ligands to assess its validity [77].

Similarly, a study from our group used a newly established FMO/GRID-DRY approach for the characterization of polar and hydrophobic interactions between PD-L1 and both PD-1 and BMS ligands. Fragment Molecular Orbital (FMO) is a powerful ab initio method particularly suited for determining the interaction energy between partner proteins or proteins and ligands. It is particularly sensitive to polar contacts, while it is less good at estimating hydrophobic interactions. The coupling with the GRID approach aims to fill this gap by calculating the PD-L1 molecular interaction fields for the DRY probe [78]. The obtained results indicate that the most important residues for hydrophobic contacts are $_{A}$Tyr56, $_{A}$Met115, $_{A}$Ala121, $_{A}$Tyr123, and $_{B}$Ile54, $_{B}$Tyr56, $_{B}$Met115, and $_{B}$Ala121 [79]. In addition, interaction with these residues appears to be responsible for dimerization. On the other hand, polar contacts also play an important role. In particular, those with the so-called G region, are delimited by Asp26, Asp122, Tyr123, Lys124, and Arg125. The results obtained in this work are in good agreement with those presented by Lim et al., who applied FMO calculations and calculated three-dimensional scattered pair interaction energies (3D-SPIEs) between PD-L1 and a series of binders. In particular, the authors examined binding to PD-1, monoclonal antibodies, macrocyclic peptides, and small molecules. After calculating all pair interaction energies (PIEs), only those at a distance of less than 5.4 Å were selected for further analysis and reported in a 3D scatter plot. The results highlighted the presence of a hot spot shared by all types of ligands and formed by Tyr56, Glu58, and Gln66. A second hot spot is characterized by Asp122 and Arg125, which are involved in interactions with PD-1 and mAbs, while small molecules interact mainly with Asp122. A special role is played by Met115, which is centrally located between the two hot spots and interacts with most ligands [80].

Sun et al. investigated the binding of the BMS compound to PD-L1, comparing the 100 ns MD trajectories of the PD-L1/PD-L1 dimer and the same dimer in the complex with BMS-8. The authors demonstrated that the presence of the ligand stabilizes the complex, increasing the number of interacting residues from eleven to thirteen, and the number of salt bridges from four to six. The molecular dynamics simulations confirm the role of key residues identified in the X-ray and in previously mentioned articles. It also highlights the conformational rearrangement of several residues at the dimer interface to host the ligand and improve the interaction network. Additionally, the binding energies of different systems, including PD-1/PD-L1, PD-L1/BMS-8, PD-L1/PD-L1, and PD-L1/PD-L1 in the complex with BMS-8, were calculated to better depict the small molecule binding mechanism. The resulting ΔG values suggest that PD-L1 prefers to bind with BMS-8 over PD-1. Additionally, the initial binding with the small molecule triggers dimerization [81].

Riccio and colleagues investigated the effect of the tumor microenvironment pH on the binding of ligands to PD-L1 in their paper. The tumor microenvironment is known to have a lower extracellular pH compared to normal tissue, which can affect processes influenced by electrostatic interactions such as protein folding and molecular recognition. The authors investigated the effect of lower pH on the binding of PD-L1 with four ligands: a macrocyclic peptide (peptide-57) and three biphenyl derivatives (BMS-202, S7911, and VIS1059). Docking and MD simulations were conducted to determine the most stable electrostatic and hydrogen bond interactions between the ligands and the protein. The pKa values of the ligands were evaluated using both Marvin and Epik. Microscale thermophoresis was used to experimentally assess pH-dependent variations in binding affinity following computational analysis. The results suggest that ligands containing a basic function that interacts with a negatively charged residue (Asp122) can increase their binding affinity at lower pH. These findings provide insight into the design of high-affinity ligands that account for pH-dependent binding [82].

The following Table 6 reports the key PD-L1 residues in the interaction with small molecules, as defined in the considered studies. The main ligand structural features affecting PD-L1 binding are schematized in Figure 8.

Table 6. Key PD-L1 residues identified in different articles studying the binding with small molecules.

Studied Ligands	Computational Approach	PD-L1 Hot Spot Residues	Reference
29 BMS derivatives	Docking Binding free energy calculation	$_A$Tyr56, $_B$Tyr56, $_B$Asp122, $_B$Lys124, $_B$Arg125	[72]
Several biphenyl derivatives from literature and patents	Docking	$_A$Tyr56, $_B$Tyr56, $_B$Asp122	[73]
35 BMS derivatives	MD AS-IE	$_A$Tyr123, $_B$Tyr56, $_B$Met155, $_A$Met115, $_B$Gln66, $_A$Asp122	[75]
6 BMS derivatives	FMO/GRID-DRY	$_A$Tyr56, $_A$Met115, $_A$Ala121, $_A$Tyr123, and $_B$Ile54, $_B$Tyr56, $_B$Met115, $_B$Ala121	[80]
4 BMS derivatives	FMO 3D-SPIE	Tyr56, Glu58, Gln66, Met115, Asp122, Arg125	[81]

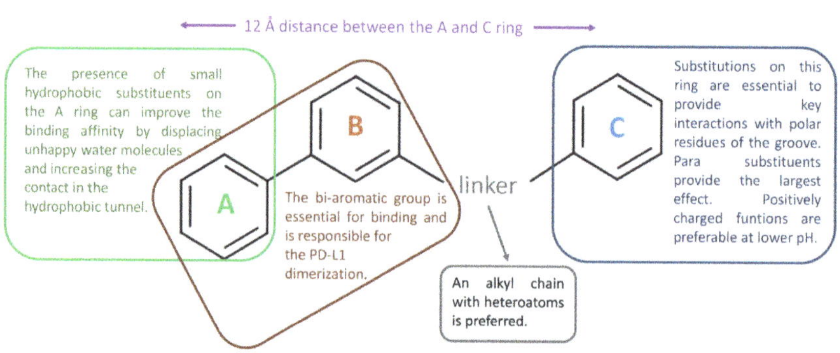

Figure 8. Schematic representation of the main structural features of the bi-aryl PD-L1 binders that are more likely to affect target binding, as suggested by the reviewed articles.

3.3. Dimerization

Closely connected to the binding of biphenyl derivatives to PD-L1, the target dimerization promoted by small molecules has been investigated by several researchers. The dimerization process needs to be elucidated in view of the potential relationship between the inhibitory activities of BMS small molecule inhibitors and the stability of the dimerized complex systems. In this context, computational approaches can represent an almost unique tool to understand this unpredicted event.

In 2019, Soremekun et al. carried out all-atom MD simulations to study the dimerization process promoted by BMS-1001 and BMS-1166. Three different systems were built: two unbound PD-L1 monomers and one PD-L1 bound to BMS-1001, and BMS-1166, in the presence of the other PD-L1 monomer in the simulation box. MD simulations of 150 ns were carried out using AMBER14 and the FF14SB force field. During the simulation, the ligand in the bound systems transitioned from the starting monomer to the second one in an intermediate phase. The final simulated state of the bound systems was represented by the dimeric form. In contrast, the unbound PD-L1 did not demonstrate any dimerization, indicating that the studied process is promoted by the ligand. Additionally, ligand binding caused an increase in residue fluctuation compared to the unbound protein. The per-residue energy decomposition analysis for both ligands was estimated using MM/PBSA. The analysis reported a high electrostatic contribution for $_{A}$Asp122, $_{A}$Tyr123, $_{A}$Lys124, and $_{A}$Arg125 outside the hydrophobic tunnel and $_{B}$Ala121 and $_{B}$Asp122 inside the tunnel, along with both $_{A}$Tyr56 and $_{B}$Tyr56 hydrophobic contacts [83].

A similar study characterizing the binding and unbinding process of BMS-8 and BMS-1166 to PD-L1 was presented by Shi et al. using MD simulations and metadynamic studies. They used the X-ray complexes with PDB ID 5J8O and 5NIX and conducted canonical MD simulations with the AMBER FF14SB force field, with a production phase of 150 ns. The authors used MM/GBSA and MM/PBSA methods to calculate the global binding free energies and the contribution of each residue through per-residue-based decomposition analysis. Additionally, they conducted metadynamics to describe the unbinding process, defining two collective variables that accounted for the ligand position in the target protein. The authors confirmed the stabilizing effect of both ligands on PD-L1 and their preferential binding to one monomer over the other. BMS-8 induced greater flexibility in the system compared to BMS-1166, which can be attributed to a larger enthalpic contribution. Metadynamics simulations suggest that the dimerization process is caused by the ligand binding to one monomer, which then recruits the second monomer, in agreement with previous predictions. The most likely dissociation mechanism involves the ligand disengaging from the dimer, which then oligomerizes after the ligand leaves [76].

In 2021, Guo et al. investigated the role of ligand chirality in PD-L1 binding. They performed docking, MD simulation, and per-residue-based decomposition analysis to study the binding of (*R*)- and (*S*)-BMS-200, along with a modified version of BMS-200, where the chiral hydroxyl is substituted by a carbonyl function. The PDB complex with BMS-200 (PDB ID 5N2F) was used as the starting point for docking with AutoDock Vina and MD simulation using GROMACS2106.4. The study confirms the dimerization mechanism induced by ligand binding to one monomer, as previously ascertained. Additionally, the authors highlight a slight difference in binding energy between the two enantiomers, with the R enantiomer showing more interactions. The residues primarily involved in ligand contacts are Ile54, Tyr56, Met115, Ala121, and Tyr123 [84]. The same research group applied a similar computational approach to investigate the binding mechanism of BMS-202 and its modified analogues where the terminal carbonyl group is substituted by a hydroxyl function, generating both enantiomers. In addition to previous results, the authors emphasized the role of the conformational rearrangement of $_{A}$Tyr56, $_{A}$Tyr123, and $_{B}$Met115 in the ligand association and dissociation process [85]. A comprehensive computational study was carried out by Ahmed et al. that explored the binding properties of four BMS ligands comprising the minimum fragment identified by Skalniak et al. [32] The authors also studied the PD-L1/PD-1 complex, the PD-L1 dimer without ligands, and the naturally occurring 'back-to-back' PD-L1 dimer (PDB ID 5JDR) through MD simulations and MM-GBSA analysis. Grid Inhomogeneous Solvation Theory (GIST) and Hydration Site Analysis (HSA) were applied to understand the role of water displacement in ligand binding and dimerization. In addition, the authors carried out a 2D-QSAR analysis using 403 ligands extracted from literature and designed a large virtual library of potential PD-L1 ligands. The MD simulations allowed for the clarification of the dimerization process, which is in line with the proposed ligand-induced mechanism. The authors confirm the biphenyl system as the minimum structural ligand requirement for PD-L1 dimerization. The computational solvent mapping suggests that BMS ligand to PD-L1 monomer can be favored by the displacement of unfavorable water molecules from their highly energetic hydration site. This result suggests that substitution of the A ring can further contribute to this effect [86].

4. Computational Studies Contributing to the Identification of New Compounds

As the final part of this review, we present computational approaches that have been applied to identify new PD-L1 ligands. As previously mentioned, this paragraph focuses on papers referring to virtual screening campaigns, while neglecting docking studies of newly synthesized compounds, which are less attractive from a computational point of view.

Only papers that include a biological assay demonstrating activity towards PD-L1, and confirming the validity of the computational results, were considered for this review. The preferred method for determining the ability of a compound to inhibit PD1/PDL1 interaction is through biological assays, with the well-established HTFR assay being the preferred option. In silico methods can also be used to virtually test a large number of compounds quickly and inexpensively, reducing the number of compounds to be tested in vitro or in vivo, speeding up the process and reducing costs.

Virtual screening (VS) techniques can be divided into two approaches: ligand- and structure-based. The ligand-based approach is useful when the 3D structure of the target is unknown and relies on the knowledge of the chemical properties of active compounds. However, this method may limit the ability to identify compounds with different structures and/or different types of interactions. On the other hand, the structure-based VS can be used when the 3D structure of the target is known. Due to the availability of numerous 3D PD-L1 structures in the Protein Data Bank, structure-based VS has become a popular method for identifying new ligands in recent years. This involves the screening of commercial and/or in-house databases to find new compounds. Several research groups conducted docking-based VS on diverse small molecule databases, including synthetic and natural compounds, as well as approved drugs. They used various methods to filter the databases and employed one or multiple 3D structures of proteins and ligands stored in the Protein Data Bank.

The section below describes the most significant structure-based VS approaches, including the most popular procedures and database filtering strategies.

Wang et al. screened the Specs database that contains more than 200,000 compounds using the 3D structure of dimeric PD-L1 protein in the complex with BMS-202 (PDB ID 5J89). They used Schrödinger programs such as Protein Preparation Wizard to refine protein structure, and LigPrep to process compounds, exploiting Epik at pH = 7.0 to predict ionized states, tautomers, and stereoisomers. The grid box was centered on the crystallographic ligand and the XP protocol was used to perform docking. The top compounds were re-docked using a flexible docking strategy (Induced Fit Docking), and the Canvas module was used to cluster the best-docked compounds and analyze the binding mode. Compound APBC (SPECS No. AG-690/11449006, Figure 9) binds, like BMS-202, to the hydrophobic site of two PD-L1 monomers with similar anchoring residues Tyr56, Met115, and Ala121. A π–π stacking interaction between the aniline group of APBC and Tyr56 was observed, and a key H-bonding to D122 stabilized the complex. The activity of compounds was valued by HTFR PD-1/PD-L1 interaction assay, with an IC_{50} of 27.82 μM [87].

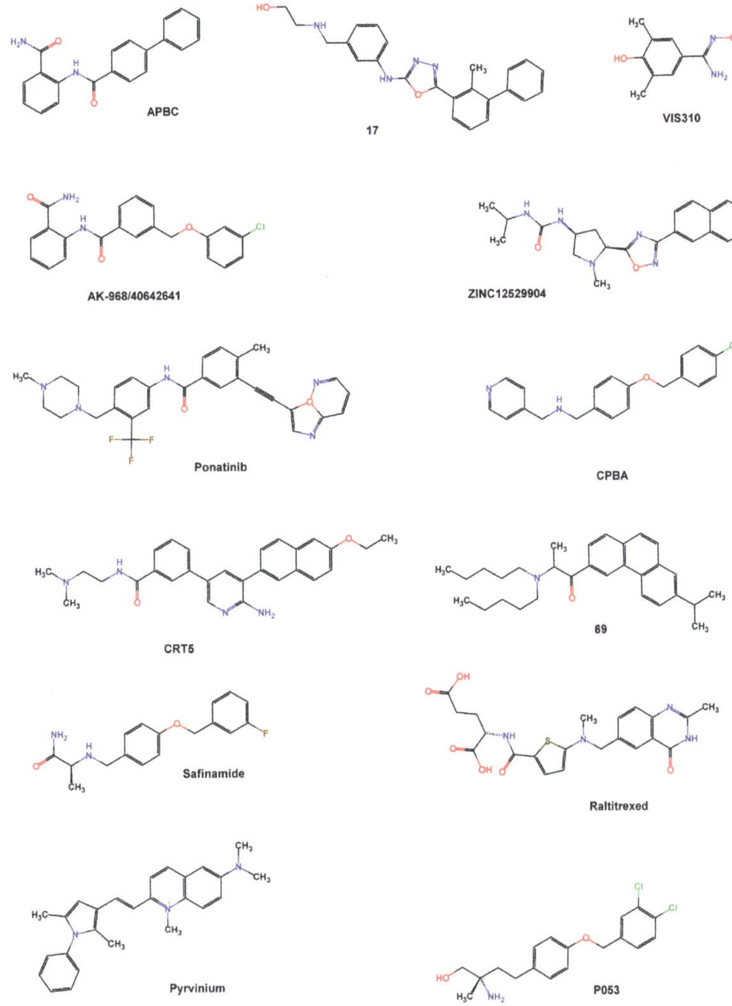

Figure 9. Chemical structures of PD-L1 inhibitors identified by structure-based virtual screening and machine learning.

Lung et al. screened the natural product dataset of the ZINC12 database (ZBC, 180,131 chemical structures) using a docking-based VS with the 3D structure of PD-L1 (PDB ID: 5J89) and the iDock program. To validate the screening protocol, four inhibitors (BMS-8, BMS-37, BMS-200, and BMS-202) with known IC_{50} values were added to the database. An arbitrary cutoff iDock score was set at the value of BMS-202 score (-9.95 kcal mol^{-1}) for further analysis. Contact fingerprint analysis was performed using the AuPosSOM web server, which automatically analyzes poses using self-organizing maps. A total of 368 compounds were clustered together with the four known BMS compounds and were filtered by drug-like properties using Data Warrior with the following criteria: molecular weight between 55 and 500 Da, no more than 5 hydrogen bond donors, no more than 10 hydrogen bond acceptors, a partition coefficient logP between -1 and 5, a net charge between -2 and 2, a topological polar surface area less than 100, and no risk of mutagenicity, tumorigenicity, irritation, or reproductive effects. The filtered 111 compounds were clustered using the FragFP descriptor with a minimum similarity of 0.8. Finally, 22 compounds with iDock scores better than BMS-202, similar contact fingerprints, and preferable drug properties were selected for in vitro evaluation. ZINC12529904 (Figure 9) inhibited approximately 40% of the PD-1/PD-L1 interaction at 100 nM in the AlphaLISA binding assay [88].

Barnwal et al. used the ParDOCK program to screen small-molecule drugs from the ZINC database, and the best-scored 11 compounds (docking score < -8 Kcal mol^{-1}) were subjected to MD simulation using the AMBER suite to determine the dynamic stability of the interaction. Among these compounds, Ponatinib (Figure 9), a tyrosine kinase inhibitor, exhibited stable binding to the active site of PD-L1, mediated by hydrophobic contacts with Glu54, Glu55, Asp56, Gln49, Val51, Tyr39, Ser100, Ile99, Met98, Ala104, and Asp105. A cell-free fluorescence-quenching study confirmed the binding of ponatinib with recombinant PD-L1 (IC_{50} 1.91 μM) [89].

Acúrcio et al. conducted a docking-based VS on several synthetic compound libraries, including NCI, Enamine, Specs, and in-house databases, comprising almost 900,000 small molecules. The crystallographic complex of BMS-202 and PD-L1 (PDB ID 5J89) was prepared and minimized using the MOE (Molecular Operating Environment (MOE), Chemical Computing Group ULC, 910-1010 Sherbrooke St. W., Montreal, QC H3A 2R7, Canada) software package with Amber 10 EHT force field. The GOLD suite was used to analyze the binding conformations, with $_{A}$Tyr56 set as the center of the binding pocket and a 10 Å radius. A preliminary screening was performed using the ChemPLP fitness scoring function and 50 genetic algorithm (GA) runs, while the top 1000 highest-ranking compounds were analyzed in a more detailed molecular docking study (GoldScore fitness scoring function with 500 GA runs). 95 compounds, resulting from filtering the docking score, fitting the active site, interacting with nearby residues, and Lipinski's rule of five criteria with the FAF-Drugs4 tool, characterized by variable chemical scaffolds, underwent testing for PD-L1 HTRF assay. Even if a total of 16 compounds, tested at 100 μM, were able to reduce the HTFR signal to 50%, the most promising compound was the **69** (IC_{50} 96 nM, Figure 9) [90].

Bianconi et al. performed a structure-based VS of 5801 small molecules (807 internal subset molecules with MW $\leq$ 500 Da and 4994 highly soluble Life Chemicals fragments with MW $\leq$ 300 Da) to identify compounds with good activity at pH 6.2, useful for overcoming drug resistance mechanisms due to an acidic tumor microenvironment. Asp122 and Lys124 residues, responsible for pH-dependent binding activity, were identified as hot spots on PD-L1 (PDB ID 5J89) and were used as key features for selecting hit compounds. Docking studies were conducted using the standard precision (SP) method of Glide and the G-score scoring function. LigPrep was used to refine the ligands, Protein Preparation Wizard to process and energetically refine the C/D PD-L1 chains, and the grid box was centered near Asp122, and Lys124 of chain C. Microscale thermophoresis (MST) experiments were used to confirm the binding to PD-L1 of the top 60 compounds based on their G-score value and interaction with the hot spot residues. The most active compound, VIS310 (Figure 9), showed a high-micromolar dissociation constant (Kd = 163.75 ± 33.61 μM), and better binding efficiency index (BEI = 21.0) than BMS-202 (Kd = 8.13 ± 1.38 μM; BEI = 12.1). Since VIS310 was the only substituted benzamidoxime of the analyzed database, the analogue-based approach to screen the REAL database of Enamine, containing about 43.8 million drug-like compounds, was used to define the structure-activity relationships of benzamidoxime compounds [91].

Wang T. et al. filtered the ChEMBL25 database (1.9 million compounds), based on Lipinski's rule of five and REOS rules. The selected compounds were submitted to structure-based VS using the 3D structure of the PD-L1 dimer co-crystallized with BMS-202 (PDB ID: 5J89) and Autodock 4.2 with a Lamarckian genetic algorithm (LGA). The first run of screening consisted of 1 round of docking while the top 50,000 molecules were docked five times for ranking purposes. From the top 100 molecules, clustered into 20 groups using FCFP_4 fingerprints, nine compounds (one from each cluster) were selected based on the drug-likeness and structural diversity of the PD-1/PD-L1 TR-FRET assay. The compound with the highest activity (IC_{50} 64.11 μM) was analyzed through a docking study in PD-L1 (PDB ID 5J89), revealing a deep insertion into the hydrophobic cleft formed at the dimer interface of PD-L1. To improve hydrophobic contacts, π–π stacking, and alkyl–π interactions with Tyr56 and Tyr123, a series of biphenyl analogues of BMS dimerizers were synthesized and tested. Compound **17** demonstrated the highest activity with an IC_{50} of 27.8 nM. The complex **17**/PD-L1 was crystallized with a resolution of 2.4 Å (PDB ID 7DY7) [57].

Another useful approach to finding new PD-L1 binders is the generation of a pharmacophore model based on the structure of the crystal ligands and using it for database screening. Wang F. et al. carried out cross-docking of four available crystal complexes (PDB ID 5J8O, 5J89, 5N2D, and 5N2F) using Glide SP and XP to select the best performing PD-L1 structure (PDB ID 5J89) based on the best RMSD in Glide docking and the best AUC in XP. Almost 200,000 compounds of the Specs database were pre-processed by PAINS filtration using the comprehensive application of the Molinspration Cheminformatics software, OpenEye, ChemAxon, RDkit, and CERTARA. A total of 120,000 compounds were analyzed by a pharmacophore model using Phase, a module of the Schrödinger Suite, on the co-crystallized inhibitors BMS-8, BMS-37, BMS-200, and BMS-202 (PDB ID 5J8O, 5N2D, 5N2F, and 5J89, respectively). The condition of mapping at least four of the six pharmacophore features of the model (one positive charge, one H-bond acceptor, one hydrophobic, and one aromatic ring feature) was satisfied by 10,125 compounds. SP and XP docking protocols, using PDB ID 5J89, in Glide, based on the ROC curves and AUC values, were used. The top 1100 molecules (XP G-scores ≤ -9.000 kcal mol^{-1}) were ranked and clustered in the Canvas module into 402 groups with Tanimoto coefficients of less than 0.5 to ensure maximum structural diversity among compounds. Best-scored compounds of each cluster were submitted to flexible docking (Induced Fit protocol) to refine the final selection. 91 compounds were purchased and tested by SPR and CBPA (SPECS No. AN-465/42833793, Figure 9) exhibited bioactivity at the molecular (Kd 48.10 μM) and cellular levels [92].

Choorakottayil Pushkaran et al. developed a 3D pharmacophore model considering the key interacting residues between BMS-202 and PD-L1 dimer (PDB ID 5J89) using the "Structure-based pharmacophore" module of Ligand Scout 4.1. The pharmacophore was characterized by seven chemical features: four hydrophobic (H), one positive ionizable (P), one H-bond acceptor (A), and one H-bond donor (D). The obtained model was validated by calculating the enrichment factor (EF) using a test dataset composed of 61 known PD-L1 inhibitors and 1425 decoys. The validated 3D pharmacophore model was used to screen all the FDA-approved drugs in the DrugBank database (1925 compounds) and small molecules in the Specs database (540,807 compounds), with the VS module of Ligand Scout software. Compounds passing the pharmacophore selection matching at list five chemical features (12 in DrugBank and 15,276 in Specs) were submitted to High Throughput VS, SP, and XP docking protocols in Glide. Ligands were prepared by LigPrep, and energy minimization was performed using the OPLS2005 force field while retaining the input structure chirality. The protein crystal structure was minimized by Protein Preparation Wizard, and the grid box was centered on BMS-202. Hits were ranked using the XP docking score, and molecular interactions between the hits and the protein were analyzed using PyMol and Biovia Discovery Studio Visualizer. The cutoff of -9 kcal mol^{-1} allowed the selection of three Drug Bank and eight Specs hits characterized by the key interactions with Tyr56, Met115, and Ala121. ADME and drug-likeness prediction were calculated using the QikProp module of Schrödinger. The in vitro toxicity and PD-1/PDL-1 inhibitory activity established that the drugs Raltitrexed, Safinamide, and the natural AK-968/40642641 (Figure 9) could be used as PDL-1 inhibitors [93].

Fattakhova et al. conducted a combined ligand- and structure-based VS to identify small molecules active on PD-L1. For the structure-based screening, ensemble docking using 7 crystal structures (PDB IDs 5N2F, 5NIU, 6R3K, 5J89, 5J8O, 5N2D, 6NM8) with approximately 10,000 approved or investigational drugs, using the AutoDock Vina algorithm, was performed. The docking protocol was validated by redocking cognate ligands and evaluating the RMSD with the crystal ligand. After merging the data of the seven proteins, the top 1000 molecules were visually inspected and twenty compounds mimicking the key ligand-PD-L1 interactions (strong hydrophobic interactions with several amino acids lining the channel-like pocket of the dimer, π–π interactions with key amino acids like Tyr56, and possible hydrogen and halogen bonds at the channel opening) were selected. ROCS 3.4.1.0 was used for ligand-based screening, which evaluates shape-similarity to the 7 crystallographic ligands. A multi-conformer database of approved and investigational drugs was screened and ranked based on the ROCS_TanimotoCombo score. The top 1000 molecules were docked in the high-resolution PD-L1 crystal structure (PDB ID 5N2F), and after combining molecules obtained by the two screening, 25 compounds were subjected to biological assays. Pyrvinium (Figure 9), an FDA-approved anthelmintic drug belonging to the phenylpyrroles class, showed comparable potency in HTRF and AlphaLISA assays, confirming its potential PD1/PD-L1 inhibitory activity (IC_{50} 29.66 μM). A post-docking optimization of the best docking pose, using the default relaxation protocol in the Desmond Molecular Dynamics v3.6 package, demonstrated that the dimethyl-phenylpyrrole moiety occupied the distal end of the PD-L1 dimer pocket similarly to the PD-L1 cognate ligand [94].

A machine learning approach was used by Patil et al. to discover bioactive PD-L1 dimerizers. They developed models based on 2D chemical descriptors using a series of small-molecule PD-L1 ligands patented by BMS. Multiple 2D fingerprint descriptors (FP1, FP2, Layered, MACCS, Morgan, RDKit) implemented in the Open Drug Discovery Toolkit (ODDT), were calculated. These descriptors were fitted by Random Forest models to 1581 "Active" molecules (BMS molecules), 50 "decoy" molecules per active compound (obtained from the DUD-E database), and 417 known inactive molecules. According to the good correlation coefficient (R) implemented in the ODDT, all fingerprints were used to screen the commercial Cayman Chemical database (16,191 bioactive molecules) and 361 compounds emerged as potentially "active" in at least 5 fingerprint models. A structure-based docking study was then realized using the highest resolution X-ray PD-L1 structure (PDB ID 5N2F) with AutoDock Vina. The binding mode of the cognate ligand (8HW) was correctly predicted (docking score -11.4 kcal mol^{-1}). The top 20 compounds, tested in the HTRF PD1-PDL1 binding assay, were selected based on the presence of typical interactions between PD-L1 and its inhibitors and their orientation relative to 8HW. MD simulations were used to predict the binding stability of the three most active compounds using Desmond, with a simulation time of 5 ns. The compound CRT5 (IC_{50} 22.35 μM, Figure 9) is the most active and stable and binds similarly to the crystal ligand. Compound P053 (IC_{50} 33.65 μM) follows a similar trend [95].

In Table 7, the main characteristics of the virtual screening campaigns reported in this paragraph are summarized.

Table 7. Principal features of the discussed virtual screening campaigns.

PDB ID	Database (# Compounds)	Program	Screening Protocol	Emerging PD-L1 Inhibitor	Activity (μM)	Ref
5J89	Specs (200,000)	Schrödinger Canva	docking study clustering	APBC	IC_{50} 27.82	[87]
5J89	natural compounds of ZINC12 (180,131)	iDock AuPosSOM Data Warrior	docking study filtering by iDock score (-9.95 kcal mol^{-1}) Contact Fingerprint Analysis filtering by drug-likeness properties clustering (FragFP descriptor)	ZINC12529904	IC_{50} 0.1	[88]
* n.r.	ZINC database	ParDOCK Amber suite	docking study filtering by docking score (-8 kcal mol^{-1}) MD simulations	Ponatinib	IC_{50} 1.91	[89]

Table 7. *Cont.*

PDB ID	Database (# Compounds)	Program	Screening Protocol	Emerging PD-L1 Inhibitor	Activity (μM)	Ref
5J89	NCI, Enamine, Specs, or in-house (900,000)	MOE FAF-Drugs4	docking-based 50 GA for rapid screening filtering by docking score re-docking 500 GA for top scoring compounds filtering by Lipinski's rule of 5	69	IC_{50} 0.096	[90]
5J89	In-house (807) Life Chemicals (4994)	Schrödinger	Docking study (Asp122 &Lys124 hot spots) filtering by docking score	VIS310	Kd 8.13	[91]
5J89	ChEMBL25 (1.9 M)	AutoDock	filtering by Lipinski's rule of 5 first docked once top-scored docked 5 times clustering by FCFP_4 fingerprints druggability	17	IC_{50} 0.0278	[57]
5J89	Specs (200,000)	Schrödinger	filtration by PAINS Pharmacophore model generation Docking study Clustering Induced fit docking	CBPA	Kd 48.10	[92]
5J89	Specs (540,807 s.m.) & DrugBank (1925 FDA-approved drug)	Schrödinger	3D pharmacophore model Docking study (3 steps: HTSV, SP, XP) Filtering by ADME and druggability	Raltitrexed Safinamide AK-968/40642641	Indirect in vitro experiments (↑ in immune cell proliferation)	[93]
5N2F, 5NIU, 6R3K, 5J89, 5J8O, 5N2D, 6NM8	10,000 approved or investigational drugs	AutoDock Vina (SB-VS) ROCS (LB-VS)	Ensemble docking (7 crystal structure) Visual inspection Filtering by shape similarity (7 crystal ligands) Docking study (top 1000 compounds)	Pyrvinium	IC_{50} 29.66	[94]
5N2F	bioactive molecules of Cayman Chemical database (16,191)	Open Drug Discovery Toolkit (ODDT) AutoDock Vina Desmond	Multiple 2D descriptors (FP1, FP2, Layered, MACCS, Morgan, RDKit) Random Forest models Docking study Molecular Dynamics	CRT5	IC_{50} 22.35 IC_{50} 33.65	[95]

* n.r. = not reported.

5. Application of AI-Based Methods to the Study of Immune Checkpoint Inhibition

Artificial intelligence (AI) has had a major impact on technologies and all fields of science, including structural biology and in silico drug discovery studies. Apart from the already discussed VS campaign by Patil et al. [95], not many other studies report the application of these approaches to the discovery of PD-L1 ligands. Most of the literature mentioning PD-L1 and AI is based on the application of these methods to aid in the diagnosis and prediction of responses to AI treatments [96].

AI is behind a breakthrough in structural biology represented by AlphaFold and similar software (RosettaFold, OpenFold, and ESMFold), which is able to predict the 3D structure of a protein from its amino acid sequence [41,97]. This specific computational method was used to predict the anti-PD-L1 antibody and antigen structures of the humanized 3D5 antibody, named h3D5-hIgG1, as well as the structure of the corresponding h3D5/PD-L1 complex [98]. Experimental results showed that h3D5-hIgG1 is characterized by an extraordinary binding affinity to the PD-L1 protein compared to the parental murine 3D5 antibody. Other Alphafold applications aimed at identifying PD-L1 inhibitors have recently been reviewed by Sobral et al. [40].

The power of AI in drug discovery has also been applied to other targets in the immune checkpoint inhibitor space, as shown in a recent study using an AI algorithm based on deep convolutional neural networks to identify small-molecule inhibitors of cytotoxic T-lymphocyte-associated protein 4 (CTLA-4) that disrupt the CTLA-4/CD80 interaction [99]. This AI approach was used to virtually screen a library of 10 million compounds. The most promising compounds were evaluated using biochemical, biophysical, immunological, and animal assays to demonstrate their ability to inhibit the CTLA-4/CD80 pathway.

Although the application of AI-based methods to the identification of PD-L1 ligands has not been widely exploited, the increasing number of available tools and technical capabilities will certainly encourage the application of these methods to this target.

6. Conclusions

The interaction between PD-1 and PD-L1 represents a well-established and valuable target for cancer immunotherapy. Several mAbs are currently in clinical use and represent a new paradigm in cancer therapy.

This protein–protein contact has been extensively investigated using experimental and computational approaches due to its significance. Our review focuses on the application of computational methods to explore the binding mechanism of PD-L1 to PD-1 and small molecule binders. The latter may represent an important goal in the field of immunotherapy, as low molecular weight compounds can overcome limitations due to mAbs administration and side effects. In this context, the resolution of the first X-ray complexes between PD-L1 and BMS derivatives has paved the way for the disclosure of a series of compounds capable of promoting the same dimerization induced by biphenyl compounds.

Most investigations in this field use the available X-ray data to perform docking, MD simulations, and other structure-based approaches. The studies presented here largely confirm the involvement of PD-L1 residues in both the binding of BMS derivatives and the dimerization process, as observed experimentally by Holak. The identification of hot spot residues in the PD-L1 binding region for both PD-1 and small molecules converged on hydrophobic contacts with Tyr56, Met115, Ala121, and Tyr123, as well as electrostatic contacts with Asp122, Lys124, and Arg125. Most of the computational studies completed the binding profile of the studied ligands; therefore, since they mostly overlap with experimental data, it is not trivial to elicit the effective contribution of these methods to the identification of new small-molecule binders.

Several MD simulations, along with other approaches such as MM/GBSA, MM/PBSA, per-residue energy decomposition, and ab initio FMO calculations, were used to gain insight into the dimerization process. The results indicate that the small-molecule ligand binds preferentially to one monomer. This binding is favored by the displacement of "unhappy" water molecules and electrostatic interactions with polar residues in the groove region. This initial contact then recruits the second PD-L1 monomer. It was found that dimerization does not occur in the absence of a ligand and that the biphenyl group (or a bi-aromatic moiety) is the minimum structural requirement for the ligand to promote this process.

The availability of X-ray complexes and activity data for biphenyl ligands, which have fed structure-based virtual screening campaigns, has a twofold implication: while it has provided an essential starting point for the development of other ligands, it has also placed a constraint on the identification of molecules with different scaffolds that can bind PD-L1 by a mechanism of action other than dimerization (assuming that compounds with this activity can exist, given the properties of PD-L1, whose surface druggable sites are unlikely to be identified). Thus, the question of the possible identification of other ligands with a novel mechanism of action is still an open one.

In addition to the need to find a good ligand, one of the most challenging aspects of ICI is the limited number of patients who respond to therapy. Better profiling of cancer protein expression and the adoption of combination therapies may increase the number of patients who can benefit from immunotherapy [100]. Recently, a series of in silico methods was applied to simulate the different intracellular signaling affecting the PD-1/PD-L1 pathway in neuroblastoma (NBM) [101]. In particular, they developed a specific network of protein kinase cascades where the corresponding Michaelis–Menten kinetics parameters were used to create a system of ordinary differential equations. The resulting computational model represents an interesting tool to predict the relation between NBM tumor phenotype and the response of anti-PD-1/PD-L1 therapy, as well as to manage the immunotherapeutic treatment of NBM patients. The latter represents an example of the even more relevant role that computational studies can play in the valuable field of immune checkpoint modulation.

Funding: This research received no external funding.

Data Availability Statement: Data sharing is not applicable.

Conflicts of Interest: The authors declare no conflict of interest.

References

1. Balkwill, F.; Mantovani, A. Inflammation and Cancer: Back to Virchow? *Lancet* **2001**, *357*, 539–545. [CrossRef] [PubMed]
2. Bickels, J.; Kollender, Y.; Merinsky, O.; Meller, I. Coley's Toxin: Historical Perspective. *IMAJ Isr. Med. Assoc. J.* **2002**, *4*, 471–472. [PubMed]
3. Burnet, F.M. Immunological Surveillance in Neoplasia. *Transplant. Rev.* **1971**, *7*, 3–25. [CrossRef] [PubMed]
4. Vesely, M.D.; Kershaw, M.H.; Schreiber, R.D.; Smyth, M.J. Natural Innate and Adaptive Immunity to Cancer. *Annu. Rev. Immunol.* **2011**, *29*, 235–271. [CrossRef] [PubMed]
5. Mahoney, K.M.; Rennert, P.D.; Freeman, G.J. Combination Cancer Immunotherapy and New Immunomodulatory Targets. *Nat. Rev. Drug Discov.* **2015**, *14*, 561–584. [CrossRef] [PubMed]
6. Adams, J.L.; Smothers, J.; Srinivasan, R.; Hoos, A. Big Opportunities for Small Molecules in Immuno-Oncology. *Nat. Rev. Drug Discov.* **2015**, *14*, 603–622. [CrossRef] [PubMed]
7. Pardoll, D.M. The Blockade of Immune Checkpoints in Cancer Immunotherapy. *Nat. Rev. Cancer* **2012**, *12*, 252–264. [CrossRef]
8. Saez-Ibanez, A.R.; Upadhaya, S.; Campbell, J. Immuno-Oncology Clinical Trials Take a Turn beyond PD1/PDL1 Inhibitors. *Nat. Rev. Drug Discov.* **2023**, *22*, 442–443. [CrossRef]
9. Ishida, Y.; Agata, Y.; Shibahara, K.; Honjo, T. Induced Expression of PD-1, a Novel Member of the Immunoglobulin Gene Superfamily, upon Programmed Cell Death. *EMBO J.* **1992**, *11*, 3887–3895. [CrossRef]
10. Nishimura, H.; Nose, M.; Hiai, H.; Minato, N.; Honjo, T. Development of Lupus-like Autoimmune Diseases by Disruption of the PD-1 Gene Encoding an ITIM Motif-Carrying Immunoreceptor. *Immunity* **1999**, *11*, 141–151. [CrossRef]
11. Shinohara, T.; Taniwaki, M.; Ishida, Y.; Kawaichi, M.; Honjo, T. Structure and Chromosomal Localization of the Human PD-1 Gene (PDCD1). *Genomics* **1994**, *23*, 704–706. [CrossRef]
12. Dong, H.; Zhu, G.; Tamada, K.; Chen, L. B7-H1, a Third Member of the B7 Family, Co-Stimulates T-Cell Proliferation and Interleukin-10 Secretion. *Nat. Med.* **1999**, *5*, 1365–1369. [CrossRef]
13. Latchman, Y.; Wood, C.R.; Chernova, T.; Chaudhary, D.; Borde, M.; Chernova, I.; Iwai, Y.; Long, A.J.; Brown, J.A.; Nunes, R.; et al. PD-L2 Is a Second Ligand for PD-1 and Inhibits T Cell Activation. *Nat. Immunol.* **2001**, *2*, 261–268. [CrossRef]
14. Butte, M.J.; Keir, M.E.; Phamduy, T.B.; Sharpe, A.H.; Freeman, G.J. Programmed Death-1 Ligand 1 Interacts Specifically with the B7-1 Costimulatory Molecule to Inhibit T Cell Responses. *Immunity* **2007**, *27*, 111–122. [CrossRef]
15. Ghiotto, M.; Gauthier, L.; Serriari, N.; Pastor, S.; Truneh, A.; Nunès, J.A.; Olive, D. PD-L1 and PD-L2 Differ in Their Molecular Mechanisms of Interaction with PD-1. *Int. Immunol.* **2010**, *22*, 651–660. [CrossRef] [PubMed]
16. Cheng, X.; Veverka, V.; Radhakrishnan, A.; Waters, L.C.; Muskett, F.W.; Morgan, S.H.; Huo, J.; Yu, C.; Evans, E.J.; Leslie, A.J.; et al. Structure and Interactions of the Human Programmed Cell Death 1 Receptor. *J. Biol. Chem.* **2013**, *288*, 11771–11785. [CrossRef] [PubMed]
17. Pabon, N.A.; Camacho, C.J. Probing Protein Flexibility Reveals a Mechanism for Selective Promiscuity. *eLife* **2017**, *6*, e22889. [CrossRef]
18. Dai, S.; Jia, R.; Zhang, X.; Fang, Q.; Huang, L. The PD-1/PD-Ls Pathway and Autoimmune Diseases. *Cell. Immunol.* **2014**, *290*, 72–79. [CrossRef] [PubMed]
19. Zuazo, M.; Gato-Cañas, M.; Llorente, N.; Ibañez-Vea, M.; Arasanz, H.; Kochan, G.; Escors, D. Molecular Mechanisms of Programmed Cell Death-1 Dependent T Cell Suppression: Relevance for Immunotherapy. *Ann. Transl. Med.* **2017**, *5*, 385. [CrossRef] [PubMed]
20. Patsoukis, N.; Li, L.; Sari, D.; Petkova, V.; Boussiotis, V.A. PD-1 Increases PTEN Phosphatase Activity While Decreasing PTEN Protein Stability by Inhibiting Casein Kinase 2. *Mol. Cell. Biol.* **2013**, *33*, 3091–3098. [CrossRef] [PubMed]
21. Patsoukis, N.; Bardhan, K.; Chatterjee, P.; Sari, D.; Liu, B.; Bell, L.N.; Karoly, E.D.; Freeman, G.J.; Petkova, V.; Seth, P.; et al. PD-1 Alters T-Cell Metabolic Reprogramming by Inhibiting Glycolysis and Promoting Lipolysis and Fatty Acid Oxidation. *Nat. Commun.* **2015**, *6*, 6692. [CrossRef]

22. Taube, J.M.; Anders, R.A.; Young, G.D.; Xu, H.; Sharma, R.; McMiller, T.L.; Chen, S.; Klein, A.P.; Pardoll, D.M.; Topalian, S.L.; et al. Colocalization of Inflammatory Response with B7-H1 Expression in Human Melanocytic Lesions Supports an Adaptive Resistance Mechanism of Immune Escape. *Sci. Transl. Med.* **2012**, *4*, 127ra37. [CrossRef] [PubMed]
23. Pedoeem, A.; Azoulay-Alfaguter, I.; Strazza, M.; Silverman, G.J.; Mor, A. Programmed Death-1 Pathway in Cancer and Autoimmunity. *Clin. Immunol.* **2014**, *153*, 145–152. [CrossRef] [PubMed]
24. Upadhaya, S.; Neftelinov, S.T.; Hodge, J.; Campbell, J. Challenges and Opportunities in the PD1/PDL1 Inhibitor Clinical Trial Landscape. *Nat. Rev. Drug Discov.* **2022**, *21*, 482–483. [CrossRef] [PubMed]
25. Sharma, P.; Hu-Lieskovan, S.; Wargo, J.A.; Ribas, A. Primary, Adaptive and Acquired Resistance to Cancer Immunotherapy. *Cell* **2017**, *168*, 707–723. [CrossRef] [PubMed]
26. Yu, J.X.; Hodge, J.P.; Oliva, C.; Neftelinov, S.T.; Hubbard-Lucey, V.M.; Tang, J. Trends in Clinical Development for PD-1/PD-L1 Inhibitors. *Nat. Rev. Drug Discov.* **2020**, *19*, 163–164. [CrossRef]
27. Cheng, J. Review of the Combination Strategies Used in Anti-PD1/PD-L1 Monoclonal Antibody Treatment. *E3S Web Conf.* **2020**, *185*, 03009. [CrossRef]
28. Chupak, L.S.; Zheng, X. Compounds Useful as Immunomodulators. WO 2015/034820A1, 12 March 2015.
29. Chupak, L.S.; Ding, M.; Martin, S.W.; Zheng, X.; Hewawasam, P.; Connoly, T.P.; Xu, N.; Yeung, K.S.; Zhu, J.; Langley, D.R.; et al. Compounds Useful as Immunomodulators. WO 2015/160641A2, 12 December 2015.
30. Zak, K.M.; Grudnik, P.; Guzik, K.; Zieba, B.J.; Musielak, B.; Dömling, A.; Dubin, G.; Holak, T.A. Structural Basis for Small Molecule Targeting of the Programmed Death Ligand 1 (PD-L1). *Oncotarget* **2016**, *7*, 30323–30335. [CrossRef]
31. Chen, S.-Y.; Zacharias, M. What Makes a Good Protein–Protein Interaction Stabilizer: Analysis and Application of the Dual-Binding Mechanism. *ACS Cent. Sci.* **2023**, *9*, 969–979. [CrossRef]
32. Skalniak, L.; Zak, K.M.; Guzik, K.; Magiera, K.; Musielak, B.; Pachota, M.; Szelazek, B.; Kocik, J.; Grudnik, P.; Tomala, M.; et al. Small-Molecule Inhibitors of PD-1/PD-L1 Immune Checkpoint Alleviate the PD-L1-Induced Exhaustion of T-Cells. *Oncotarget* **2017**, *8*, 72167. [CrossRef]
33. Sasikumar, P.G.; Ramachandra, M. Small Molecule Agents Targeting PD-1 Checkpoint Pathway for Cancer Immunotherapy: Mechanisms of Action and Other Considerations for Their Advanced Development. *Front. Immunol.* **2022**, *13*, 752065. [CrossRef]
34. Lin, X.; Lu, X.; Luo, G.; Xiang, H. Progress in PD-1/PD-L1 Pathway Inhibitors: From Biomacromolecules to Small Molecules. *Eur. J. Med. Chem.* **2020**, *186*, 111876. [CrossRef]
35. Wu, X.; Meng, Y.; Liu, L.; Gong, G.; Zhang, H.; Hou, Y.; Liu, C.; Wu, D.; Qin, M. Insights into Non-Peptide Small-Molecule Inhibitors of the PD-1/PD-L1 Interaction: Development and Perspective. *Bioorg. Med. Chem.* **2021**, *33*, 116038. [CrossRef] [PubMed]
36. Islam, M.K.; Stanslas, J. Peptide-Based and Small Molecule PD-1 and PD-L1 Pharmacological Modulators in the Treatment of Cancer. *Pharmacol. Ther.* **2021**, *227*, 107870. [CrossRef] [PubMed]
37. Chen, R.; Yuan, D.; Ma, J. Advances of Biphenyl Small-Molecule Inhibitors Targeting PD-1/PD-L1 Interaction in Cancer Immunotherapy. *Future Med. Chem.* **2022**, *14*, 97–113. [CrossRef] [PubMed]
38. Deng, J.; Cheng, Z.; Long, J.; Dömling, A.; Tortorella, M.; Wang, Y. Small Molecule Inhibitors of Programmed Cell Death Ligand 1 (PD-L1): A Patent Review (2019–2021). *Expert Opin. Ther. Pat.* **2022**, *32*, 575–589. [CrossRef]
39. Wang, S.; Wang, Y.; Yan, H. Progress on Biphenyl Derivatives as PD-1/PD-L1 Inhibitors. *Med. Chem. Res.* **2023**, *32*, 2089–2115. [CrossRef]
40. Sobral, P.S.; Luz, V.C.C.; Almeida, J.M.G.C.F.; Videira, P.A.; Pereira, F. Computational Approaches Drive Developments in Immune-Oncology Therapies for PD-1/PD-L1 Immune Checkpoint Inhibitors. *Int. J. Mol. Sci.* **2023**, *24*, 5908. [CrossRef] [PubMed]
41. Jumper, J.; Evans, R.; Pritzel, A.; Green, T.; Figurnov, M.; Ronneberger, O.; Tunyasuvunakool, K.; Bates, R.; Žídek, A.; Potapenko, A.; et al. Highly Accurate Protein Structure Prediction with AlphaFold. *Nature* **2021**, *596*, 583–589. [CrossRef]
42. Sadybekov, A.V.; Katritch, V. Computational Approaches Streamlining Drug Discovery. *Nature* **2023**, *616*, 673–685. [CrossRef]
43. Lin, D.Y.-W.; Tanaka, Y.; Iwasaki, M.; Gittis, A.G.; Su, H.-P.; Mikami, B.; Okazaki, T.; Honjo, T.; Minato, N.; Garboczi, D.N. The PD-1/PD-L1 Complex Resembles the Antigen-Binding Fv Domains of Antibodies and T Cell Receptors. *Proc. Natl. Acad. Sci. USA* **2008**, *105*, 3011–3016. [CrossRef]
44. Zak, K.M.; Kitel, R.; Przetocka, S.; Golik, P.; Guzik, K.; Musielak, B.; Dömling, A.; Dubin, G.; Holak, T.A. Structure of the Complex of Human Programmed Death 1, PD-1, and Its Ligand PD-L1. *Structure* **2015**, *23*, 2341–2348. [CrossRef] [PubMed]
45. Boisgerault, N.; Bertrand, P. Inside PD-1/PD-L1,2 with Their Inhibitors. *Eur. J. Med. Chem.* **2023**, *256*, 115465. [CrossRef] [PubMed]
46. Zhang, F.; Wei, H.; Wang, X.; Bai, Y.; Wang, P.; Wu, J.; Jiang, X.; Wang, Y.; Cai, H.; Xu, T.; et al. Structural Basis of a Novel PD-L1 Nanobody for Immune Checkpoint Blockade. *Cell Discov.* **2017**, *3*, 17004. [CrossRef]
47. Perry, E.; Mills, J.J.; Zhao, B.; Wang, F.; Sun, Q.; Christov, P.P.; Tarr, J.C.; Rietz, T.A.; Olejniczak, E.T.; Lee, T.; et al. Fragment-Based Screening of Programmed Death Ligand 1 (PD-L1). *Bioorg. Med. Chem. Lett.* **2019**, *29*, 786–790. [CrossRef] [PubMed]
48. Chen, Y.; Liu, P.; Gao, F.; Cheng, H.; Qi, J.; Gao, G.F. A Dimeric Structure of PD-L1: Functional Units or Evolutionary Relics? *Protein Cell* **2010**, *1*, 153–160. [CrossRef] [PubMed]
49. Wen, M.; Cao, Y.; Wu, B.; Xiao, T.; Cao, R.; Wang, Q.; Liu, X.; Xue, H.; Yu, Y.; Lin, J.; et al. PD-L1 Degradation Is Regulated by Electrostatic Membrane Association of Its Cytoplasmic Domain. *Nat. Commun.* **2021**, *12*, 5106. [CrossRef]

50. Wang, Q.; Cao, Y.; Shen, L.; Xiao, T.; Cao, R.; Wei, S.; Tang, M.; Du, L.; Wu, H.; Wu, B.; et al. Regulation of PD-L1 through Direct Binding of Cholesterol to CRAC Motifs. *Sci. Adv.* **2022**, *8*, eabq4722. [CrossRef]
51. Pascolutti, R.; Sun, X.; Kao, J.; Maute, R.L.; Ring, A.M.; Bowman, G.R.; Kruse, A.C. Structure and Dynamics of PD-L1 and an Ultra-High-Affinity PD-1 Receptor Mutant. *Structure* **2016**, *24*, 1719–1728. [CrossRef]
52. Guzik, K.; Zak, K.M.; Grudnik, P.; Magiera, K.; Musielak, B.; Törner, R.; Skalniak, L.; Dömling, A.; Dubin, G.; Holak, T.A. Small-Molecule Inhibitors of the Programmed Cell Death-1/Programmed Death-Ligand 1 (PD-1/PD-L1) Interaction via Transiently-Induced Protein States and Dimerization of PD-L1. *J. Med. Chem.* **2017**, *60*, 5857–5867. [CrossRef]
53. Muszak, D.; Surmiak, E.; Plewka, J.; Magiera-Mularz, K.; Kocik-Krol, J.; Musielak, B.; Sala, D.; Kitel, R.; Stec, M.; Weglarczyk, K.; et al. Terphenyl-Based Small-Molecule Inhibitors of Programmed Cell Death-1/Programmed Death-Ligand 1 Protein-Protein Interaction. *J. Med. Chem.* **2021**, *64*, 11614–11636. [CrossRef]
54. Basu, S.; Yang, J.; Xu, B.; Magiera-Mularz, K.; Skalniak, L.; Musielak, B.; Kholodovych, V.; Holak, T.A.; Hu, L. Design, Synthesis, Evaluation, and Structural Studies of C2-Symmetric Small Molecule Inhibitors of Programmed Cell Death-1/Programmed Death-Ligand 1 Protein-Protein Interaction. *J. Med. Chem.* **2019**, *62*, 7250–7263. [CrossRef]
55. Park, J.J.; Thi, E.P.; Carpio, V.H.; Bi, Y.; Cole, A.G.; Dorsey, B.D.; Fan, K.; Harasym, T.; Iott, C.L.; Kadhim, S.; et al. Checkpoint Inhibition through Small Molecule-Induced Internalization of Programmed Death-Ligand 1. *Nat. Commun.* **2021**, *12*, 1222. [CrossRef]
56. Butera, R.; Ważyńska, M.; Magiera-Mularz, K.; Plewka, J.; Musielak, B.; Surmiak, E.; Sala, D.; Kitel, R.; De Bruyn, M.; Nijman, H.W.; et al. Design, Synthesis, and Biological Evaluation of Imidazopyridines as PD-1/PD-L1 Antagonists. *ACS Med. Chem. Lett.* **2021**, *12*, 768–773. [CrossRef]
57. Wang, T.; Cai, S.; Cheng, Y.; Zhang, W.; Wang, M.; Sun, H.; Guo, B.; Li, Z.; Xiao, Y.; Jiang, S. Discovery of Small-Molecule Inhibitors of the PD-1/PD-L1 Axis That Promote PD-L1 Internalization and Degradation. *J. Med. Chem.* **2022**, *65*, 3879–3893. [CrossRef] [PubMed]
58. Sun, C.; Cheng, Y.; Liu, X.; Wang, G.; Min, W.; Wang, X.; Yuan, K.; Hou, Y.; Li, J.; Zhang, H.; et al. Novel Phthalimides Regulating PD-1/PD-L1 Interaction as Potential Immunotherapy Agents. *Acta Pharm. Sin. B* **2022**, *12*, 4446–4457. [CrossRef] [PubMed]
59. Zhang, H.; Zhou, S.; Plewka, J.; Wu, C.; Zhu, M.; Yu, Q.; Musielak, B.; Wang, X.; Awadasseid, A.; Magiera-Mularz, K.; et al. Design, Synthesis, and Antitumor Activity Evaluation of 2-Arylmethoxy-4-(2,2′-Dihalogen-Substituted Biphenyl-3-Ylmethoxy) Benzylamine Derivatives as Potent PD-1/PD-L1 Inhibitors. *J. Med. Chem.* **2023**, *66*, 10579–10603. [CrossRef] [PubMed]
60. Wang, K.; Zhang, X.; Cheng, Y.; Qi, Z.; Ye, K.; Zhang, K.; Jiang, S.; Liu, Y.; Xiao, Y.; Wang, T. Discovery of Novel PD-L1 Inhibitors That Induce the Dimerization, Internalization, and Degradation of PD-L1 Based on the Fragment Coupling Strategy. *J. Med. Chem.* **2023**, *66*, 16807–16827. [CrossRef] [PubMed]
61. Surmiak, E.; Ząber, J.; Plewka, J.; Wojtanowicz, G.; Kocik-Krol, J.; Kruc, O.; Muszak, D.; Rodríguez, I.; Musielak, B.; Viviano, M.; et al. Solubilizer Tag Effect on PD-L1/Inhibitor Binding Properties for m-Terphenyl Derivatives. *ACS Med. Chem. Lett.* **2023**, *15*, 36–44. [CrossRef]
62. Hollingsworth, S.A.; Dror, R.O. Molecular Dynamics Simulation for All. *Neuron* **2018**, *99*, 1129–1143. [CrossRef]
63. Karplus, M.; Petsko, G.A. Molecular Dynamics Simulations in Biology. *Nature* **1990**, *347*, 631–639. [CrossRef]
64. Ahmed, M.; Barakat, K. The Too Many Faces of PD-L1: A Comprehensive Conformational Analysis Study. *Biochemistry* **2017**, *56*, 5428–5439. [CrossRef]
65. Liu, W.; Huang, B.; Kuang, Y.; Liu, G. Molecular Dynamics Simulations Elucidate Conformational Selection and Induced Fit Mechanisms in the Binding of PD-1 and PD-L1. *Mol. BioSyst.* **2017**, *13*, 892–900. [CrossRef]
66. Kenn, M.; Karch, R.; Tomasiak, L.; Cibena, M.; Pfeiler, G.; Koelbl, H.; Schreiner, W. Molecular Dynamics Identifies Semi-Rigid Domains in the PD-1 Checkpoint Receptor Bound to Its Natural Ligand PD-L1. *Front. Bioeng. Biotechnol.* **2022**, *10*, 838129. [CrossRef]
67. Shi, D.; Zhou, S.; Liu, X.; Zhao, C.; Liu, H.; Yao, X. Understanding the Structural and Energetic Basis of PD-1 and Monoclonal Antibodies Bound to PD-L1: A Molecular Modeling Perspective. *Biochim. Et Biophys. Acta—Gen. Subj.* **2018**, *1862*, 576–588. [CrossRef]
68. Huang, D.; Wen, W.; Liu, X.; Li, Y.; Zhang, J.Z.H. Computational Analysis of Hot Spots and Binding Mechanism in the PD-1/PD-L1 Interaction. *RSC Adv.* **2019**, *9*, 14944–14956. [CrossRef]
69. Du, J.; Qin, Y.; Wu, Y.; Zhao, W.; Zhai, W.; Qi, Y.; Wang, C.; Gao, Y. The Design of High Affinity Human PD-1 Mutants by Using Molecular Dynamics Simulations (MD). *Cell Commun. Signal.* **2018**, *16*, 25. [CrossRef] [PubMed]
70. Klyukin, K.; Alexandrov, V. Kinetics of pH-Dependent Interactions between PD-1 and PD-L1 Immune Checkpoint Proteins from Molecular Dynamics. *Proteins Struct. Funct. Bioinform.* **2020**, *88*, 1162–1168. [CrossRef] [PubMed]
71. Almahmoud, S.; Zhong, H.A. Molecular Modeling Studies on the Binding Mode of the PD-1/PD-L1 Complex Inhibitors. *Int. J. Mol. Sci.* **2019**, *20*, 4654. [CrossRef]
72. Sasmal, P.; Kumar Babasahib, S.; Prashantha Kumar, B.R.; Manjunathaiah Raghavendra, N. Biphenyl-Based Small Molecule Inhibitors: Novel Cancer Immunotherapeutic Agents Targeting PD-1/PD-L1 Interaction. *Bioorg. Med. Chem.* **2022**, *73*, 117001. [CrossRef] [PubMed]
73. Liu, X.; Peng, L.; Zhou, Y.; Zhang, Y.; Zhang, J.Z.H. Computational Alanine Scanning with Interaction Entropy for Protein–Ligand Binding Free Energies. *J. Chem. Theory Comput.* **2018**, *14*, 1772–1780. [CrossRef] [PubMed]

74. Xia, W.; He, L.; Bao, J.; Qi, Y.; Zhang, J.Z.H. Insights into Small Molecule Inhibitor Bindings to PD-L1 with Residue-Specific Binding Free Energy Calculation. *J. Biomol. Struct. Dyn.* **2022**, *40*, 12277–12285. [CrossRef] [PubMed]
75. Kuang, Z.; Heng, Y.; Huang, S.; Shi, T.; Chen, L.; Xu, L.; Mei, H. Partial Least-Squares Discriminant Analysis and Ensemble-Based Flexible Docking of PD-1/PD-L1 Inhibitors: A Pilot Study. *ACS Omega* **2020**, *5*, 26914–26923. [CrossRef] [PubMed]
76. Shi, D.; An, X.; Bai, Q.; Bing, Z.; Zhou, S.; Liu, H.; Yao, X. Computational Insight into the Small Molecule Intervening PD-L1 Dimerization and the Potential Structure-Activity Relationship. *Front. Chem.* **2019**, *7*, 764. [CrossRef] [PubMed]
77. Mejías, C.; Guirola, O. Pharmacophore Model of Immunocheckpoint Protein PD-L1 by Cosolvent Molecular Dynamics Simulations. *J. Mol. Graph. Model.* **2019**, *91*, 105–111. [CrossRef] [PubMed]
78. Goodford, P.J. A Computational Procedure for Determining Energetically Favorable Binding Sites on Biologically Important Macromolecules. *J. Med. Chem.* **1985**, *28*, 849–857. [CrossRef]
79. Paciotti, R.; Agamennone, M.; Coletti, C.; Storchi, L. Characterization of PD-L1 Binding Sites by a Combined FMO/GRID-DRY Approach. *J. Comput.-Aided Mol. Des.* **2020**, *34*, 897–914. [CrossRef]
80. Lim, H.; Chun, J.; Jin, X.; Kim, J.; Yoon, J.H.; No, K.T. Investigation of Protein-Protein Interactions and Hot Spot Region between PD-1 and PD-L1 by Fragment Molecular Orbital Method. *Sci. Rep.* **2019**, *9*, 16727. [CrossRef]
81. Sun, X.; Liang, L.; Gu, J.; Zhuo, W.; Yan, X.; Xie, T.; Wu, Z.; Liu, X.; Gou, X.; Liu, W.; et al. Inhibition of PD-L1 by Benzyl Ether Derivatives: Analyses of Conformational Change, Molecular Recognition and Binding Free Enrgy. *J. Biomol. Struct. Dyn.* **2019**, *11*, 1–14.
82. Riccio, A.; Coletti, A.; Dolciami, D.; Mammoli, A.; Cerra, B.; Moretti, S.; Gioiello, A.; Ferlin, S.; Puxeddu, E.; Macchiarulo, A. The Stone Guest: How Does pH Affect Binding Properties of PD-1/PD-L1 Inhibitors? *ChemMedChem* **2021**, *16*, 568–577. [CrossRef]
83. Soremekun, O.S.; Olotu, F.A.; Agoni, C.; Soliman, M.E.S. Recruiting Monomer for Dimer Formation: Resolving the Antagonistic Mechanisms of Novel Immune Check Point Inhibitors against Programmed Death Ligand-1 in Cancer Immunotherapy. *Mol. Simul.* **2019**, *45*, 777–789. [CrossRef]
84. Guo, Y.; Jin, Y.; Wang, B.; Liu, B. Molecular Mechanism of Small-Molecule Inhibitors in Blocking the PD-1/PD-L1 Pathway through PD-L1 Dimerization. *Int. J. Mol. Sci.* **2021**, *22*, 4766. [CrossRef]
85. Liang, J.; Wang, B.; Yang, Y.; Liu, B.; Jin, Y. Approaching the Dimerization Mechanism of Small Molecule Inhibitors Targeting PD-L1 with Molecular Simulation. *Int. J. Mol. Sci.* **2023**, *24*, 1280. [CrossRef]
86. Ahmed, M.; Ganesan, A.; Barakat, K. Leveraging Structural and 2D-QSAR to Investigate the Role of Functional Group Substitutions, Conserved Surface Residues and Desolvation in Triggering the Small Molecule-Induced Dimerization of hPD-L1. *BMC Chem.* **2022**, *16*, 49. [CrossRef] [PubMed]
87. Wang, F.; Ye, W.; Wang, S.; He, Y.; Zhong, H.; Wang, Y.; Zhu, Y.; Han, J.; Bing, Z.; Ji, S.; et al. Discovery of a New Inhibitor Targeting PD-L1 for Cancer Immunotherapy. *Neoplasia* **2021**, *23*, 281–293. [CrossRef] [PubMed]
88. Lung, J.; Hung, M.-S.; Lin, Y.-C.; Hung, C.-H.; Chen, C.-C.; Lee, K.-D.; Tsai, Y.H. Virtual Screening and In Vitro Evaluation of PD-L1 Dimer Stabilizers for Uncoupling PD-1/PD-L1 Interaction from Natural Products. *Molecules* **2020**, *25*, 5293. [CrossRef]
89. Barnwal, A.; Das, S.; Bhattacharyya, J. Repurposing Ponatinib as a PD-L1 Inhibitor Revealed by Drug Repurposing Screening and Validation by In Vitro and In Vivo Experiments. *ACS Pharmacol. Transl. Sci.* **2023**, *6*, 281–289. [CrossRef]
90. Acúrcio, R.C.; Pozzi, S.; Carreira, B.; Pojo, M.; Gómez-Cebrián, N.; Casimiro, S.; Fernandes, A.; Barateiro, A.; Farricha, V.; Brito, J.; et al. Therapeutic Targeting of PD-1/PD-L1 Blockade by Novel Small-Molecule Inhibitors Recruits Cytotoxic T Cells into Solid Tumor Microenvironment. *J. Immunother. Cancer* **2022**, *10*, e004695. [CrossRef]
91. Bianconi, E.; Riccio, A.; Ruta, L.; Bigiotti, C.; Carotti, A.; Moretti, S.; Cerra, B.; Gioiello, A.; Ferlin, S.; Puxeddu, E.; et al. Turning a Tumor Microenvironment Pitfall into Opportunity: Discovery of Benzamidoxime as PD-L1 Ligand with pH-Dependent Potency. *Int. J. Mol. Sci.* **2023**, *24*, 5535. [CrossRef]
92. Wang, F.; Ye, W.; He, Y.; Zhong, H.; Zhu, Y.; Han, J.; Gong, X.; Tian, Y.; Wang, Y.; Wang, S.; et al. Identification of CBPA as a New Inhibitor of PD-1/PD-L1 Interaction. *Int. J. Mol. Sci.* **2023**, *24*, 3971. [CrossRef] [PubMed]
93. Choorakottayil Pushkaran, A.; Kumaran, K.; Ann Maria, T.; Biswas, R.; Mohan, C.G. Identification of a PD1/PD-L1 Inhibitor by Structure-Based Pharmacophore Modelling, Virtual Screening, Molecular Docking and Biological Evaluation. *Mol. Inform.* **2023**, *42*, e2200254. [CrossRef]
94. Fattakhova, E.; Hofer, J.; DiFlumeri, J.; Cobb, M.; Dando, T.; Romisher, Z.; Wellington, J.; Oravic, M.; Radnoff, M.; Patil, S.P. Identification of the FDA-Approved Drug Pyrvinium as a Small-Molecule Inhibitor of the PD-1/PD-L1 Interaction. *ChemMedChem* **2021**, *16*, 2769–2774. [CrossRef] [PubMed]
95. Patil, S.P.; Fattakhova, E.; Hofer, J.; Oravic, M.; Bender, A.; Brearey, J.; Parker, D.; Radnoff, M.; Smith, Z. Machine-Learning Guided Discovery of Bioactive Inhibitors of PD1-PDL1 Interaction. *Pharmaceuticals* **2022**, *15*, 613. [CrossRef] [PubMed]
96. Wang, Y.; Huang, S.; Feng, X.; Xu, W.; Luo, R.; Zhu, Z.; Zeng, Q.; He, Z. Advances in Efficacy Prediction and Monitoring of Neoadjuvant Immunotherapy for Non-Small Cell Lung Cancer. *Front. Oncol.* **2023**, *13*, 1145128. [CrossRef]
97. Varadi, M.; Anyango, S.; Deshpande, M.; Nair, S.; Natassia, C.; Yordanova, G.; Yuan, D.; Stroe, O.; Wood, G.; Laydon, A.; et al. AlphaFold Protein Structure Database: Massively Expanding the Structural Coverage of Protein-Sequence Space with High-Accuracy Models. *Nucleic Acids Res.* **2022**, *50*, D439–D444. [CrossRef] [PubMed]
98. Du, K.; Huang, H. Development of Anti-PD-L1 Antibody Based on Structure Prediction of AlphaFold2. *Front. Immunol.* **2023**, *14*, 1275999. [CrossRef] [PubMed]

99. Sobhani, N.; Tardiel-Cyril, D.R.; Chai, D.; Generali, D.; Li, J.-R.; Vazquez-Perez, J.; Lim, J.M.; Morris, R.; Bullock, Z.N.; Davtyan, A.; et al. Artificial Intelligence-Powered Discovery of Small Molecules Inhibiting CTLA-4 in Cancer. *BJC Rep.* **2024**, *2*, 4. [CrossRef]
100. Anderson, K.G.; Braun, D.A.; Buqué, A.; Gitto, S.B.; Guerriero, J.L.; Horton, B.; Keenan, B.P.; Kim, T.S.; Overacre-Delgoffe, A.; Ruella, M.; et al. Leveraging Immune Resistance Archetypes in Solid Cancer to Inform Next-Generation Anticancer Therapies. *J. Immunother. Cancer* **2023**, *11*, e006533. [CrossRef]
101. Lombardo, S.D.; Presti, M.; Mangano, K.; Petralia, M.C.; Basile, M.S.; Libra, M.; Candido, S.; Fagone, P.; Mazzon, E.; Nicoletti, F.; et al. Prediction of PD-L1 Expression in Neuroblastoma via Computational Modeling. *Brain Sci.* **2019**, *9*, 221. [CrossRef]

MDPI
St. Alban-Anlage 66
4052 Basel
Switzerland
www.mdpi.com

Pharmaceuticals Editorial Office
E-mail: pharmaceuticals@mdpi.com
www.mdpi.com/journal/pharmaceuticals

www.ingramcontent.com/pod-product-compliance
Lightning Source LLC
Chambersburg PA
CBHW040828250926
51850CB00007BA/1288
9783725808144